Science: a second level cour

S269 Earth and Life

The Dynamic Earth

Prepared for the Course Team by Angela Colling (book editor), Nancy Dise, Peter Francis, Nigel Harris and Chris Wilson

The Open University

S269 Course Team

Course Team Chair	Peter Francis
Book Editor	Angela Colling
Course Team	Angela Colling
	Nancy Dise
	Steve Drury
	Peter Francis
	Iain Gilmour
	Nigel Harris
	Allister Rees
	Peter Skelton
	Bob Spicer
	Charles Turner
	Chris Wilson
	Ian Wright
	John Wright
Course Managers	Kevin Church
	Annemarie Hedges
Secretaries	Anita Chhabra
	Janet Dryden
	Marilyn Leggett
	Jo Morris
	Rita Quill
	Denise Swann
Series Producer	David Jackson
Editors	Gerry Bearman
	Rebecca Graham
	Kate Richenburg
Graphic Design	Sue Dobson
	Ray Munns
	Pam Owen
	Rob Williams
Liaison Librarian	John Greenwood
Course Assessor	Professor W. G. Chaloner (FRS)

The Open University, Walton Hall, Milton Keynes MK7 6AA

First published 1997. Reprinted with corrections 2001.

Copyright © 1997 The Open University

All rights reserved; no part of this publication may be reproduced, stored in a retrieval system, or transmitted in any form or by any means, electronic, mechanical, photocopying, recording, or otherwise without either the prior written permission of the Publishers or a licence permitting restricted copying issued by the Copyright Licensing Agency, 90 Tottenham Court Road, London W1P 9HE. This book may not be lent, resold, hired out or otherwise disposed of by way of trade in any form of binding or cover other than that in which it is published, without the prior consent of the Publishers.

Edited, designed and typeset by The Open University.

Printed and Bound in Singapore under the supervision of MRM Graphics Ltd, Winslow, Bucks.

ISBN 0 7492 8183 9

This text forms part of an Open University Second Level Course. Details of this and other Open University courses can be obtained from the Call Centre, PO Box 724, The Open University, Milton Keynes MK7 6ZS, United Kingdom: tel. +44 (0)1908 653231, e-mail ces-gen@open.ac.uk

Alternatively, you may visit the Open University website at http://www.open.ac.uk where you can learn more about the wide range of courses and packs offered at all levels by the Open University.

To purchase this publication or other components of Open University courses, contact Open University Worldwide Ltd, The Berrill Building, Walton Hall, Milton Keynes MK7 6AA, United Kingdom: tel. +44 (0)1908 858785; fax +44 (0)1908 858787; e-mail ouwenq@open.ac.uk; website http://www.ouw.co.uk

1.2

S269b2i1.2

Contents

Preface 5

Chapter 1
An hospitable planet 7
1.1 How is the Earth different? 7
1.2 Energy from the Sun 10
1.3 Summary of Chapter 1 24

Chapter 2
Keeping the Earth habitable 26
2.1 The distribution of temperature over the Earth's surface ... 26
2.2 The Earth's air-conditioning and heating systems 28
    2.2.1 Transport of heat and water by the atmosphere 30
    2.2.2 Heat transport by the ocean 44
    2.2.3 Atmosphere–ocean coupling 46
    2.2.4 The deep circulation 52
2.3 A closer look at climate 55
    2.3.1 The atmosphere – bringing in chemistry 55
    2.3.2 Atmosphere–ocean–Earth: the support system for life ... 58
2.4 Sunspots and climatic change: the importance of keeping an open mind 66
2.5 Summary of Chapter 2 68

Chapter 3
The carbon cycle 73
3.1 Carbon and life 73
3.2 Carbon and climate 75
3.3 The natural carbon cycle: a question of time-scale 76
    3.3.1 Short time-scales: the terrestrial carbon cycle 77
    3.3.2 Intermediate time-scales: the marine carbon cycle 83
    3.3.3 Long time-scales: the geological carbon cycle 100
3.4 A system in balance? 103
    3.4.1 Short-circuiting the geological carbon cycle 105
3.5 Summary of Chapter 3 109

Chapter 4
Volcanism and the Earth system 112
4.1 Introduction 112
4.2 Volcanic aerosols and climatic change 119
    4.2.1 Climatic effects of the eruption of Mount Pinatubo ... 122
    4.2.2 Historic eruptions 124
    4.2.3 The Toba eruption 127
4.3 Flood basalts and their climatic effects 129
    4.3.1 Lessons from the Laki Fissure eruption 130
    4.3.2 Are flood basalts implicated in mass extinctions? 134

4.4 CO$_2$ emissions from the Deccan Traps: did it cause greenhouse warming? ... 136
4.5 Taking a global view ... 137
4.6 Summary of Chapter 4 ... 139

Chapter 5
Plate tectonics, carbon and climate ... 141

5.1 Continental drift and climate ... 141
    5.1.1 Rearranging the continents ... 144
    5.1.2 Continental drift, ocean currents and climate change ... 151
5.2 Glacials and interglacials ... 158
5.3 Sea-level change: causes and consequences ... 161
    5.3.1 Contributions to sea-level changes ... 167
    5.3.2 Causes of eustatic sea-level change ... 167
    5.3.3 Sea-level, climate and atmospheric CO$_2$ ... 168
5.4 Plate tectonics, mountain-building and climate ... 175
    5.4.1 Mountain-building ... 175
    5.4.2 Mountain climate ... 176
    5.4.3 Subduction, mountains and atmospheric CO$_2$ ... 177
5.5 Summary of Chapter 5 ... 180

Chapter 6
Tibet, the Himalayas and the Arabian Sea ... 182

6.1 Raising the roof of the world ... 182
    6.1.1 Why are the Himalayas and Tibet so high? ... 183
    6.1.2 When was the Tibetan Plateau uplifted? ... 185
6.2 The uplift of Tibet and climate change ... 188
    6.2.1 Present-day climate ... 188
    6.2.2 Evidence for climate change ... 191
6.3 The Himalayas, Tibet and atmospheric CO$_2$ ... 197
    6.3.1 Strontium isotopes and the rate of chemical weathering ... 203
6.4 Cycles in the sediments: climate records in the Arabian Sea ... 209
6.5 Summary of Chapter 6 ... 214

Objectives ... 216

Answers to Questions ... 218

Comments on Activities ... 234

Acknowledgements ... 239

Index ... 241

Preface

Frontispiece
The Earth and Moon in space, photographed on the *Galileo* mission.

This book is about the Earth's climate and the various factors that cause it to change. We begin, of course, with the influence of the Sun, and look at how the solar radiation reaching the Earth's surface varies over three different time-scales: over the course of a year (i.e. seasonally); over tens of thousands of years (in response to the Milankovich cycles); and over decades (in response to changes in solar activity). This idea of change occurring simultaneously over a number of different time-scales, in response to a number of different driving mechanisms or 'forcing factors', is one that recurs throughout.

In the second chapter, we begin to bring in the other factors which modify the Sun's influence — winds and currents which redistribute heat over the surface of the globe, the chemical compositions of the atmosphere and ocean, and the activities of living organisms. Chapter 3 builds on these latter themes in its discussion of the carbon cycle. Life on Earth is based on the chemistry of carbon, and the CO_2 concentration in the atmosphere is an important influence on global temperatures; it should not be surprising, therefore, that the cycling of carbon between atmosphere, biosphere, hydrosphere and lithosphere is an important theme of this book, as well as others in the *Earth and Life* series.

Chapter 4 begins with a brief look at volcanism on the Earth at the present day, including volcanic emission of CO_2 (which tends to cause global warming) and SO_2 (which, by forming atmospheric aerosols, tends to cause global cooling). By examining the impacts of relatively recent eruptions, we attempt to quantify the effects of eruptions recorded in the geological record – eruptions which could be implicated in mass extinctions, including that which saw the demise of the dinosaurs.

The next chapter brings together various themes introduced in earlier chapters – including the influence of ocean currents, the cycling of carbon, and volcanism – in the context of sea-level change, plate tectonics and continental drift. Climatic changes related to plate tectonics (including, probably, the coming and going of Ice Ages) take place over tens to hundreds of millions of years; however, in the context of changing sea-level, we also discuss the glacial–interglacial fluctuations that occur *within* Ice Ages.

The Dynamic Earth concludes with a case-study centred on the uplift of the Tibetan Plateau and the Himalayas – an event that resulted from lithospheric plate movements, yet eventually involved all components of the Earth system – the atmosphere, ocean and biosphere, as well as the lithosphere. The climatic changes that resulted are hard to disentangle: it seems clear that there was a strengthening of the Asian monsoon; it is also possible that the uplift was responsible for the decline in temperatures that has affected the whole globe over the last 50 million years or so.

Cover photograph:
View of the Himalayas in the Khumbu region of eastern Nepal.
(Courtesy of Roy Lawrance)

Chapter 1
An hospitable planet

We do not know whether life flourishes on planets orbiting stars even more distant than our own Sun. Despite the excitement in 1996 about possible fossils in martian meteorites, we *still* do not know whether life exists, or has ever existed, on Mars. But we do know that within the Solar System, the Earth is unique in actually being *hospitable* to life. Why is this?

1.1 How is the Earth different?

If we could observe the Solar System from far off, we would see the Sun and its orbiting planets, all virtually in the same plane – a legacy of their common origin from the same spinning nebular disc. Each planet is spinning about its own axis, and at any one time one side is lit by the Sun and the other is in darkness. Though separated from the Sun by the blackness of space, the planets remain tied to it through its enormous gravitational pull, and also influence one another through their own smaller gravitational fields.

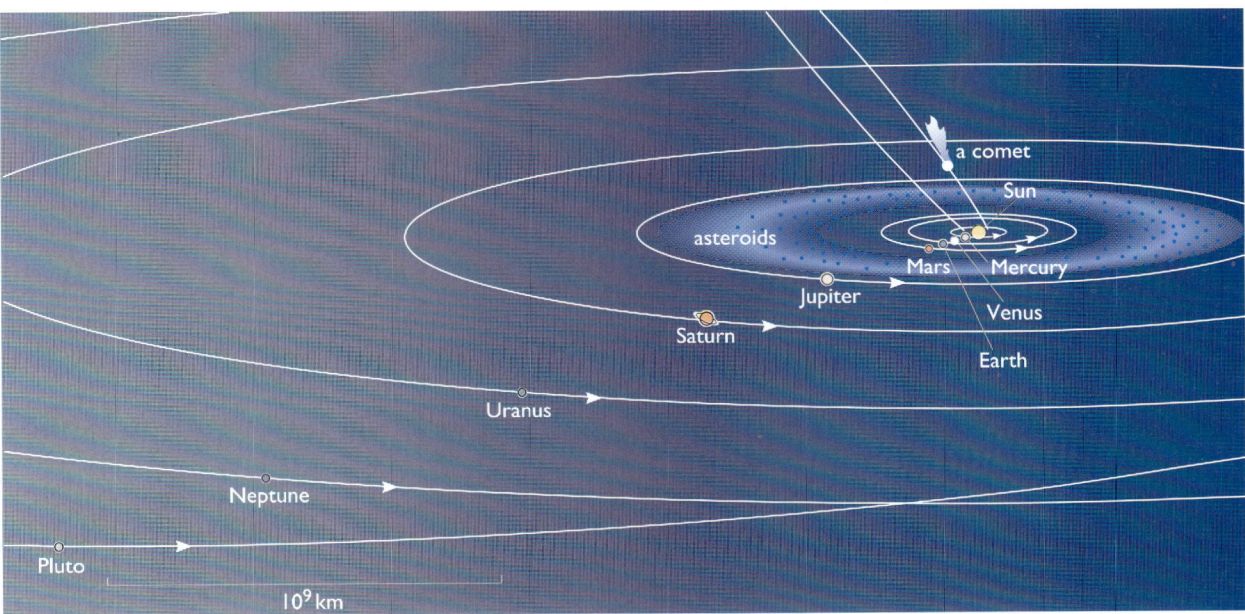

Figure 1.1
A schematic representation of the Solar System, viewed obliquely. Most planetary orbits are in the same plane as the Earth's orbit, and most planets (including the Earth) spin so that their direction of rotation is consistent with their orbital motion about the Sun. (Note that planetary orbits are in reality nearly circular.)

If we were to look more closely at the four planets nearest to the Sun, we would see a small one, Mercury, then a very bright one – Venus; then our own blue Earth with its swirls of white cloud; and finally the reddish globe of Mars. Four planets with very different surface environments, yet formed in more-or-less the same part of the solar nebula, and so likely to be made up of the same elements, in very similar proportions. Looking closer still at the blue planet, we see that it has a thin envelope of hazy atmosphere; through this, below the clouds, we see not only the blue of oceans and seas and the bright white of the ice-caps, but the greens, greys and browns of land. And in the night hemisphere we see cities defined by billions of tiny light sources, flashes of bluish lightning, the red glow of fires, and the occasional glow of hot magma.

Environmental conditions at the Earth's surface, in particular those we think of as constituting 'climate', are a result of the complex interplay of many processes – physical, chemical, geological *and* biological. These act over a wide range of time-scales, and have spatial scales ranging from chemical reactions between atoms and molecules to gravitational interaction with other planets.

Looked at simply, in the final analysis the Earth is hospitable to life as a result of its distance from – or perhaps closeness to – the Sun. Earth is half as far again from the Sun as Venus, where the average surface temperature is 460 °C, and about two-thirds of the distance of Mars, where the average surface temperature is −50 °C; by contrast, the average temperature at the surface of the Earth is a moderate 15 °C. Averages can, of course, conceal enormous ranges, but on the Earth at the present time surface temperatures rarely rise above 50 °C, and rarely fall below −50 °C (Figure 1.2). This relatively small range is a result of the form and content of the Earth's fluid envelopes – its atmosphere and ocean. As you will see, the full story is very complicated (and by no means completely understood), but the two most important volatile constituents in this context are water and carbon dioxide.

On the Earth, surface temperatures are such that most of the planet's water is in liquid form, with the remainder in the ice-caps and in the atmosphere; the atmosphere contains a small amount of CO_2, the oceans a good deal more, and large amounts are effectively 'locked up' in crustal rocks. By contrast, on Venus the atmosphere is largely CO_2; there is a minute amount of atmospheric water vapour (i.e. H_2O gas) and, at the prevailing temperatures, none of it can condense on the planet's surface. The martian atmosphere, like the venusian one, is largely CO_2, but some CO_2 and most of the planet's water is in the form of ice. Were it not for the presence of liquid water on the surface of the Earth, life – at least in the form we know it – could not exist here.

Look at Figure 1.3. The cloud cover, which obscures so much of the Earth's surface in the Frontispiece photograph, has been removed. Immediately obvious are bright areas of ice cover, the enormous area of ocean in the Southern Hemisphere, and the green of forests and crops.

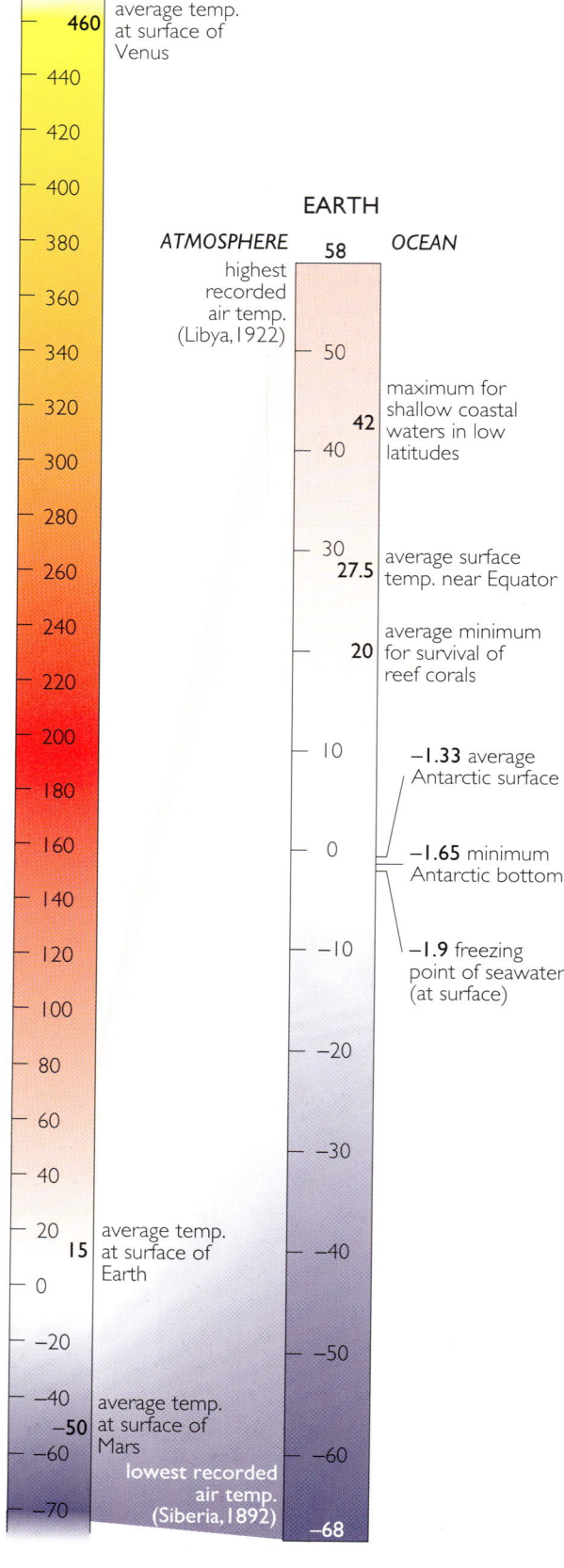

Figure 1.2
Characteristic temperatures at the Earth's surface (left, in the atmosphere; right, in the ocean), in comparison with temperatures on Mars and Venus.

An hospitable planet

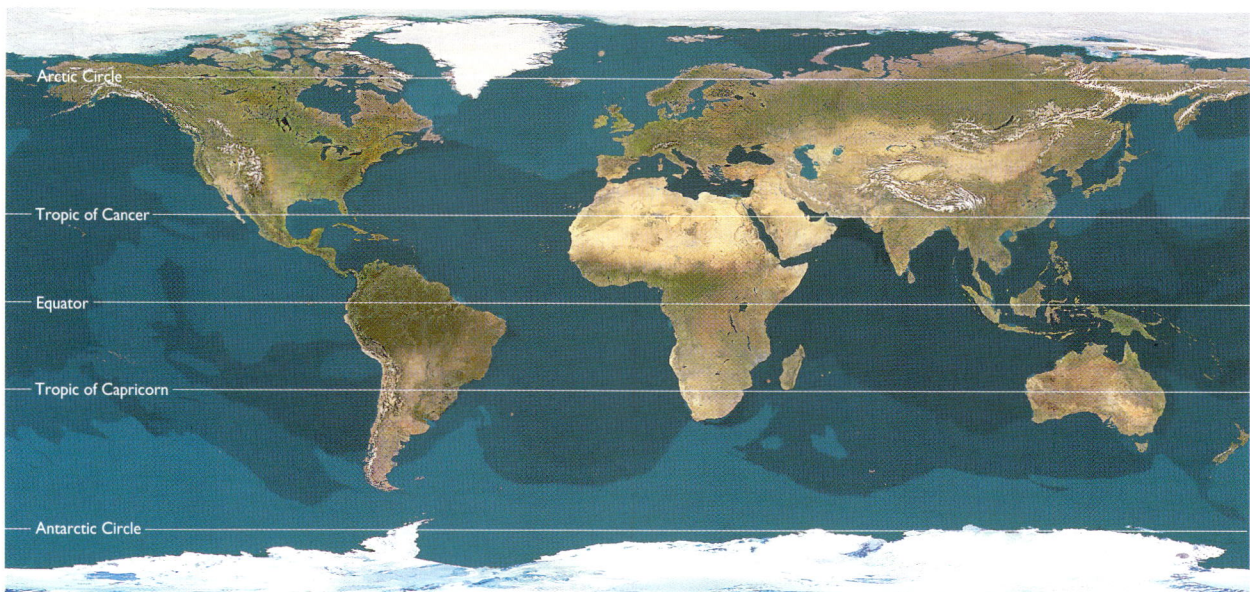

Figure 1.3
Composite satellite image of the Earth's surface. The green areas are forests and crops; the yellow areas are savannah, semi-arid scrub and desert; and the brown areas at high northern latitudes are tundra. At high latitudes, and at lower latitudes where there are mountains (e.g. the Rockies, Andes and Himalayas), white corresponds to ice and snow. Note that the area of ice at high latitudes is grossly exaggerated: at the North and South Poles, single points are expanded to the width of the map.

■ To what extent does this map tell us about life on Earth?

To a significant, but limited, extent. This map shows life on Earth because, from many points of view, life means *plants* – organisms that can build their own organic material by harnessing the energy of sunlight using **photosynthesis**. Expressed simply, the chemical equation for photosynthesis can be written as follows:

$$\underbrace{nCO_2(g)}_{\text{carbon dioxide from atmosphere}} + nH_2O \xrightarrow{\text{light energy}} \underbrace{(CH_2O)_n}_{\text{organic matter}} + \underbrace{nO_2(g)}_{\text{oxygen}} \qquad \text{(Equation 1.1)}$$

where n can have various values, and $(CH_2O)_n$ represents a range of carbohydrate materials of which glucose ($C_6H_{12}O_6$) is the simplest; (g) is gas. For convenience, from now on we will write this equation with $C_6H_{12}O_6$ representing organic matter in general:

$$\underbrace{6CO_2(g)}_{\text{carbon dioxide from atmosphere}} + 6H_2O \longrightarrow \underbrace{C_6H_{12}O_6}_{\text{organic matter}} + \underbrace{6O_2(g)}_{\text{oxygen}} \qquad \text{(Equation 1.2)}$$

This process of taking 'free' carbon from the atmosphere and combining it into living organic material is referred to as 'fixing' carbon, and the process of building living material by fixing carbon is known as **primary production**. Animals cannot fix carbon and so can live only by consuming primary producers, either directly or indirectly. (Although at the present time most primary production is by plants, it's important to remember that some bacteria are also primary producers.)

But Figure 1.3 tells only a small part of the story. For one thing, the oceans, too, support abundant plant life – Figure 1.4 shows not only the geographical variation of the potential for primary production on land (i.e. the potential for carbon to be incorporated into terrestrial living material) but also the average concentration of chlorophyll in algae living in surface waters. Furthermore, although they are composites of many satellite images, both Figures 1.3 and 1.4 are essentially snapshots in time. They provide information about the 'standing stock' of plant material, but by themselves they do not tell us anything about the *rates* at which plant material is being made – the primary *productivity* – in different environments.

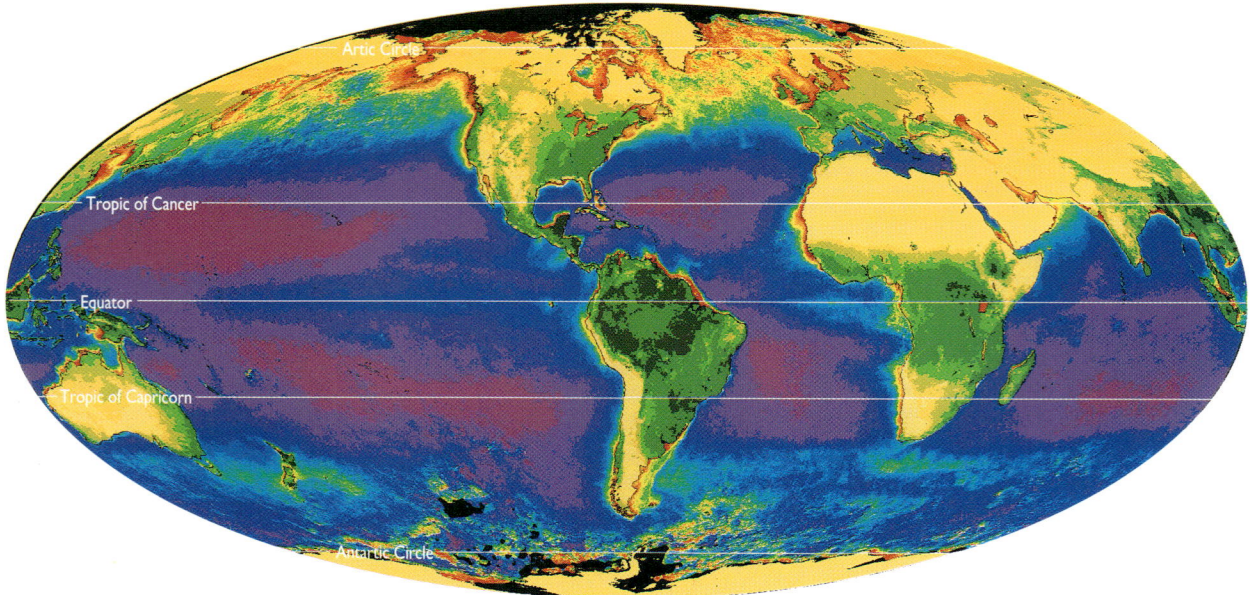

Figure 1.4
Global distribution of the potential for primary production on land and in surface waters, as indicated by chlorophyll concentration (determined using satellite-borne sensors). On land, darkest green areas correspond to the greatest potential for production of new plant material; decreasing production potential is indicated by increasingly paler greens. Least productive of all are deserts, high mountains and arctic regions shown in yellow. In surface waters, regions of highest productivity are bright red, followed by yellow, green and blue. Least productive oceanic regions are shown in purplish-red. As terrestrial and marine plants make up more than 99% of living matter on Earth, the map effectively represents the global **biosphere** – the totality of living organisms on Earth.

Nor do they tell us anything about the rates at which the plant material is being eaten by animals, decomposing or being recycled. As you will see, because organic material is essentially carbon, almost all of which can eventually find its way back into the atmosphere as gaseous CO_2, all these processes are important influences on the Earth's climate.

- Returning to Figure 1.3 for a moment, can you suggest why the vegetation patterns shown are not wholly reliable as indicators of local climatic conditions?

- Because of the influence of humans and their domesticated animals – the loss of forests (particularly in temperate latitudes), irrigation of land in arid regions, grazing and so on.

Nevertheless, the patterns of primary production seen in Figures 1.3 and 1.4 are to a large extent determined by the movement of air and water over the surface of the Earth – including the swirling clouds seen in the Frontispiece. We will see *how* shortly, but first we must look at what drives the continual motion of the Earth's fluid envelopes – energy from the Sun.

1.2 Energy from the Sun

The Earth is, on average, about 150×10^6 km from the Sun, and at this distance the average amount of solar energy falling upon unit area (1 m^2) per second at the top of the atmosphere is effectively ~343 J, so that the *effective solar flux* is ~343 W m^{-2}. If you want to see how that value is arrived at, look at Figure 1.5; note that watts (W) are joules per second (J s^{-1}). Not all of this incoming solar energy is available to heat the Earth–atmosphere system: about 30% of it is reflected back into space, mainly from the tops of clouds. In other words, the **albedo** of the Earth as a whole – the percentage of incoming solar radiation that is reflected from it – is about 30%. This means that the Earth–atmosphere system receives ~240 W of solar energy per square metre, i.e. 240 W m^{-2}. But this figure of 240 W m^{-2} is an average – and in the final analysis it is the uneven heating of the Earth that drives the Earth's climate 'engine'.

An hospitable planet

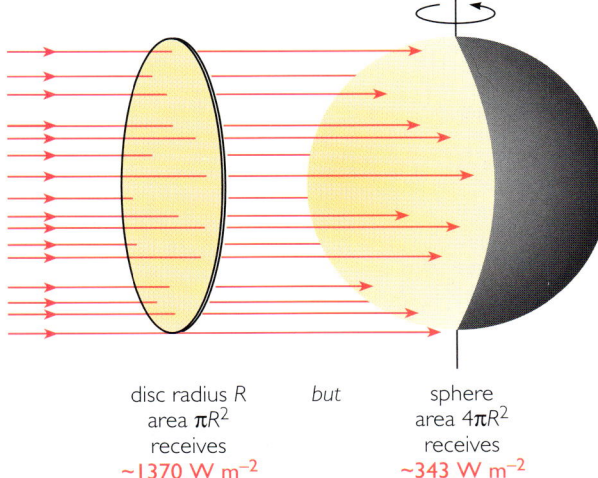

incoming solar radiation spread over surface of rotating Earth

disc radius R
area πR^2
receives
~1370 W m^{-2}

but

sphere
area $4\pi R^2$
receives
~343 W m^{-2}

Figure 1.5
Diagram to illustrate the calculation of the average flux of solar energy reaching the Earth. The amount of solar energy that would fall on a surface at right-angles to the Sun's rays, known as the solar flux (or solar irradiance, or solar constant), would be ~1370 W m^{-2}. This means that the amount of solar radiation that the Earth 'intercepts' is ~1370 × πR^2 W, where πR^2 is the area of a disc with the same radius as the Earth. However, the Earth is *spherical* so the area presented to the incoming solar radiation by the rotating Earth (over any period longer than a day) is $4\pi R^2$, i.e. four times as great. The average flux of solar energy is therefore effectively only a quarter of the solar flux, i.e. 1370/4 ≈ 343 W m^{-2} (or 343 J s^{-1} m^{-2}). This is the effective solar flux.

- Look at Figure 1.6. Bearing in mind that the atmosphere absorbs a proportion of incoming solar energy, suggest *two* reasons why the intensity of solar radiation at the Earth's surface, and hence the surface temperature, is generally lower at high latitudes than at low latitudes.

- The intensity of solar radiation at the Earth's surface (on the diagram, the number of 'rays' per unit area) depends on the angle of the rays with respect to the surface: the more oblique the angle, the larger the area over which the solar energy will be spread. Furthermore, the more oblique the rays, the greater the thickness of atmosphere through which the rays will have to travel.

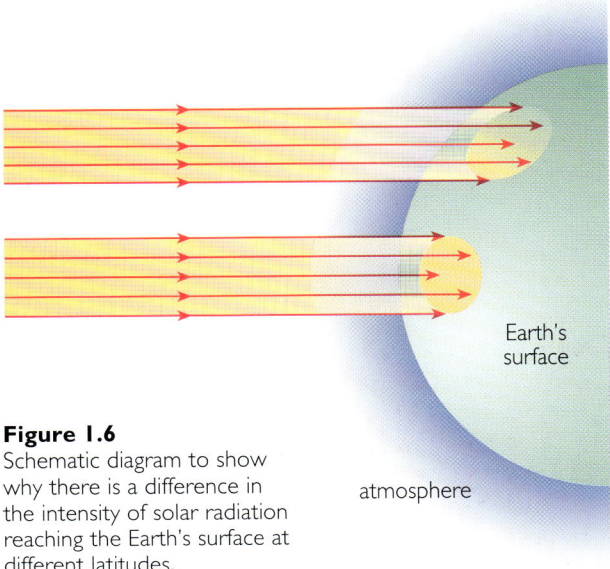

Figure 1.6
Schematic diagram to show why there is a difference in the intensity of solar radiation reaching the Earth's surface at different latitudes.

The relationship between the angle the Sun's rays make with the ground and their ability to warm it is, in fact, the origin of the word 'climate'. It derives from the Greek 'κλιμα' (pronounced *kleema*) which originally meant 'slope', and incorporated the long-observed fact that south-facing (i.e. equatorward-facing) slopes – where the Sun's rays meet the ground at a steeper angle – are warmer.

If the Earth's axis of rotation were at right-angles to the plane of its orbit, for any given latitude the angle at which the rays of the noonday Sun fell upon the surface would remain constant, with higher latitudes in both hemispheres always receiving less solar radiation than lower latitudes (Figure 1.7, overleaf). In other words, the Earth's surface at (say) 10° N and 10° S would always receive the same amount of solar radiation, and always more than that received at (say) 40° N and 40° S. The noonday Sun would be overhead at the Equator, and nowhere else.

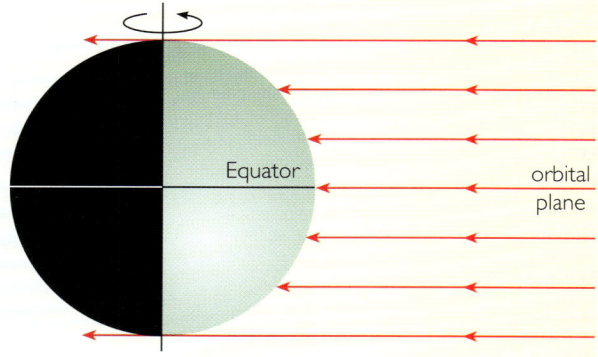

Figure 1.7
Hypothetical situation in which the Earth's axis of rotation is at right-angles to the plane of its orbit. (Note that at the poles the Sun's rays graze the Earth's surface.)

- Look at Figure 1.7. What would be the implications of such a situation for the lengths of night and day?

- Except at the poles, which would have perpetual twilight, night and day would always be the same length (i.e. each 12 hours long), everywhere on the globe.

But the Earth's axis of rotation is *tilted* with respect to the plane of its orbit, currently at an angle of 23.4° (Figure 1.8). As a result, the noonday Sun is overhead at the Equator only twice a year, at the equinoxes (when the lengths of night and day are equal). At other times, the latitude at which the noonday Sun is overhead is migrating between 23.4° N (the Tropic of Cancer) and 23.4° S (the Tropic of Capricorn) and back again. This change in the position of the noonday Sun throughout the year is the cause of the seasons – if you need to be reminded why this is so, read Box 1.1.

Box 1.1 The cause of seasons

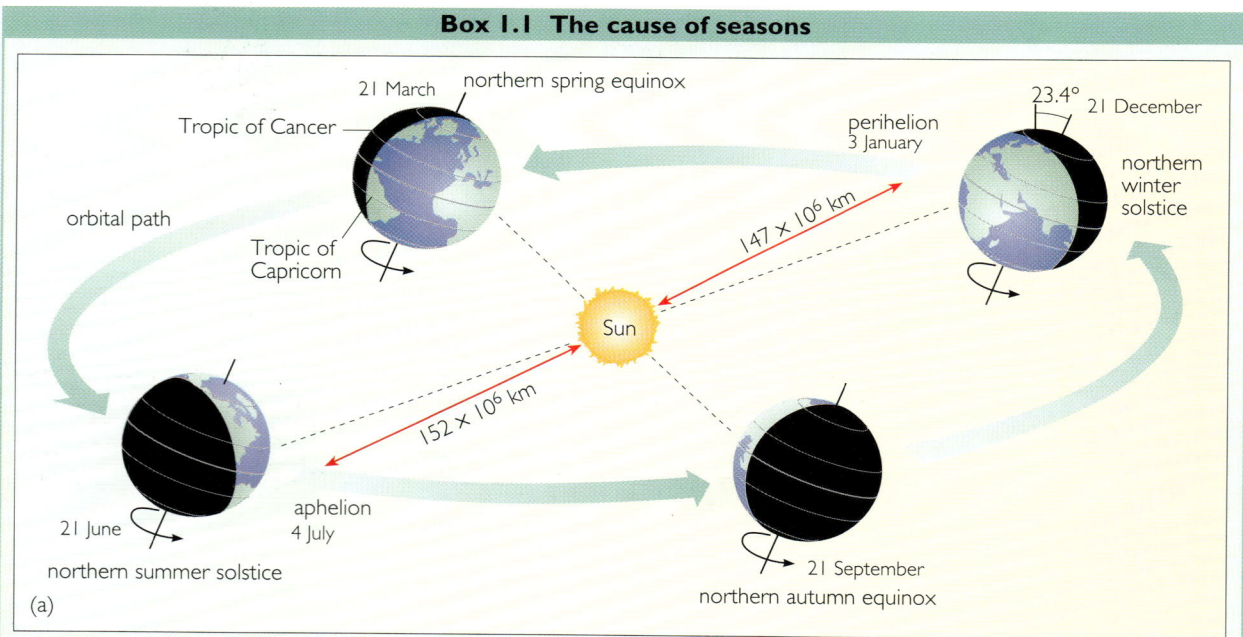

Figure 1.8
(a) The four seasons (here given for the Northern Hemisphere) related to the Earth's orbit around the Sun. The Earth's axis is tilted at approximately 23.4° to the normal to the plane of its orbit around the Sun, with the result that between latitudes 23.4° N and 23.4° S (the Tropics of Cancer and Capricorn, respectively), the noonday Sun may be directly overhead for at least part of the year. When the Sun is overhead one or other of the Tropics, it is the summer solstice (the longest day) in the hemisphere experiencing summer, and the winter solstice (the shortest day) in the other. Poleward of the Tropics, the Sun is never directly overhead.

An hospitable planet

Figure 1.8a shows the passage of the seasons for the Northern Hemisphere. Along the Tropic of Cancer (23.4° N), the noonday Sun is overhead, and maximum solar radiation received, during the summer solstice: the longest day, on 21 June. After that, days begin to shorten, until at the autumn equinox, on 21 September, day and night are of equal length. Day-lengths continue to shorten, until the shortest day – the winter solstice, on 21 December – after which days begin to lengthen again. It's not hard to see why this day was of such significance for our ancestors, whose activities were ruled by the seasonal changes in light and warmth from the Sun.

At the Equator, maximum solar radiation is received at the March and September equinoxes, when the noonday Sun is overhead, and day and night are of equal length. Poleward of the Tropics, the Sun is *never* overhead, although it is at its highest at the summer solstice. The poles themselves are wholly illuminated in summer and wholly dark in winter.

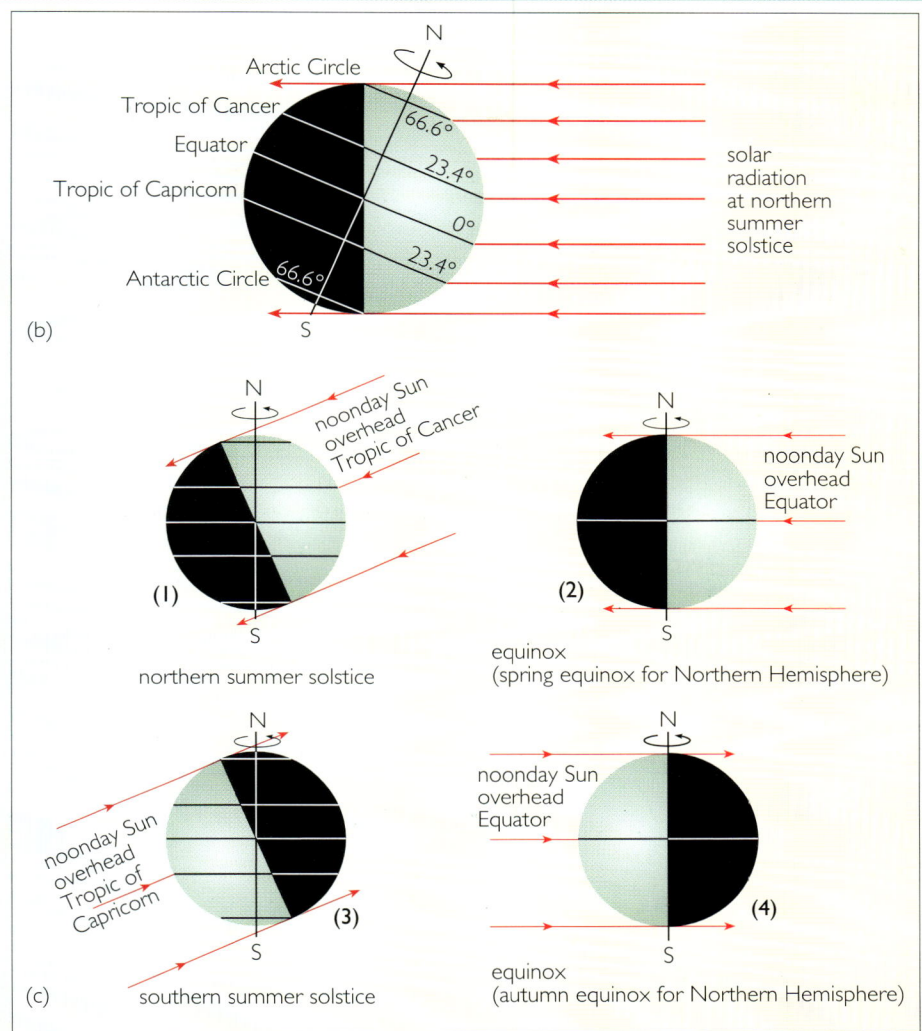

Figure 1.8
(b) The Earth's angle of tilt (at present 23.4°) determines the latitude of the Tropics (where the Sun is overhead at one of the solstices) and of the Arctic and Antarctic Circles (66.6° = 90° − 23.4°), poleward of which there is total darkness for at least part of the year.

(c) The passage of the seasons shown in terms of the position of the noonday Sun with respect to the Earth. (1) The noonday Sun overhead the Tropic of Cancer, i.e. the northern summer solstice (cf. (b)) and (3) the noonday Sun overhead the Tropic of Capricorn, i.e. the southern summer solstice. (2), (4) At the equinoxes, by contrast, the Sun is overhead the Equator, and Northern and Southern Hemispheres are illuminated equally; days and nights are the same duration at all latitudes, except at the poles which are grazed by the Sun's rays for 24 hours.

Figure 1.9 shows the seasonal variation in the amount of solar radiation received daily at the Earth's surface (i.e. taking into account absorption in the atmosphere). The zero contour corresponds to 24-hour darkness. At the North Pole (90° N), it encompasses the period between 21 September and 21 March, and at the South Pole it encompasses the period between 21 March and 21 September. In each case, the first of these dates is the autumn equinox and the second is the spring equinox (cf. Figure 1.8). Don't worry if this type of diagram seems strange – you should be able to see how it works when you have tried Question 1.1.

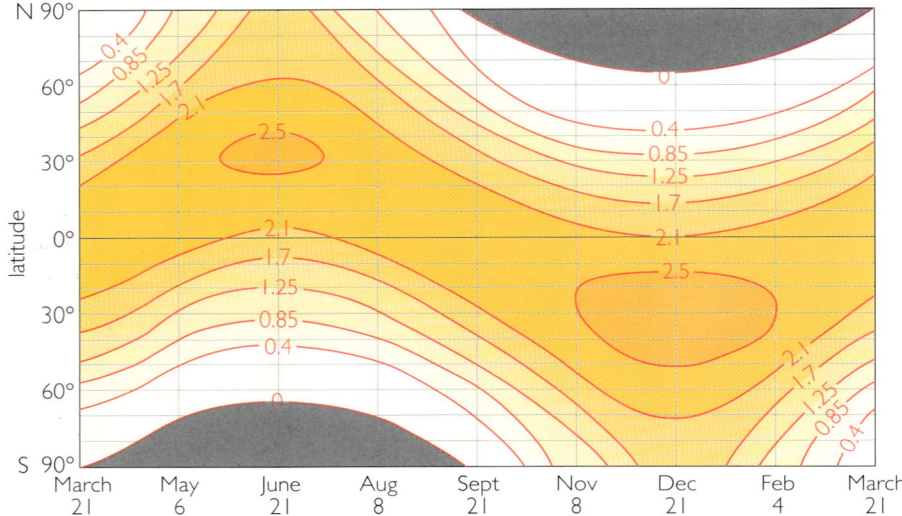

Figure 1.9
Seasonal variation of daily incoming solar radiation (in $10^7\,J\,m^{-2}$) at the Earth's surface, taking account of absorption by the atmosphere but ignoring the effect of topography. Note that this is not a map, but a plot of incoming solar energy against latitude on the one hand and time of year on the other.

Question 1.1
(a) (i) Describe briefly how the incoming solar radiation changes over the course of the year at 50° N (the British Isles lie mainly between 50° and 60° N). (ii) The units used in Figure 1.9 are $J\,m^{-2}$ because it shows values of incoming solar radiation *per day*. How would you convert the contour values so that they were for *average* incoming solar radiation in $W\,m^{-2}$?

(b) Figure 1.9 shows that on average, over the year as a whole, the Equator receives the most solar radiation. But which part(s) receive the most at any one time? Can you suggest the reason for this?

An aspect of Figure 1.9 that is at first sight puzzling is that the maximum amount of solar energy received by southern mid-latitudes in the southern summer is *greater* than the maximum amount received by northern mid-latitudes in the northern summer – compare, for example, the areas enclosed by the $2.5 \times 10^7\,J\,m^{-2}$ contour. Furthermore, if you study Figure 1.9 carefully you will see that in the southern summer *all* latitudes receive more energy than the corresponding latitudes in the other hemisphere in the northern summer. This is because the Earth's orbit is elliptical, and at the present time the Earth comes closest to the Sun – i.e. is at **perihelion** – during the southern summer (on 3 January), and is furthest from the Sun – i.e. is at **aphelion** – during the northern summer (on 4 July). It is because the Sun is at one of the two foci of the ellipse, rather than at its geometric centre, that perihelion and aphelion only occur once a year, rather than twice.

An hospitable planet

Because of the varying gravitational attraction of the Sun and of the other planets (notably Jupiter and Saturn), the degree of ellipticity (*eccentricity*, or off-centredness) of the Earth's orbit varies with time, and over a period of about 110 000 years changes from its most elliptical (maximum eccentricity) to nearly circular and back again (Figure 1.10a). This 110 000-year cycle is the longest of three astronomical cycles which affect the amount and distribution of solar radiation reaching the Earth's surface, and it is the only one that affects the *total* amount of solar radiation reaching the Earth. The two shorter cycles (Figure 1.10b) involve the orientation of the Earth's axis, and so affect the *distribution* of solar radiation over the Earth's surface.

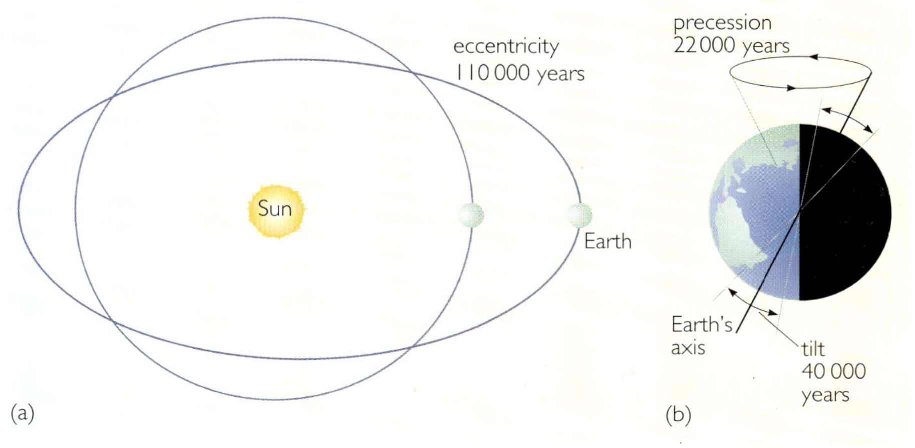

Figure 1.10
The component Milankovich cycles. (a) Plan view of the Earth's orbit to show how it changes shape from circular to most elliptical and back again, over the course of the 110 000-year eccentricity cycle. (b) The Earth showing the 40 000-year tilt cycle and the 22 000-year precession cycle.

First, over the course of about 22 000 years, the direction in which the Earth's axis points traces a circle in the sky.

- Sketch a version of Figure 1.8a, showing what the tilted Earth in its orbit will look like in ~11 000 years' time (ignore changes in the shape of the orbit). Indicate the summer and winter solstices and the spring and autumn equinoxes (for the Northern Hemisphere), and the orbital positions of perihelion and aphelion. Why can you ignore changes in the shape of the orbit?

Figure 1.11 overleaf (plus, hopefully, your own sketch) show that ~11 000 years from now, the positions in the orbit of the northern and southern summer will be reversed; the seasons – the solstices and the equinoxes – will have moved clockwise round the orbit, and in another 11 000 years would be back in their current positions. This phenomenon is often referred to as the *precession of the equinoxes*. (Of course, our calendars will continually have to be adjusted to take account of this so that, for example, the northern summer solstice will remain in June and not gradually drift to December.)

Now to the second of the two shorter cycles. At the same time that the Earth's axis traces a circle in the sky, the angle it makes with the normal to the orbital plane varies between about 21.8° and 24.4°, and back again, with a periodicity of about 40 000 years; at the moment, the angle of *tilt* is about 23.4°.

- Bearing in mind Figure 1.8, what do you think the latitude of the Tropics would be if the angle of tilt increased so that it was 24.4°, rather than 23.4°? What effect would that have on a diagram like Figure 1.9?

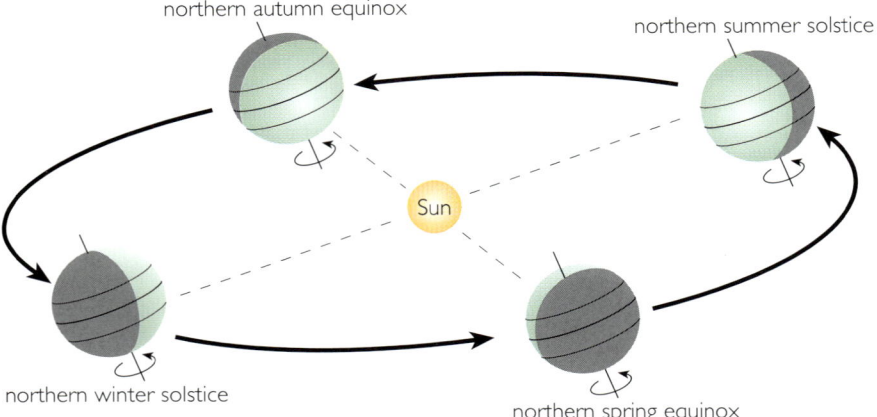

Figure 1.11
The tilted Earth in its orbit in ~11 000 years' time; the axis points in the opposite direction to what it does at present, with the result that summer and winter occupy diametrically opposite positions in the orbit to those they occupy at present. We can ignore changes in the shape of the orbit because they will have been small over a time equivalent to only about one-tenth of the period of the eccentricity cycle.

■ If the angle of tilt increased to 24.4°, the Tropics (over which the Sun would be directly overhead during the summer solstice) would be at 24.4° N and 24.4° S. This would mean that on a diagram like Figure 1.9, the areas of maximum incoming solar radiation corresponding to summer months would be shifted polewards slightly, and for the winter hemisphere, the zero contour (for example) would extend a little further towards the Equator.

So the greater the angle of tilt, the greater the difference between winter and summer. At present the angle of tilt is in fact *decreasing*, so summers are very gradually becoming cooler and winters are very gradually becoming warmer.

These three astronomical cycles are usually known as **Milankovich cycles**, after Milutin Milankovich, a Serbian astronomer. However, as long ago as the 1860s, the Scotsman James Croll recognized that these astronomical cycles could affect the Earth's climate; Figure 1.12 is taken from his book, *Climate and Time*, published in 1875. Milankovich's work in the 1930s and 1940s was an improvement and refinement of Croll's work and for this reason the cycles are sometimes referred to as Milankovich–Croll cycles.

The form of the three cycles over the past 800 000 years can be seen in Figure 1.13; in (b) you can see that the angle of tilt is currently decreasing, as mentioned above. At present the eccentricity of the Earth's orbit is such that the Earth–Sun distance is about 147×10^6 km at perihelion and about 152×10^6 km at aphelion (Figure 1.8a).

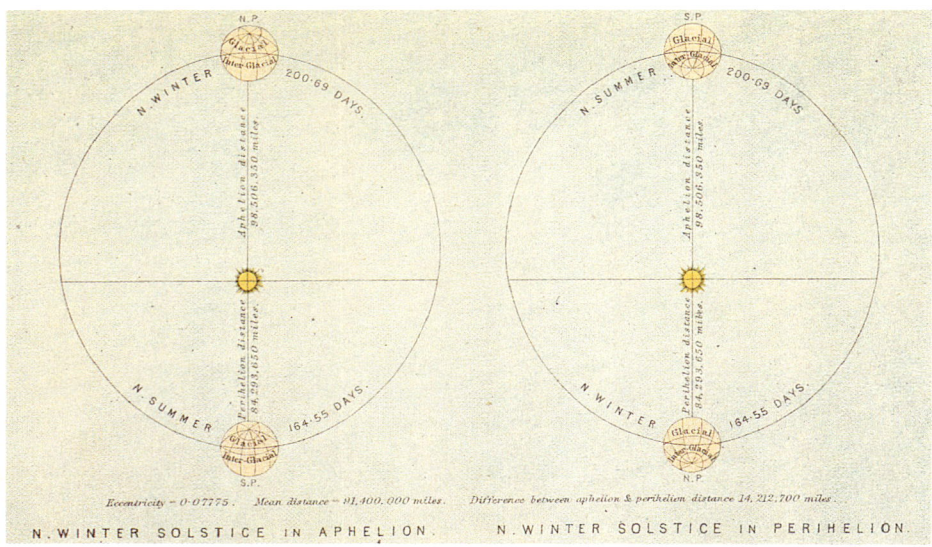

Figure 1.12
The Frontispiece to Croll's book, *Climate and Time*, published in 1875. The orbits of the Earth are seen from above. If you look carefully at the globes, you can see the direction of tilt of the Earth's axis.
N.P. = North Pole;
S.P. = South Pole. (See also Question 1.6, later.)

An hospitable planet

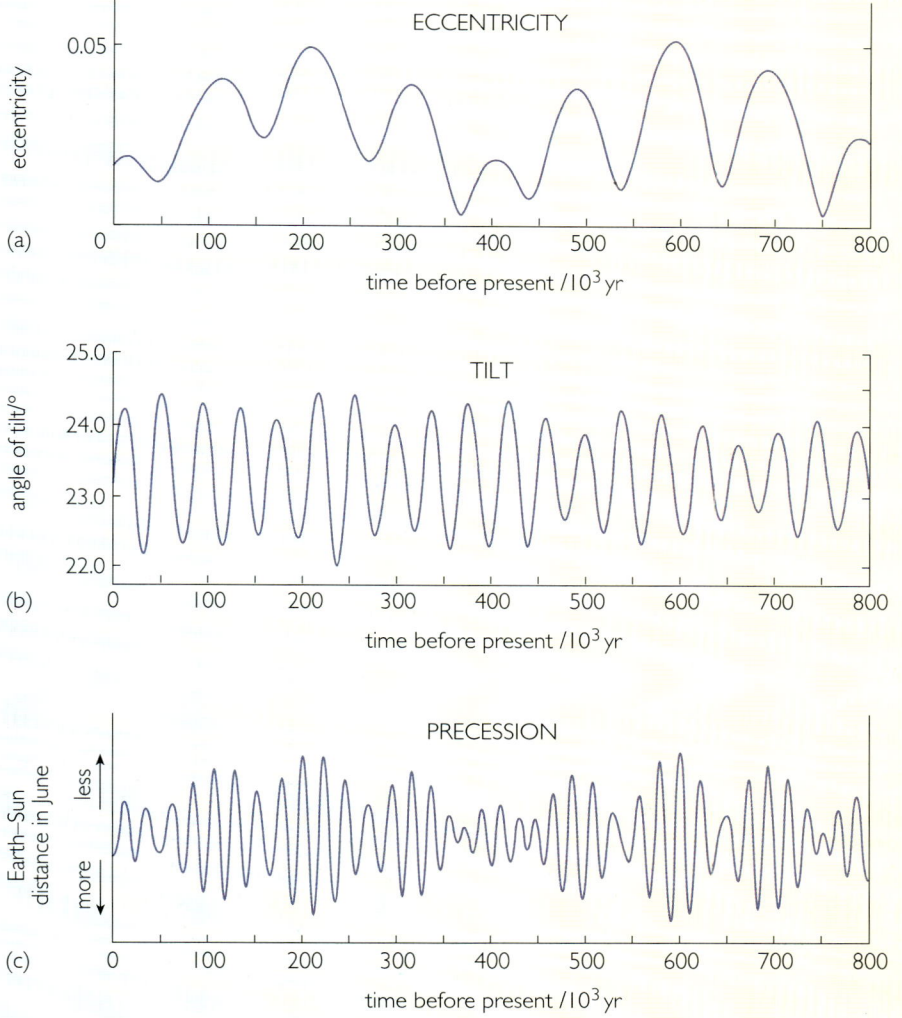

Figure 1.13
Milankovich–Croll orbital changes over the past 800 000 years, based on astronomical data.
(a) Variations in orbit eccentricity (the higher the value, the more elliptical the orbit: an eccentricity of zero corresponds to circular); note that this 110 000-year cycle is actually a combination of a 100 000-year cycle with a much weaker 413 000-year cycle, and is sometimes referred to as the 100 000-year cycle. (b) Variations in the angle of tilt, which you may see referred to as the 40 000-year or 41 000-year cycle. (c) Precession expressed in terms of the Earth–Sun distance in June; this 22 000-year cycle may be further broken down into a 19 000-year cycle and a more dominant 23 000-year cycle.

■ According to Figure 1.13, is the Earth's orbit currently becoming more or less elliptical? Approximately when will it next be most nearly circular?

■ It is currently becoming less elliptical. At the present, the eccentricity curve, (a), is tending towards smaller values, and if you look at the shape of the eccentricity cycle as a whole you can see that it will next reach its minimum – i.e. be most nearly circular – in about 50 000 years' time.

Note that an elliptical orbit tends to exaggerate the seasons in one hemisphere – the one for which winter occurs during aphelion and summer occurs during perihelion – and to moderate them in the other. The more elliptical the orbit, the more extreme this effect is.

Figure 1.14 (overleaf) is a curve computed by combining the periodic variations shown in Figure 1.13, to show how the intensity of summer sunshine at high northern latitudes (65° N) has varied in response to Milankovich–Croll orbital changes over the past 600 000 years.

Activity 1.1

(a) Compare the contour and axis values in Figures 1.14 and 1.9. Does Figure 1.14 show the variation in incoming solar radiation at the top of the atmosphere or at ground level?

(b) Explain very briefly whether you would expect the corresponding curve for high southern latitudes to be the same as that in Figure 1.14, a mirror image of it, or neither of these.

(c) At 50° N (the southern limit of the British Isles; cf. Question 1.1), what approximately is the difference between the solar radiation received during mid-summer and mid-winter, according to Figure 1.9? Bearing this in mind, do you think that changes in incoming solar radiation related to the Milankovich cycles (Figure 1.14) are likely to result directly in climatic changes detectable on human time-scales?

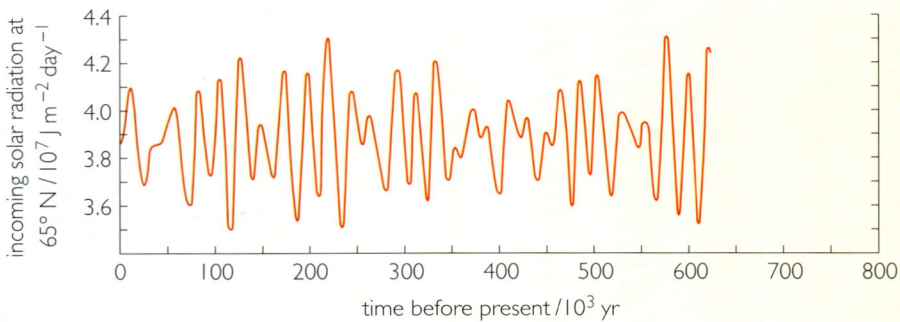

Figure 1.14 The variation in incoming solar radiation in summer at northern latitudes (65° N) over the past 600 000 years.

However, the complexity of the climate system and interactions between the different components – including feedback effects, both positive and negative – mean that it is difficult to predict the effect of a given change in incoming solar radiation. There is good geological evidence (of which more later) that Milankovich cycles have been important agents of climatic variation for at least the past 2.5 million years. Furthermore, it is possible that climatic changes influenced by, or even triggered by, changes in incoming solar radiation associated with the Milankovich cycles could occur over a fairly short time-span, that is, hundreds or even tens of years.

One way in which astronomical variations may affect the Earth's climate is through the influence they have on the growth and decay of the polar ice-caps and hence the amount of solar radiation reflected by the Earth. As mentioned earlier, the percentage of incoming solar radiation reflected from the Earth – its planetary albedo – is about 30%. Albedos for various surfaces are given in Table 1.1; note that it is not possible to provide precise values because a surface's reflectivity depends not only on what it is made of, its colour and its roughness, but also the angle of the incoming radiation, and the wavelength.

Table 1.1 Some typical albedos.

Type of surface	Albedo
clouds	average ~55% (~15–80%, depending on type and thickness)
fresh snow and sea-ice	80–90%
thawing snow	~45%
desert	35%
grassland	25–33%
forest, bare soil, rock, cities	~10–20%
water:	
moderate–high Sun (elevation > 40°)	< 5%
low Sun (elevation ~10°)	> 50%

An hospitable planet

■ Look at the Frontispiece and Figure 1.3. Ice cover at high latitudes, and the extensive areas of cloud, both reflect strongly, looking bright from space. Given all this, does a planetary albedo of 30% seem reasonable?

■ Obviously, it's not possible to make an accurate estimate on the basis of these two images, but the moderate albedo of most land surfaces and the low albedo (mostly) of the great areas of ocean (particularly in the Southern Hemisphere) must compensate for the effect of clouds and ice. Overall, a planetary albedo of 30% does seem reasonable.

So far, we have been discussing illumination and heating by solar radiation almost interchangeably. However, visible and thermal radiation occupy different parts of the Sun's spectrum, i.e. they have different frequencies and wavelengths. Because frequency and wavelength of radiation have very important climatic implications, you should now read Box 1.2 if you need to be reminded about the **electromagnetic spectrum** and thermal energy.

Box 1.2 The electromagnetic spectrum and thermal energy

The Sun emits electromagnetic radiation over a wide range of the electromagnetic spectrum, from gamma-rays to radio waves (see Figure 1.15), but ~99.9% of the energy is in the wavelength range 0.15–5 µm (see Figure 1.16), which includes ultraviolet radiation, visible radiation (violet to red), and infrared (heat) radiation. As with any propagating waves, the shorter the wavelength (λ), the higher the frequency (f), and the two multiplied together give the speed (c):

$$c = f\lambda$$

So the term 'ultraviolet' refers to radiation with a higher frequency than that of violet, and the term 'infrared' to radiation with a lower frequency than that of red.

It is often convenient to consider the Sun (and indeed the Earth) as a 'black body', i.e. a body that is radiating energy at the maximum possible rate for its temperature. If we make this assumption, we can say that:

energy radiated per unit area per unit time is proportional to T^4

where T is the absolute temperature in kelvins (K). This expression is known as the *Stefan–Boltzmann Law*, and it tells us (1) that *any*

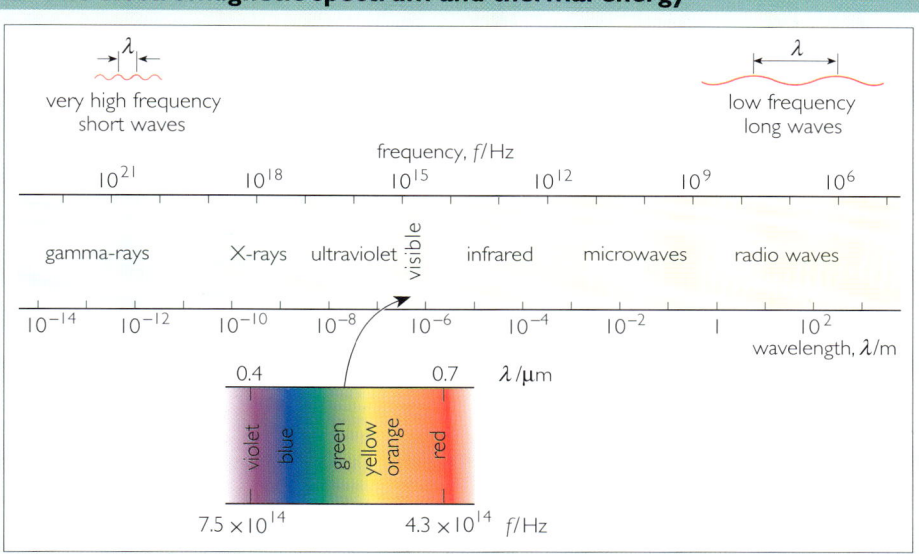

Figure 1.15
The electromagnetic spectrum. Wavelength is given in metres. For the expanded visible spectrum, wavelength is given in micrometres (µm) = 10^{-6} m; frequency is given in Hz (hertz = cycles s^{-1}).

body above absolute zero emits radiant energy, and (2) that the radiation energy emitted per unit area per unit time is proportional to the fourth power of the temperature in kelvins; remember that the kelvin scale starts at absolute zero, or −273 °C, so 1 °C is (1 + 273) = 274 K.

Figure 1.16 (in the Box continued overleaf) shows the spectral curves for the Sun and the Earth, assuming that they radiate like black bodies of 6000 K and 255 K, respectively.

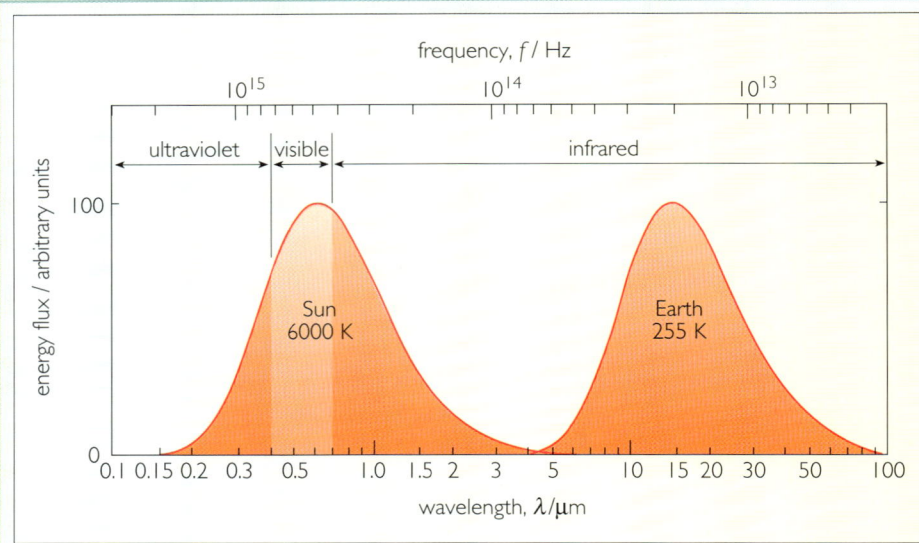

Figure 1.16
Spectral curves for radiation emitted by the Sun, assuming that it radiates as a black body of surface temperature 6000 K, and for the Earth, assuming that it radiates as a black body of surface temperature 255 K. This latter value (255 K = −18 °C) is the Earth's *effective* planetary temperature, consistent with the radiation it emits to space. As you will see, because the Earth has an atmosphere, the average temperature *at its surface* is about 288 K (= 15 °C). *Note*: Both horizontal scales are logarithmic. Wavelength and frequency are inversely related, so while the wavelength axis increases from left to right, frequency increases from right to left.

These curves demonstrate a general principle: the higher the temperature of a surface, the more the maximum in the spectrum of energy it radiates is shifted towards shorter wavelengths and higher frequencies. If you find this hard to imagine, think of a lump of coke (as it happens, a reasonable approximation to a black body) in a furnace. As it starts to get hot, the coke glows red; as it gets even hotter and hotter, shorter and shorter wavelengths are also emitted, until all of the visible part of the spectrum is being emitted and there is a 'white heat'.

■ To what extent do the ranges of wavelengths emitted by the Sun and by the Earth overlap? Is it reasonable to use 'short-wave' and 'long-wave' as a shorthand for solar radiation and thermal energy radiated by the Earth, respectively?

□ The curves overlap only very slightly. Incoming solar radiation is in the ultraviolet, visible and infrared (thermal) bands; outgoing radiation is all in the infrared, and there is a small overlap between the two curves at wavelengths of 4.0–4.5 μm. In the context of radiation incident on and emitted by the Earth, it is reasonable to think of solar radiation as short-wave and terrestrial radiation as long-wave.

Note that the process of radiation is completely different from reflection, in which energy is not absorbed and re-radiated, but 'bounced off' with its frequency and wavelength unaffected (think of light reflected by a mirror).

All molecules vibrate, and when a molecule absorbs electromagnetic energy, the amplitude of its vibration increases. Once 'excited' in this way, molecules can lose the energy again, either by re-emitting infrared radiation or by converting it into kinetic energy, by colliding with other molecules. The effect of such collisions is to raise the internal energy of the material, i.e. to raise its temperature. In the case of a solid or liquid, the vibrations of the excited atoms/molecules not only cause the surface to radiate long-wave radiation but the collisions of excited atoms (particularly electrons) with adjacent, less energetic, atoms results in the transfer of energy 'down the temperature gradient', by **conduction**.

Now let's look closer at the radiation balance of the Earth–atmosphere system as a whole. The solid curve in Figure 1.17 shows the average daily amount of solar energy absorbed by the Earth and atmosphere, as a function of latitude. Much of this incoming solar radiation is *short-wave* radiation; a proportion of this energy is then re-radiated, at longer wavelengths (see Box 1.1). Little of this long-wave radiation is radiated directly into space – most is absorbed in the atmosphere, particularly by carbon dioxide, water vapour and cloud droplets. The atmosphere is therefore heated from below and itself re-emits long-wave radiation into space, mostly from the top of the cloud cover. As temperatures at the top of the cloud cover do not vary much with latitude, neither does the intensity of long-wave radiation emitted to space: this can be seen from the dashed curve in Figure 1.17.

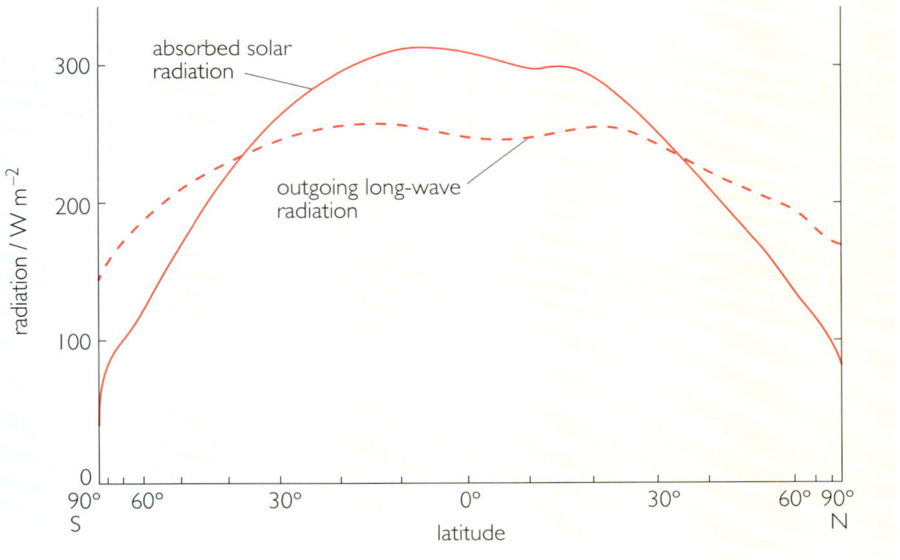

An hospitable planet

Figure 1.17
The variation with latitude of the solar radiation absorbed by the Earth–atmosphere system (solid curve) and the outgoing long-wave radiation lost to space (dashed curve). Values are averaged over the year, and are scaled according to the area of the Earth's surface in different latitude bands.

Question 1.2

(a) Looking first at the full curve for the solar radiation absorbed by the Earth, can you suggest why it has the general shape it does? (*Give two reasons.*) *Hint*: Refer to the Frontispiece and Figures 1.3 and 1.6, and Table 1.1.

(b) Over what latitudes does the Earth–atmosphere system have a net gain of heat, and over which does it have a net loss?

Note that together the two regions in Figure A1 marked 'net loss' are equivalent in area to the region marked 'net gain', demonstrating that – at least over short time-scales – the overall Earth–atmosphere radiation budget is in balance, so the Earth is neither cooling down nor heating up. Whether this is true over time-scales of decades is, of course, hotly debated.

Despite the positive radiation balance at low latitudes and the negative one at high latitudes, there is no evidence that low-latitude regions are steadily heating up while high-latitude regions are steadily cooling. The reason for this, of course, is the continual redistribution of heat over the globe by winds in the atmosphere and currents in the ocean – the subject of the next chapter.

It may have struck you that throughout we have been making an implicit assumption: namely that the intensity of radiation emitted by the Sun remains constant. Is this a valid assumption? The answer is no, it isn't. For one thing, according to theories of stellar evolution, the amount of radiation emitted by the Sun at the formation of the Solar System 4600 million years ago would have been only 70–75% of what it is now. On a much shorter time-scale, there are variations in solar activity and hence luminosity on an 11-year cycle (cf. Figure 1.18a). These variations have been directly observed only recently, by means of satellite-borne instruments (Figure 1.18b), but they have been occurring for hundreds of years, at least. We know this because one manifestation of high solar activity is a relatively large number of sunspots (cf. Figure 1.19a): sunspots are actually darker, cooler areas of the Sun's surface, but maxima in the number of sunspots are accompanied by maxima in the numbers of 'faculae' – bright, hot areas of the Sun – whose effect outweighs that of sunspots. These become visible when the Sun is photographed using its X-ray emission (Figure 1.19b).

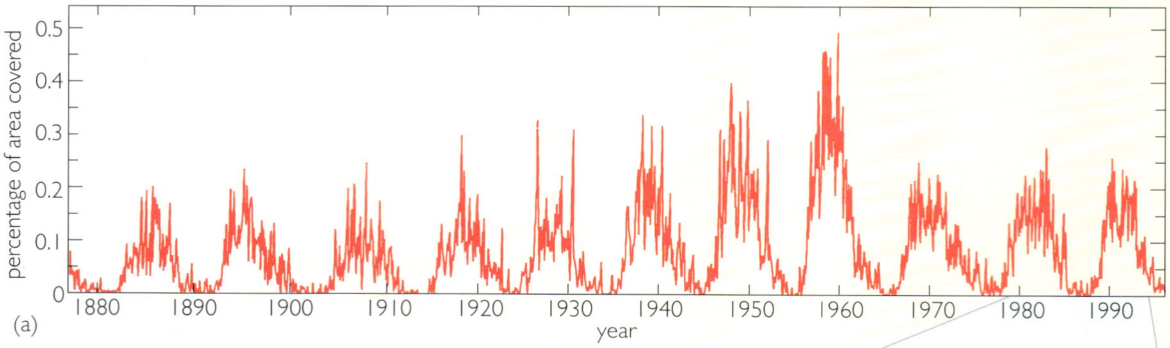

Figure 1.18
(a) The ~11-year solar activity (sunspot) cycle. The vertical scale gives the percentage of the Sun's surface area covered by sunspots. (b) The variation in solar flux at the top of the atmosphere, as measured by satellite-borne radiometer between 1979 and 1995. Note the rough correlation between the two traces, with minimum values around 1985–87 and maximum values ~5 years either side. (In this diagram, we are showing variation in the total solar flux, not the effective flux at the Earth's surface; cf. Figure 1.5.)

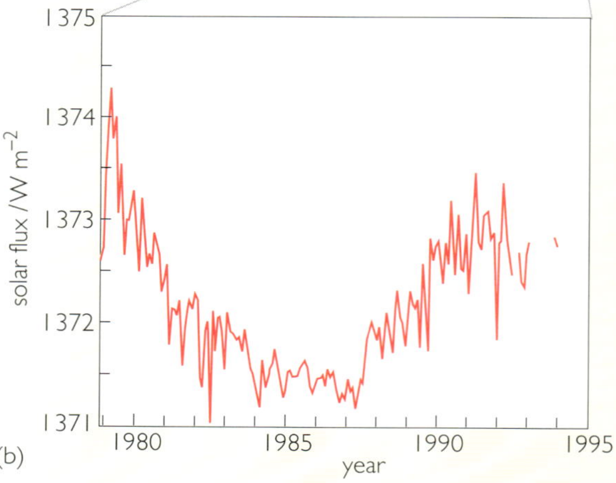

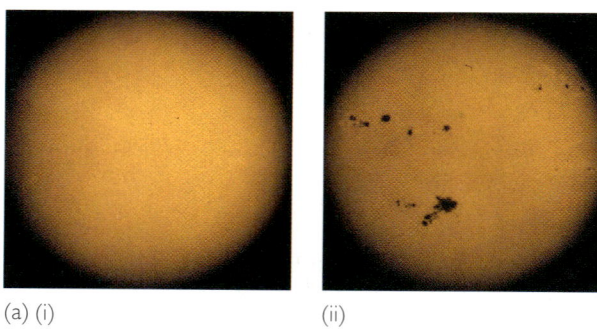

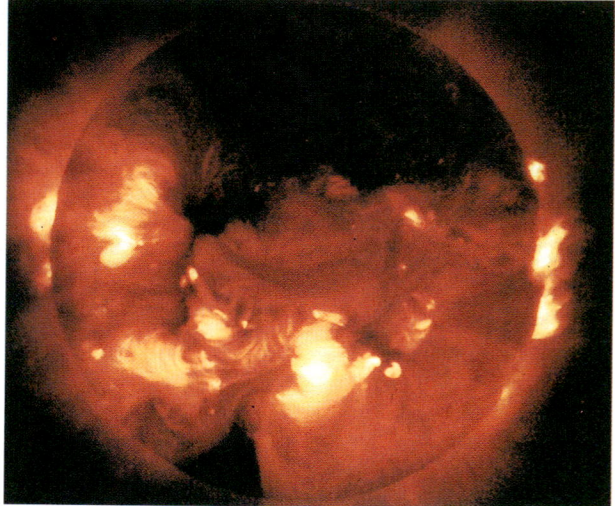

Figure 1.19
(a) The Sun (i) near a time of minimum solar activity and (ii) near a time of maximum activity, both photographed using the visible spectrum. (b) The Sun in an active period, photographed using X-rays which are generated by energetic activity and high temperatures in the Sun's atmosphere.

■ According to Figure 1.18b, by approximately how much did the solar flux vary between 1979 (near the peak of a sunspot cycle) and the minimum value in mid-1983? (Express your answer as a percentage of the 1983 flux.)

■ The peak flux in 1979 was ~1374.3 W m^{-2} and the minimum flux value in 1983 was just over 1371 W m^{-2}. The difference between these two is ~3.3 W m^{-2}, which is $\frac{3.3}{1371} \times 100\% \approx 0.2\%$ of the 1983 flux.

The changes in the amount of solar radiation reaching the Earth *within* an 11-year cycle are therefore relatively small; furthermore, the fluctuations occur on such a short time-scale that it seems unlikely that the components of the climate system that respond on long time-scales – the oceans or the polar ice-caps, for example – could be affected by them.

However, Figure 1.18a shows clearly that the individual cycles vary slightly one to another, having different numbers of sunspots and being of very slightly different lengths. It seems that 'waxing and waning' of sunspot peaks like those in Figure 1.18a occurs with a periodicity of 80–90 years; in other words, the high sunspot peaks of the middle decades of the 20th century mark the maximum of a cycle that began around 1905. Assuming that the approximate correlation between sunspot numbers and solar flux holds good *between* cycles as well as *within* them (cf. Figure 1.18), solar flux must on average be greater during 'more active' cycles than during 'less active' cycles.

Changes in incoming solar radiation such as those shown in Figure 1.18b, or any other factor that has the potential to cause changes in the Earth's climate system, is known as a **forcing function**. The term comes from computer modelling studies of the climate, particularly climate change, and 'function' here means the mathematical expression of the 'forcing' or driving mechanism. In general, you can think of a forcing function as something that causes a system to respond in a way other than it would do if left alone, forcing it to change; for researchers working on computer models of the climate system, particularly in the context of global warming, 'forcing function' often means something that alters the Earth's radiation balance, either locally or globally (sometimes referred to as 'radiative forcing').

Climate modellers describe the size of the response of the climate system to a forcing of given magnitude as the climate's **sensitivity**. An analogy can be drawn with human diet. If we maintain a steady regime of exercise and eating, our weight stabilizes at a certain figure, characteristic of our own metabolism. If we change our eating pattern, and consistently eat 10% more, our weight will increase until it stabilizes at a different, higher, level. This will almost certainly not be 10% more than our initial body weight, nor will different individuals show the same change: we each have a different sensitivity to the same forcing.

Unfortunately, the Earth's sensitivity to climate forcing is hard to measure, and (as implied in our discussion of the effect of Milankovich cycles) hard to predict. Theoretical modelling suggests that an increase in the average amount of incoming solar radiation of 4.0–4.5 W m^{-2} could cause a change in temperature of anything from 1.5 to 5.5 °C, which means that the climate's sensitivity is of the order of 0.5–1.0 °C for 1 W m^{-2} of radiative forcing (say 0.75 °C per W m^{-2} of forcing).

Notice the uncertainty in the estimate. One reason for this is the time it takes for the Earth to respond to a change: as you will see, its systems have huge inertia. If the Sun were to be completely extinguished for a few minutes (as it is locally during eclipses), the effects would be negligible. But if the solar flux were to decrease abruptly and permanently by 0.1%, this change would ultimately show up in small changes in all aspects of the climate system – the temperature distribution within the atmosphere and ocean, the extents of polar ice-caps and of tropical rainforests, and so on. Each of these changes would have its own feedback effects – so it would take a long time, perhaps a century, before the climate system settled into a new equilibrium. Thus, in trying to assess the sensitivity of the Earth's climate to changes in forcing, we have to take into account all of the various feedback processes that may come into play, and their characteristic time-scales. An important implication of this is that the climate system may not have the same sensitivity to different types of climatic forcing.

> **Question 1.3**
> During the 'Little Ice Age', which altogether lasted from the 15th century until well into the 19th century, different parts of the Earth experienced unusually cold decades: icebergs became common off Norway, glaciers advanced down valleys and the Thames froze in winter. It has been suggested that these cold decades were caused by a prolonged period of low solar activity, known as the 'Maunder minimum', which was most marked from 1640 to 1720. Given that average global temperatures were about 1 °C cooler than today, by how much (as a percentage) would incoming solar radiation (i.e. the effective solar flux) have to have been below its present value (~343 W m^{-2})?
>
> *Note*: Assume a climate sensitivity of 0.75 °C per W m^{-2} of solar forcing; both theoretical modelling (see above) and studies of the Earth's glacial history indicate that this is a reasonable average value.

In fact, various other factors have been proposed as causing (or contributing to) the Little Ice Age, including natural changes in concentrations of greenhouse gases in the atmosphere and changes in patterns of ocean circulation. Such influences, which we will be looking at more closely later, are sometimes referred to as 'internal forcing functions'.

- What other forcing functions have been discussed or alluded to so far?

- The Milankovich cycles (sometimes described as 'astronomical forcing' or 'orbital forcing') and variations in solar irradiance, not only over very long time-scales but also over ~11-year and ~80–90-year time-scales.

Both solar forcing and astronomical forcing act on the Earth from outside, and so are known as 'external forcing functions'. Examples of other forcing functions will be discussed later, but first we will take a look at how, together, winds and currents form the Earth's air-conditioning cum central heating/cooling system, which in some cases moderates, and in others mediates, the effects of the various forcing factors.

1.3 Summary of Chapter 1

1   In the final analysis, the Earth is hospitable to life because of its particular orbit around the Sun, which determines the amount of solar radiation that reaches it. The amount of solar radiation actually available to warm the Earth's surface is determined by how much is reflected rather than absorbed, i.e. by its albedo. At the temperatures that obtain at the Earth's surface, water can exist as solid, liquid or gas: were it not for the presence of liquid water, life could not exist on Earth.

2   The fixation of carbon by primary producers (mainly plants) is the basis of all life on Earth.

3   Seasonal changes in incoming solar radiation are a result of the tilt of the Earth's axis in relation to the orbital plane (i.e. if the Earth's axis were at right-angles to the orbital plane, there would be no seasons). Over time-scales of tens of thousands to hundreds of thousands of years, there are cyclical variations in incoming solar radiation caused by changes in the degree of eccentricity of the Earth's orbit, in the *angle* of tilt of the Earth's axis, and in the *direction* in which the axis points. These cycles have periodicities of ~110 000 years, ~40 000 years and ~22 000 years, respectively, and are known as the Milankovich cycles or the Milankovich–Croll cycles.

4   Most of the radiation energy emitted by the Sun is in the wavelength range 0.15–5 µm, which includes ultraviolet radiation, visible radiation and infrared radiation. Solar radiation is often referred to as short-wave radiation, to distinguish it from longer-wavelength (thermal) radiation emitted by the Earth (and any other body) as a consequence of having been warmed.

5   The luminosity of the Sun itself varies with time. It is thought that the amount of radiation emitted by the early Sun was only 70–75% of what it is now. There is evidence for variation in solar luminosity over an ~11-year cycle and over an ~80–90-year cycle (and there may well be longer-term variations, so far undetected).

6   Although the Earth's surface is heated unevenly by the Sun, the redistribution of heat by winds and currents ensures that low latitudes do not continually heat up and high latitudes continually cool down.

Now try the following questions to consolidate your understanding of this chapter.

Question 1.4
Mars is an arid planet where strong winds sometimes whip up dust clouds, but clouds of condensed water (as ice because of the low pressure) are much less common than on Earth. Venus is totally covered in a thick layer of cloud (which, incidentally, seems to be ~75% sulfuric acid and ~25% water). How would you expect the planetary albedos of Mars and Venus to compare with that of the Earth?

Question 1.5
The Tropics of Cancer and Capricorn are not shown on Figure 1.7. Why is this?

Question 1.6
(a) Croll's Frontispiece to *Climate and Time* (Figure 1.12) shows a way in which the interaction of two of the three astronomical cycles might possibly lead to glaciations in one or other hemisphere. (i) Which two cycles are involved? (ii) Can you suggest what Croll's argument might have been?

(b) One of the two diagrams in Figure 1.12 is intended to represent the present-day situation. Which is it? And how far apart in time are the two orbital configurations?

(c) To what extent does the situation illustrated for the present day actually correspond to that seen on the Earth now?

Chapter 2
Keeping the Earth habitable

As discussed in Chapter 1, the Earth is hospitable to life because temperatures at its surface are moderate (Figure 1.2) and allow the existence of large amounts of liquid water. The distribution of temperature and freshwater supply over the surface of the globe are influenced by winds and current flow, leading to various climatic conditions which in turn determine the distribution of primary production (Figure 1.4) and the types of species that thrive. This is not the whole story, however, because survival and growth of living organisms are also intimately linked to the chemical compositions of the atmosphere, ocean and solid Earth. Nevertheless, let's begin our discussion of the present-day climate with the geographical variation of surface temperature.

So far, we have tended to think of the Earth simply as a spherical planetary body, with no features on its surface except polar ice-caps. If the Earth were really like that, we would expect its surface temperature to decrease smoothly poleward, away from a zone of maximum temperature, the latitude of which would seasonally shift northwards or southwards, depending on which hemisphere was experiencing summer (cf. Figure 1.9). Not surprisingly, for the real Earth – with its oceans and continents, forests and deserts – things are more complicated.

2.1 The distribution of temperature over the Earth's surface

Parts (a) and (b) of Figure 2.1 show daytime surface temperatures for January and July, as measured by satellite-borne radiometer (which measures thermal radiation emitted by the Earth), and part (c) shows differences in surface temperature between January and July. In Figure 2.1 (a and b), surface temperatures generally decrease from low to high latitudes, but the **isotherms** – contours of equal temperature – do not run simply east–west.

- With what are the most extreme departures from a simple east–west trend associated?

- With the distribution of continents and oceans. In the hemisphere experiencing summer, at a given latitude continental temperatures are generally *higher* than sea-surface temperatures; in the hemisphere experiencing winter, continental temperatures are generally *lower* than sea-surface temperatures.

This point is even more forcefully borne out in Figure 2.1c. The large areas of red and brown in the Northern Hemisphere and the great areas of blue and green in the Southern Hemisphere show that the greatest warming and cooling occurs over the continents. Seasonal changes of up to 30 °C occur over land in both hemispheres; by contrast, seasonal changes in sea-surface temperature, which are greatest in mid-latitudes, rarely exceed 8–10 °C.

- Look at Box 2.1. Which of the properties of water listed explains the contrast in seasonal temperature changes on land and sea?

Figure 2.1 (opposite) Daytime surface temperatures as measured by satellite-borne radiometer. For the two upper pictures, temperatures below 0 °C are green and blue; the highest temperatures are shown by red and dark brown. Contours of equal temperature – in this case unlabelled – are known as isotherms. The distributions of sea-surface temperature shown are discussed in the text.
(a) In January, temperatures at high northern latitudes are very low, having fallen below zero in eastern Europe and the northern USA, and approaching −30 °C over Siberia and most of Canada. In the Southern Hemisphere it is summer, with mid-latitude temperatures of 20–30 °C.
(b) By July, areas of the Northern Hemisphere have warmed by 10–20 °C. The Greenland ice-cap remains frozen, but temperatures are considerably lower in the Antarctic where there is now a large area of sea-ice.
(c) Temperature differences between January and July. Areas of greatest increase in temperature are red and dark brown; areas of greatest decrease in temperature are bright blue and dark blue. The greatest changes are over land in mid-latitudes; low latitudes are more stable.

Keeping the Earth habitable

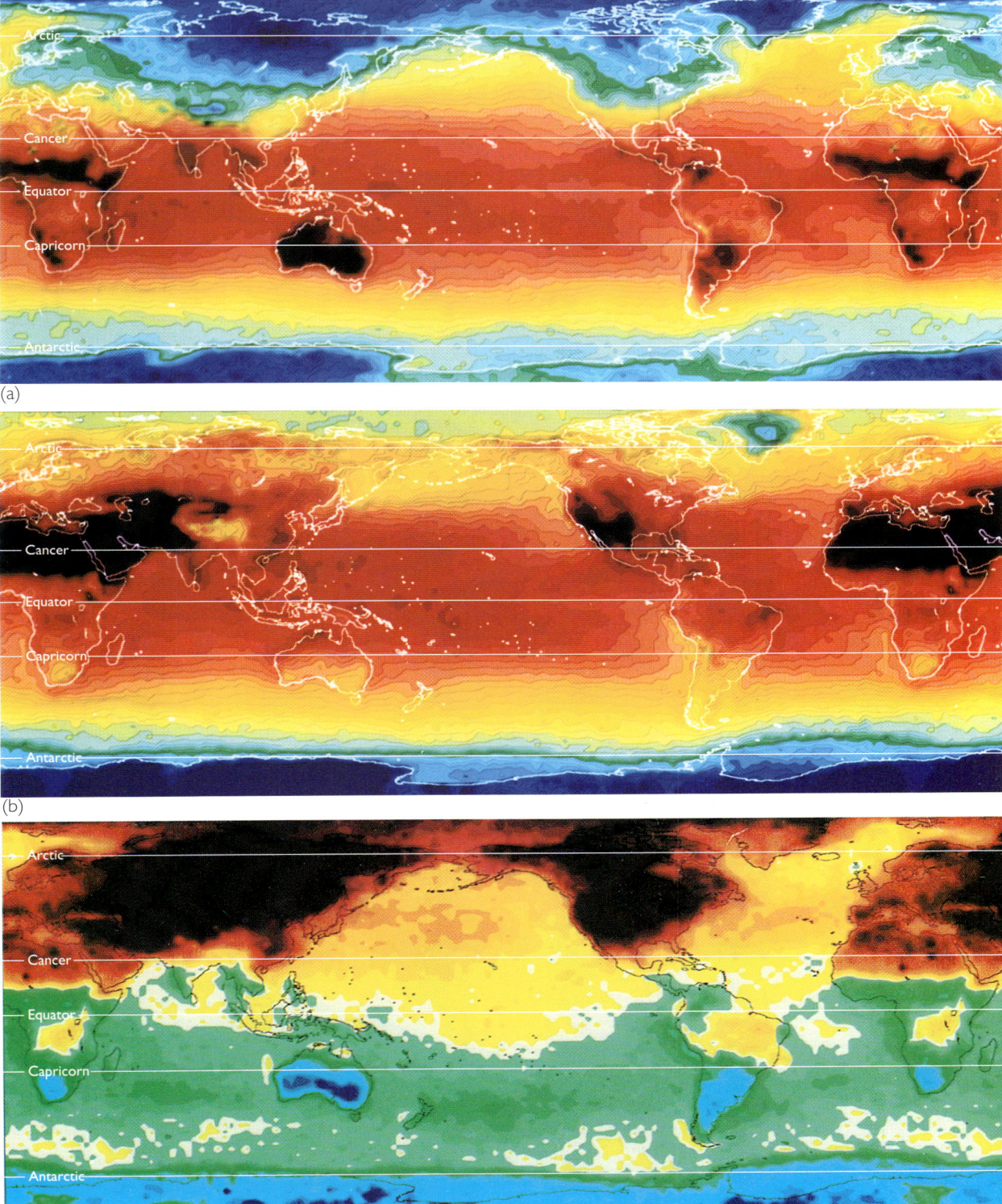

(a)

(b)

(c)

> **Box 2.1 Properties of water of importance for climate**
>
> The relevance of some of these properties of water to the Earth's climate may already be obvious; the importance of others will become clear as you read further.
>
> - At the range of temperatures found at the surface of the Earth, water can exist as a gas (water vapour), as a liquid, and as a solid (ice).
> - Its specific heat, i.e. the amount of heat needed to raise the temperature of 1 kg by 1 °C (or 1 K) is $4.18 \times 10^3 \, \text{J kg}^{-1} \, °C^{-1}$. This is the highest specific heat of all solids and liquids except ammonia.
> - Its latent heat of fusion, i.e. the amount of heat needed to convert 1 kg of ice to water at the same temperature (and the amount of heat *given up* to the surrounding environment when 1 kg of ice forms) is $3.3 \times 10^5 \, \text{J kg}^{-1}$. This is the highest latent heat of fusion (or freezing) of all solids and liquids except ammonia.
> - Its latent heat of evaporation, i.e. the amount of heat needed to convert 1 kg of liquid water to water vapour at the same temperature (and the amount of heat *released* to the surrounding environment when 1 kg of water vapour condenses) is $2.25 \times 10^6 \, \text{J kg}^{-1}$. This is the highest latent heat of evaporation (condensation) of all substances.
> - It dissolves more substances, and in greater quantities, than any other liquid.
> - It conducts heat more efficiently than most other liquids naturally occurring on the Earth.
> - Its temperature of maximum density decreases with increasing salt content; pure water has its maximum density at 4 °C, but the density of seawater increases right down to the freezing point at about −1.9 °C.
> - The density of ice is less than that of water, with the result that: (1) ice occupies more space than the water from which it formed, and (2) ice floats on water. (For most substances, the solid phase is denser than the liquid phase.)
> - Compared with other liquids, it is relatively transparent.

■ It is the high specific heat of water, which means that the thermal capacity of the oceans is much greater than that of the continents. A much greater heat input is needed to raise the temperature of a mass of ocean by 1 °C than is needed to raise the temperature of the same mass of continental rock by 1 °C. As a result, continental areas heat up and cool down more quickly than oceanic areas.

The property that the oceans have of heating up and cooling down slowly is sometimes referred to as their 'thermal inertia'. It is part of the reason why the range of temperature found in the oceans is less than half that which occurs on land (see Figure 1.2). The temperature contrasts between continents and oceans is one of the reasons why the distribution of temperature over the surface of the Earth is not simply zonal*, in latitudinal bands. In the next section, we will look at the other reasons.

2.2 The Earth's air-conditioning and heating systems

In Chapter 1, we saw that the Earth's radiation budget has an excess at low latitudes and a deficit at high latitudes (Figure 1.17). Figure 2.2 shows, very schematically, how heat is redistributed over the surface of the Earth. Put simply, the three principal processes involved are:

1. Air moving over the surface of the oceans and continents takes up heat from them. Warm air rising at low pressure regions like the Equator, and moving polewards in the upper troposphere, transports heat from low to high latitudes, as does any warm air moving polewards. Air moving equatorward, which has been cooled at high latitudes by contact with ice and the cold surface of land and sea, also contributes to the redistribution of heat.

---

* The corresponding term for north–south is *meridional*.

Keeping the Earth habitable

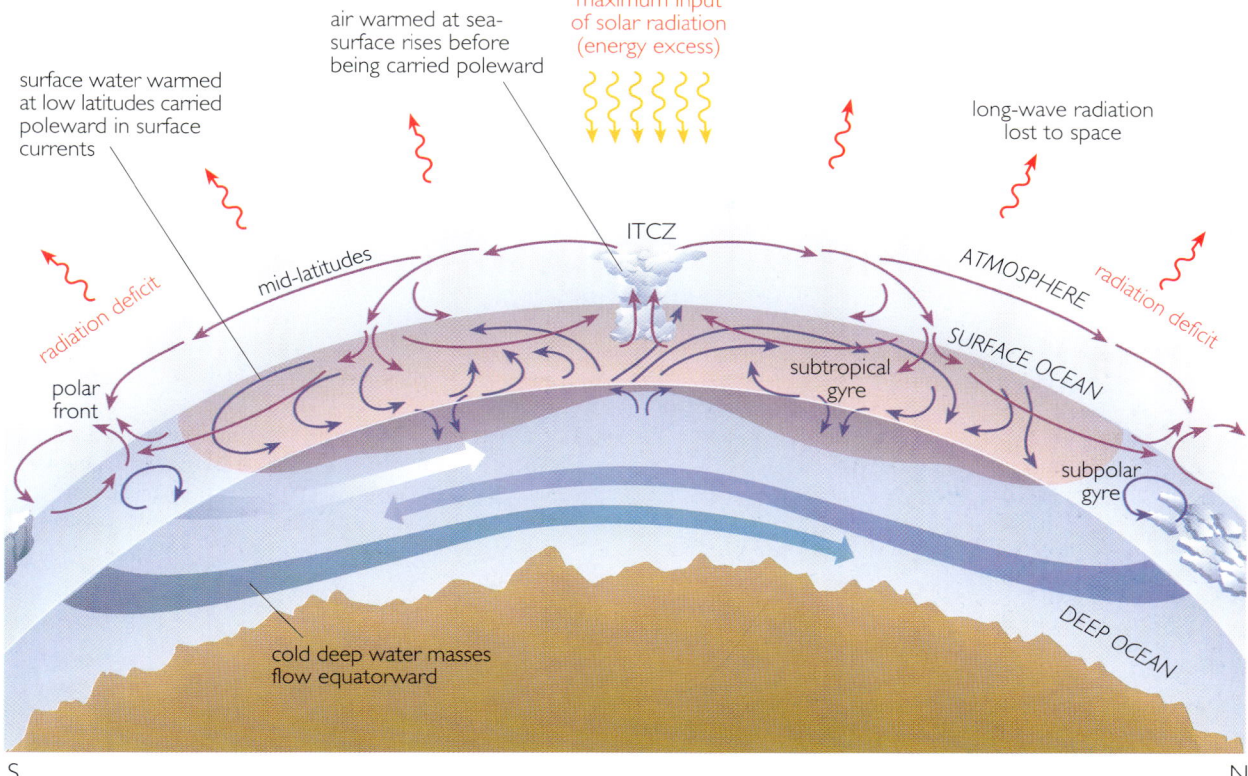

Figure 2.2 Schematic diagram of the Earth's heat-redistribution system (not to scale). The three interlinked circulatory systems are: winds in the atmosphere; wind-driven surface currents; and density-driven currents in the deep ocean. ITCZ = Intertropical Convergence Zone. (See text for more details.)

2   Under the influence of winds, surface ocean water warmed at low latitudes flows polewards in surface currents, while that cooled at high latitudes flows equatorwards.

3   Surface ocean water cooled at high latitudes increases in density, sinks and flows equatorwards as deep currents.

Another way of looking at processes 1–3 above is as interrelated circulatory systems, largely driven by **convection**. We are all familiar with what happens when a pan of water is heated on an electric ring. Heat is transmitted from the element, through the pan to the water at the bottom of the pan by conduction (cf. Box 1.2). Being heated, this water expands, becomes less dense and rises, to be replaced by cooler, denser water sinking from above. On reaching the surface, the warmed water begins to lose heat to the air; it cools, becomes denser and sinks, then is heated again and rises, and so on (Figure 2.3), so forming convection 'cells'. (You can sometimes see such cells when cooking spaghetti, because the spaghetti strands tend to become aligned with the flow.)

Figure 2.3 The circulatory pattern in a pan of water heated on an electric ring.

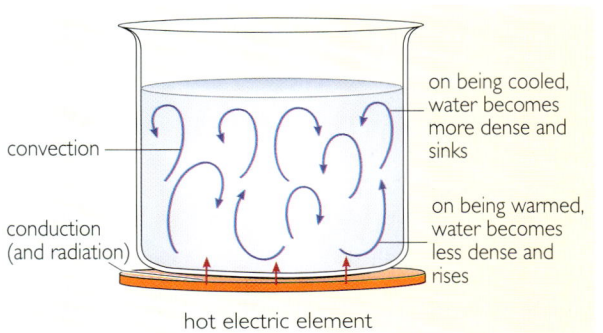

- Does the convection occurring (i) in the atmosphere, and (ii) in the ocean (see Figure 2.2) resemble that occurring in a pan of water being heated (Figure 2.3)?

- Convection occurring in the atmosphere does resemble that occurring in a pan of water, because it is driven by heating from below. However, that occurring within the ocean is different *because it is driven by cooling from above.*

2.2.1 Transport of heat and water by the atmosphere

We will return to the ocean later, but first we need to look more closely at *how* the atmosphere is warmed. We all know that 'warm air rises' or, to be more precise, that 'air that is warmer than its surroundings (and is therefore less dense) rises'. As the analogy with the pan of water demonstrates, it is the convective 'bulk mixing' of water that distributes the heat supplied at the bottom of the pan so that eventually all of the water becomes warm. So it is for the atmosphere: when air is warmed by contact with a warm sea or land surface and rises, it is replaced by cooler air which is warmed in turn. By contrast, the transfer of energy by conduction occurs at the atomic level (see Box 1.2) – if we had to rely on conduction to heat a pan of water, we would have to wait a very long time indeed.

What happens in practice is complicated by two things. The first is that air, like all fluids, is compressible; the second is that it contains variable amounts of water vapour. First compression: when a fluid is compressed, the internal energy that it possesses per unit volume by virtue of the motions of its constituent atoms, and which determines its temperature, is increased. Thus, a fluid heats up when compressed (a well-known example is the compression of air in a bicycle pump), and cools – i.e. undergoes a decrease in energy per unit volume – when it expands (this is what occurs in the cooling system of a refrigerator). Changes in temperature that occur in this way, and not as a result of a gain or loss of heat from the surroundings, are described as **adiabatic**. When warmed air rises, the atmospheric pressure it is subjected to decreases (see black curve in Figure 2.4), and so it expands and becomes less dense; however, because it is expanding, it is undergoing adiabatic cooling. Whether the air continues to rise depends on the relative sizes of these two effects.

Imagine a parcel of air heated by contact with the ground and beginning to move upwards in random, turbulent eddies. Temperature decreases with height in the lower atmosphere (see red curve in Figure 2.4), but as long as the adiabatic decrease in temperature of a rising parcel of air is *less* than the decrease of temperature with height in the lower atmosphere, the rising parcel of air will be warmer and less dense than its surroundings and will continue to rise: the situation will be unstable and conducive to convection. On the other hand, if the adiabatic cooling of the rising parcel of air is sufficient to reduce its temperature to below that of the surrounding air, it will sink back to its original level, and convection will be inhibited.

So far, we have been assuming that the rising parcel of air is dry. Rising air, particularly over the ocean, may be saturated with water vapour or become saturated as a result of adiabatic cooling (cool air being able to contain less water vapour than warm air). Continued rise and associated adiabatic cooling result in cloud formation – the condensation of water vapour onto 'nuclei' such as dust particles, to produce tiny water droplets or, at higher levels in the atmosphere, ice crystals. This condensation releases latent heat to the rising air (cf. Box 2.1),

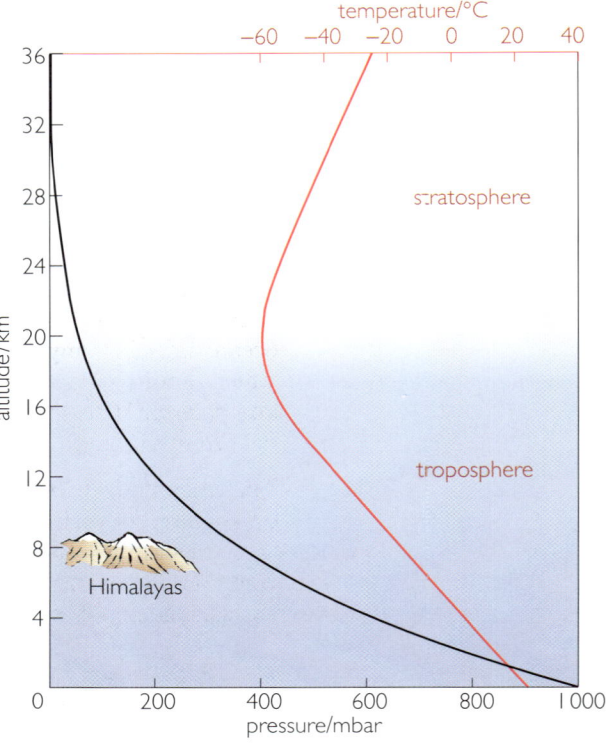

Figure 2.4
Schematic diagram to show how both pressure (black curve) and temperature (red curve) decrease with increasing height in the lowermost atmosphere, or troposphere. Above that, in the stratosphere, temperature increases again (this is discussed later). The curves are generalized, and intended to illustrate the general principle only. The thickness of the troposphere and actual values of temperature and pressure near the Earth's surface vary with latitude and location; the temperature curve shown is appropriate for a latitude of about 30°. One mbar is a thousandth of a bar, where 1 bar is 1 atm (atmosphere).

offsetting the effect of adiabatic cooling. In other words, humid air convects much more easily than dry air because condensation of water vapour releases additional heat energy, keeping the rising air less dense than the surrounding air for longer than would be the case for dry air.

Rising air warmed locally by conduction/convection becomes part of the global scale atmospheric circulatory system shown schematically in Figure 2.2 (and Figure 2.7), whose horizontal flows over the surface of the Earth are the surface winds. Wind systems redistribute heat partly by the **advection** (bulk transport) of warm air masses into cooler regions (and vice versa), and partly by the transfer of latent heat bound up in water vapour, which is released when the water vapour condenses to form cloud in a cooler environment, perhaps thousands of kilometres from the site of evaporation. Most of this moisture comes from the surface of the ocean; indeed, at any one time a large proportion of the water in the atmosphere – water vapour and clouds – has only recently evaporated from the tropical ocean, and poleward transport of warm humid air is the most important way in which heat from the ocean at low latitudes is transferred to higher latitudes.

The amount of evaporation from land depends on the moisture content of the exposed soil or rock. Vegetation is also a source of atmospheric moisture, both through simple evaporation from surfaces and through *transpiration*, whereby water drawn up from the soil by roots is lost to the atmosphere through pores in leaves; together, these two processes are known as **evapotranspiration**.

- From which types of land areas would you expect transfer of latent heat to the atmosphere to be greatest?
- From tropical rainforests, where evapotranspiration releases large amounts of water vapour to the atmosphere.

Indeed, being both warm and a good source of atmospheric moisture, rainforests behave climatically rather like the tropical ocean.

As mentioned earlier, heat is also transferred from the Earth's surface to the atmosphere by simple conduction and convection. This is referred to as transfer of **sensible heat** (heat which can be felt, or sensed) and it increases as the temperature difference between the Earth's surface and the overlying atmosphere increases.

- By reference to Figure 2.1a–c, would you expect loss of sensible heat to the atmosphere to be greatest from land surfaces or from the ocean?
- From land surfaces, especially at mid- and low latitudes in summer when the landmasses have heated up.

At low latitudes, sensible heat loss from land surfaces to the atmosphere is an order of magnitude greater than that by evaporation. It is true that there is also a loss of sensible heat from the sea to the air, if the sea-surface is warmer than the air above it (which is the case more often than not), but as far as the ocean is concerned, *an order of magnitude more* heat is transferred to the atmosphere via evaporation and subsequent condensation than is transferred by conduction, even when enhanced by convection.

Cumulus and cumulonimbus clouds, over both land and sea, are visible evidence of convection in the atmosphere (see Box 2.2). Cloud formation, the turbulence of atmospheric convection and the winds which redistribute heat over the Earth's

Box 2.2 Clouds

Clouds form when water vapour in the atmosphere condenses around solid particles or 'nuclei' (e.g. pollen grains, dust or salt from sea spray) to form tiny water droplets, or at greater heights and lower temperatures, ice crystals. Condensation can also occur on aerosols, minute droplets, often of sulfate compounds, notably sulfuric acid formed from water vapour and sulfur dioxide (SO_2) emitted from volcanoes or produced by industrial processes.

As pressure decreases with height in the atmosphere (black curve in Figure 2.4), water droplets can exist in clouds at temperatures down to −12 °C; between −12 and −30 °C, there can be a mixture of water droplets and ice crystals; at temperatures less than −30 °C, ice crystals predominate, and at temperatures below −40 °C, clouds consist entirely of ice crystals.

As shown in Figure 2.5, clouds can have a variety of forms: for example, cumulus and cumulonimbus, being associated with convection, have flat bottoms (marking the height at which condensation began to occur) and bubbly tops; stratus clouds, formed by more gentle uplift of air at a front or as a result of the presence of mountains, are layered. Wispy cirrus-type clouds are made of ice and are usually associated with flow in jet streams; cumulus-type clouds consist of water droplets, though the tops of towering cumulonimbus clouds consist of ice crystals. Stratus-type clouds – which at ground level are what we refer to as fog – can be made of either.

All clouds are highly reflective, being white in colour when illuminated (they only appear grey if their undersides are in shadow). However, the veil-like structure of cirrus clouds means that they have a fairly low albedo (<20%), while stratocumulus and cumulonimbus, which are thicker and consist of more densely packed droplets and/or ice crystals, have fairly high albedos of about 50% and 70%, respectively.

The role of clouds in the climate system is still unclear. Because of their high albedo, they may greatly reduce the amount of solar radiation reaching the Earth's surface, particularly at high and low latitudes (see discussion in connection with Table 1.1). However, as you will see, they absorb radiation as well as reflect it. Last but not least, their complex three-dimensional shapes make their effect on the radiation budget much more difficult to quantify than, say, land ice. We will come back to the role of clouds in the climate system later in this chapter.

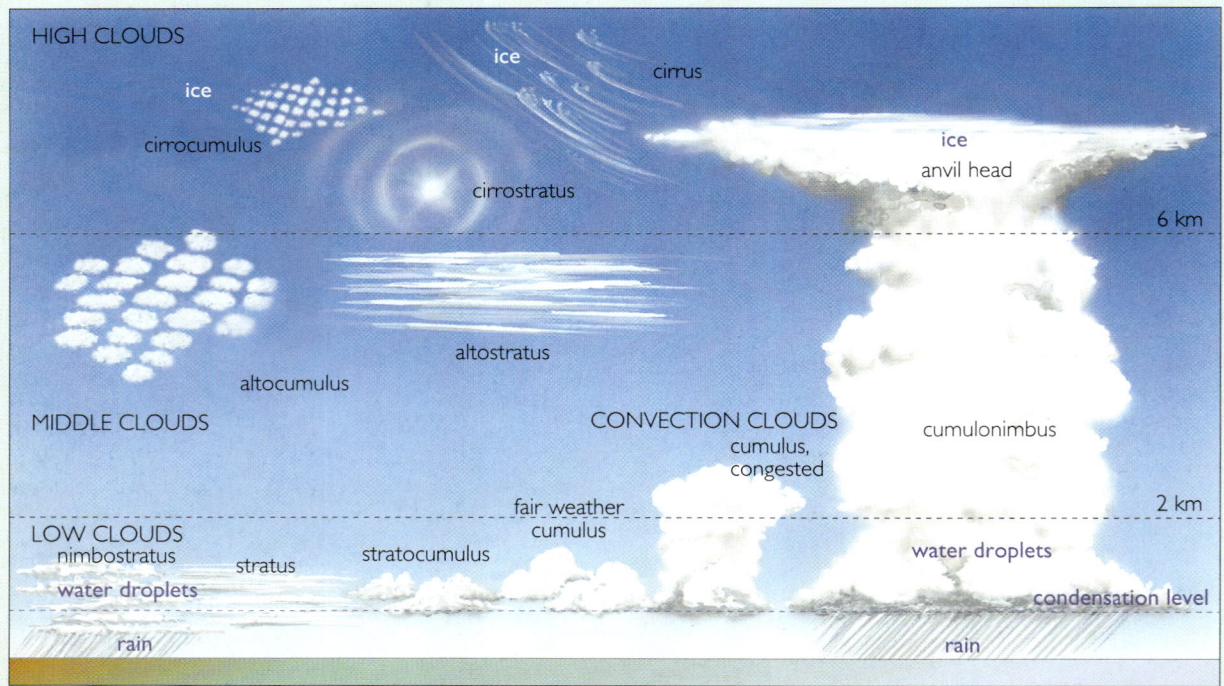

Figure 2.5
Various types of clouds and their characteristic levels and/or extents in the atmosphere.

Keeping the Earth habitable

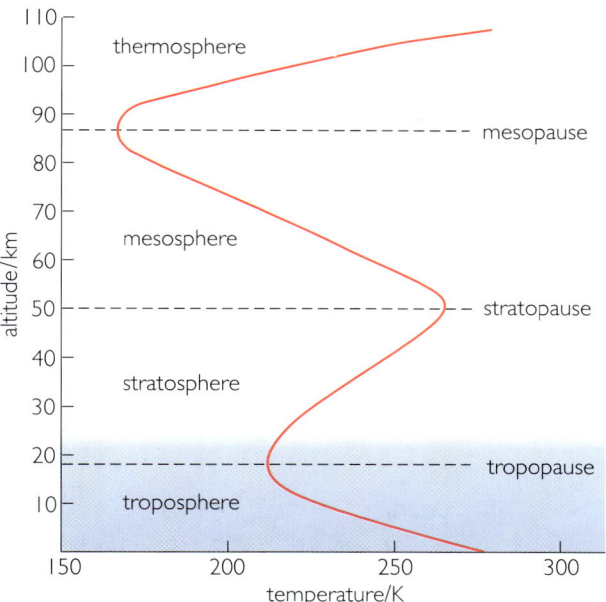

Figure 2.6
The vertical temperature 'structure' of the atmosphere, defined by the way that temperature varies with altitude. In each successive 'sphere', the temperature gradient is reversed: temperature decreases with height in the troposphere, so that conditions are conducive to convection; by contrast, temperature increases with height in the stratosphere, and vertical motions are suppressed. Note that, as indicated by the vertical air movements in Figure 2.7 (below), the tropopause is higher at low latitudes (where it is at ~ 17 km) than at high latitudes (where it is at ~ 8–10 km).

surface, are mostly confined to the lower atmosphere. This is known as the **troposphere**, after the Greek word *tropos* meaning 'turn' (causing mixing), and makes up ~ 80% of the total mass of the atmosphere. Temperature decreases with height within the troposphere (cf. Figure 2.4) up to its upper 'boundary' – the *tropopause* – then begins to increase again in the overlying stratosphere because of absorption of radiation in the ozone layer. It is this temperature *inversion* that generally prevents convection in the lower atmosphere from reaching any higher. There is some interchange of air between the troposphere and stratosphere, particularly in mid-latitudes, and in certain circumstances rapidly rising air masses can overshoot the tropopause. In general, though, stratospheric winds do not interact strongly with the winds of the lower atmosphere and so do not directly affect conditions at the Earth's surface. It is within the troposphere that the Earth's weather occurs.

Figure 2.6 shows how temperature varies with height in the atmosphere as a whole (cf. Figure 2.4 for the troposphere), and you can see that above the troposphere and the stratosphere there are two more changes in temperature gradient within two more 'spheres'. However, it is generally thought that these outer 'spheres' are not important influences on the Earth's climate system. Part of the reason for this is that the atmosphere is so rarefied at these altitudes that the heat content of the air is very low, even when the temperature (determined by the vibrations of individual molecules, Box 1.2) is high.

Figure 2.7 shows what the wind system and surface atmospheric pressure pattern would be if the Earth were completely covered by ocean.

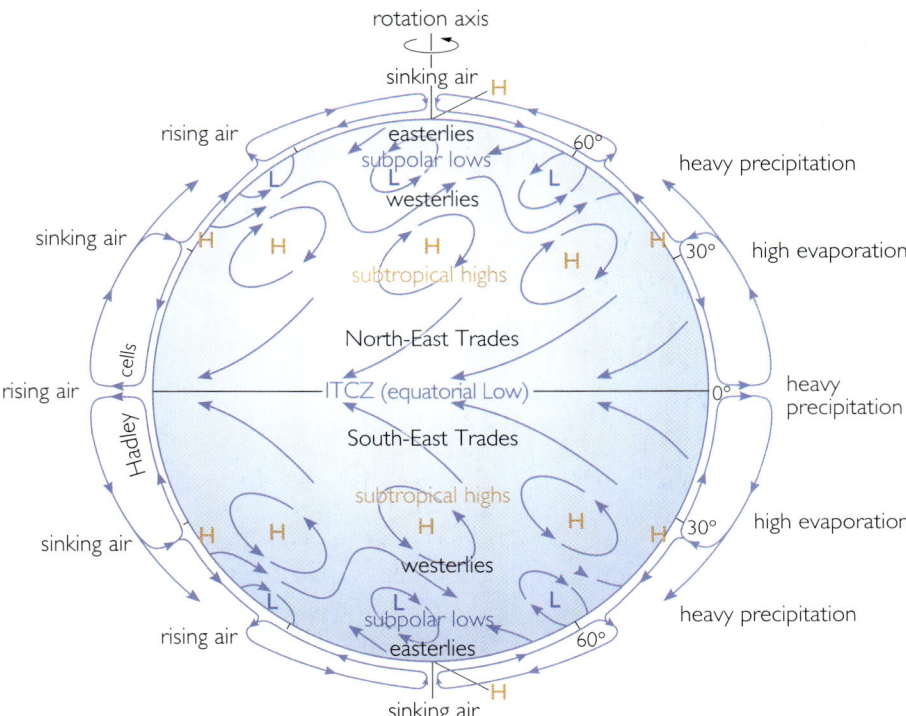

Figure 2.7
Wind system for a hypothetical water-covered Earth, showing the major surface winds and the zones of low and high pressure. Vertical air movements are indicated on the left-hand side of the diagram; characteristic surface conditions are given on the right-hand side.

Question 2.1

(a) By reference to Figure 2.7, describe how *surface* wind directions relate to regions of high and low surface pressure.

(b) By reference to the vertical air motions shown on Figure 2.7, describe how vertical air movements relate to regions of high and low surface pressure.

So, air sinks and flows anticyclonically *outwards* from regions of high surface pressure, and flows cyclonically *inwards* and rises at regions of low surface pressure (Figure 2.8).

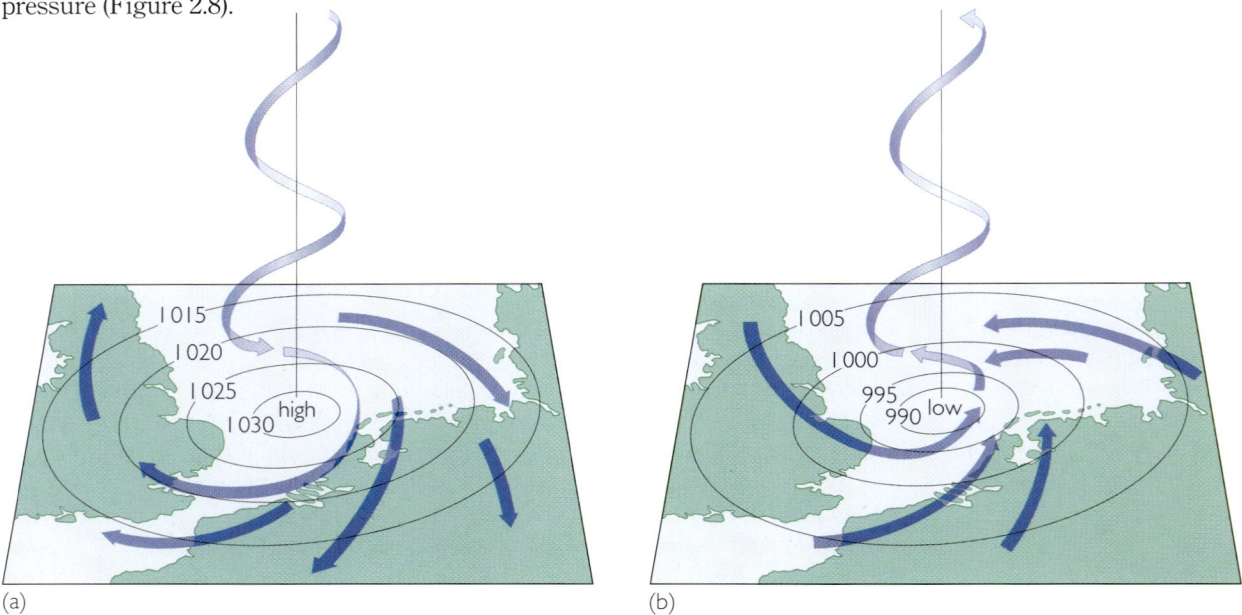

Figure 2.8
(a) Air spiralling downwards and outwards from an atmospheric high (an anticyclone).
(b) Air spiralling inwards and upwards towards an atmospheric low (a cyclone or depression). Note that the flow directions shown are for the Northern Hemisphere; in the Southern Hemisphere, flow around anticyclones is anticlockwise, and flow around cyclones is clockwise (cf. Figure 2.7). Contour values are typical atmospheric pressures at sea-level, expressed in millibars (10^{-3} bar). Surface wind speeds typically reach a few tens of m s^{-1}.

But why are the paths of winds curved? And why isn't the global wind system a simple convection system in which air sinks only at the poles, and surface winds blow directly from the poles to the Equator, as illustrated in Figure 2.9? The answer lies in the rotation of the Earth. Air masses moving above the surface of the Earth are bound to it by friction extremely weakly, and as they move, the Earth turns beneath them. The deflection of winds and currents relative to the surface of the Earth is known as the **Coriolis effect**, and it increases with latitude, from no deflection at the Equator to a maximum at very high latitudes. The Coriolis effect can be quantified in terms of a Coriolis *force*, acting at right-angles to the direction of flow. In the Northern Hemisphere, the Coriolis force acts to the right of a flow and, in the absence of balancing forces, turns it clockwise; in the Southern Hemisphere, the Coriolis force acts to the left of the flow, and in the absence of balancing forces turns it anticlockwise. Figure 2.10 shows typical paths of air or water moving over the Earth's surface under the influence of the Coriolis force. Further information about how such paths arise, *which you need not remember*, is given in Box 2.3.

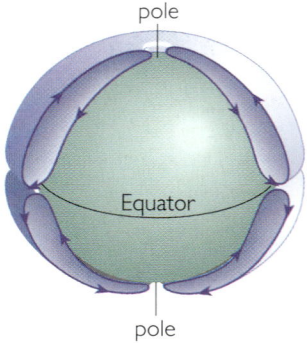

Figure 2.9
Simple hypothetical wind system for a non-rotating Earth.

Keeping the Earth habitable

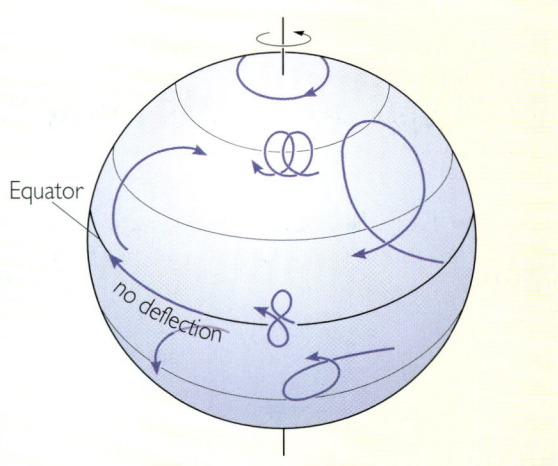

Figure 2.10
Various paths for air or water moving over the surface of the Earth under the influence of the Coriolis force, assuming that other forces, including friction, are negligible.

Box 2.3 The Coriolis effect

Air or water moving over the surface of the Earth is bound to it by friction only loosely. This means that – unlike the motion of (say) a train or animal – its motion *relative to the Earth* is a combination of the motion it has on its own account *and* the motion of the Earth spinning about its axis. From the point of view of moving air and water, what matters is the rotation of the Earth's surface about a vertical local axis, i.e. an axis at right-angles to the Earth's surface (Figure 2.11). It's fairly straightforward to see that the surface of the Earth is rotating about a vertical axis at the poles; it's also fairly easy to see that it is *not* rotating about a vertical axis at the Equator, because here such an axis is at right-angles to the Earth's axis of rotation. It may take a leap of imagination to realize that, apart from at the Equator, every part of the Earth's surface is rotating about a local vertical axis, at a rate that increases from zero at the Equator to a maximum at the poles.

- Imagine a flow of air moving over the (rotating) surface of the Earth in the Northern Hemisphere. Will the air flow move clockwise or anticlockwise in relation to the Earth's surface?

- In the Northern Hemisphere, the surface of the Earth is turning anticlockwise so, *in relation to the Earth*, the air flow will turn clockwise.

(You can see this for yourself by visiting the Foucault Pendulum exhibit at the Science Museum in London.)

If you still need convincing, and find rotatory motion hard to imagine, here is a simpler, but less complete, explanation. Think of a missile being fired northwards from a launcher positioned on the Equator (Figure 2.12a, overleaf). When it leaves the launcher, the missile is moving eastwards at the same velocity as the launcher, as well as moving northwards at its firing velocity. As it moves northwards, the Earth is turning beneath it. Initially, because it has the same eastwards velocity as the Earth, the missile appears to be travelling in a straight line. However, the eastwards velocity of the surface of the Earth is greatest at the Equator and decreases polewards, so as the missile travels progressively

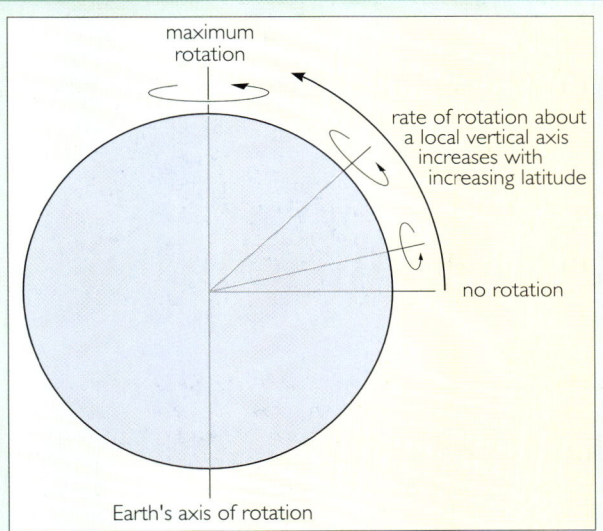

Figure 2.11
Schematic diagram to illustrate the reason for the Coriolis effect. The angular velocity of the surface of the Earth about a local vertical axis increases from zero on the Equator to a maximum at the poles. (You can think of this in terms of the angular velocity of any part of the Earth's surface being a *component* of the angular velocity about the Earth's axis, the size of the component depending on the latitude.) A ⊗ marked on the Earth, and viewed from a satellite in space positioned above it, would be seen to rotate. This is easiest to imagine for a location close to the poles – you can try it yourself using a globe.

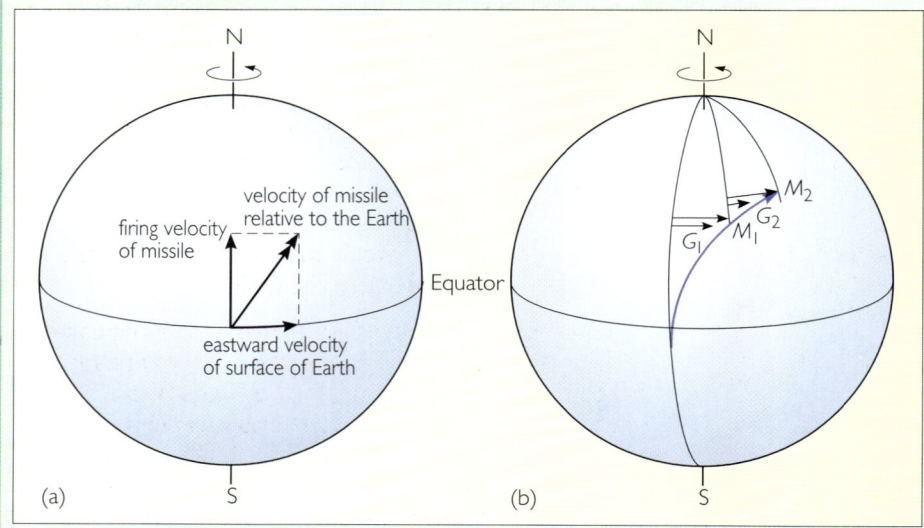

Figure 2.12
(a) A missile launched from the Equator has not only its northward firing velocity but also the same eastward velocity as the surface of the Earth at the Equator. The resultant firing velocity is therefore a combination of the two, as shown by the double-headed arrow.
(b) The path taken by the missile. In time interval T_1, the missile has moved eastwards to M_1 but the Earth has moved only to G_1; in time interval T_2, the missile has moved to M_2 and the Earth to G_2. Note that the deflection attributed to the Coriolis force (the difference between M_1 and G_1, and between M_2 and G_2 etc.) increases with increasing latitude.

Figure 2.13
Hypothetical cylindrical Earth, spinning about its axis.

northwards, the eastward velocity of the Earth turning beneath it becomes less and less. As a result, in relation to the Earth, the missile is moving not only northwards but also eastwards, at a progressively greater rate, and so travels in a curved path as shown in Figure 2.12b). A similar argument can be used to show that a missile fired from the North Pole will be deflected towards the west.

■ So, what can you say about the Coriolis effect for motion over a hypothetical *cylindrical* Earth, rotating on its axis as shown in Figure 2.13?

■ For a rotating cylindrical Earth, there would be no Coriolis effect.

The eastwards velocity of the surface of a cylindrical Earth would be the same all along its length, so a missile fired polewards (with the eastwards velocity of its launcher) would not be deflected. In contrast to the situation illustrated in Figure 2.11, for a rotating cylinder, there is no rotation about axes at right-angles to the curved surface, as these are also at right-angles to the axis of rotation. In other words, the Coriolis effect is *a result of the fact that the spinning Earth is round.*

The characteristic conditions indicated at the right-hand side of Figure 2.7 are a result of the fact that while sinking air is generally dry, rising air often has a high moisture content. The most dramatic common manifestations of rising (i.e. convecting) moist air are cumulonimbus clouds, the tallest of which form at low latitudes in the **Intertropical Convergence Zone** (or **ITCZ**), where the wind systems of the two hemispheres meet (Figures 2.2 and 2.7). Here, moist air carried in the Trade Winds converges and rises, resulting in the formation of towering thunderheads (cf. Figure 2.5).

Because of the Coriolis effect, the Trade Winds blow towards the Equator not from the north and the south, but from the north-east and the south-east. The convention for naming winds is to use the direction *from* which they blow, and so these winds are known as the North-East Trades and the South-East Trades. As mentioned above, in the lower atmosphere, pressure is low along the Equator (or, more accurately, along the ITCZ, which corresponds more or less to the region of

Keeping the Earth habitable

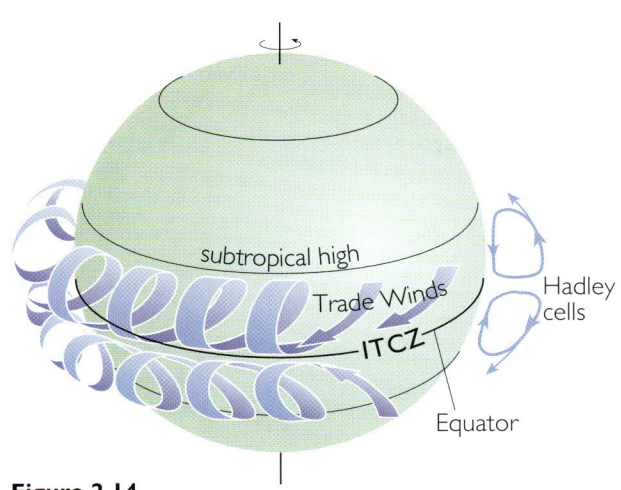

Figure 2.14
The helical circulation patterns of which the Trade Winds form the surface expression; the north–south component of this helical circulation is known as the Hadley circulation; the two 'Hadley cells' can be seen on either side of the Equator (cf. left-hand side of Figure 2.7). (*Note*: Some texts also use the term 'Hadley cell' to refer to the north–south components of the vertical atmospheric circulation at higher latitudes.)

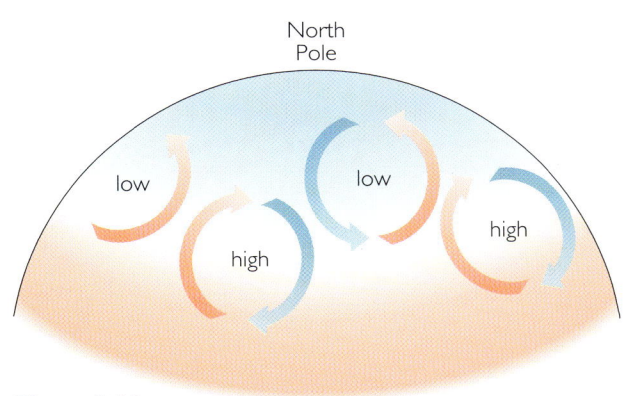

Figure 2.15
Highly schematic diagram to show the poleward transport of heat through the action of mid-latitude cyclones and anticyclones.

highest surface temperature), and here air converges and rises; it then moves poleward in the upper troposphere, with much of it sinking in mid-latitudes where, as a result, the pressure at the Earth's surface is high. Thus, the Trade Winds are the surface expression of a helical circulatory system known as the **Hadley circulation** or Hadley cells (Figure 2.14); strictly speaking, only the north–south component of the circulation, as shown on Figure 2.7, is named the Hadley circulation.

Because the Coriolis effect increases with increasing latitude, there is a change in the *style* of atmospheric circulation with latitude. As the Trade Winds blow in tropical latitudes where the Coriolis effect is relatively small, they are deflected laterally only slightly and the overall pattern of circulation is the predominantly vertical Hadley circulation. At higher latitudes, where the degree of deflection is much greater, vortices tend to form in the lower atmosphere and have a predominantly horizontal, or slantwise, circulation. As shown in Figure 2.8, these are the anticyclonic and cyclonic winds familiar from weather charts.

- How can mid-latitude cyclones and anticyclones contribute to the poleward transport of heat?

Moving air masses mix with adjacent air masses along *atmospheric fronts*, and heat is exchanged between them. Air moving northwards in a Northern Hemisphere cyclone or anticyclone will be transporting relatively warm air polewards, while air returning equatorwards will have been cooled. This is, in effect, a kind of large-scale stirring, and is shown schematically in Figure 2.15.

The paths taken by mid-latitude depressions and anticyclones are determined by the behaviour of the **polar jet stream** (Figure 2.16, overleaf). This is a high-level, fast air current that in each hemisphere flows around the Earth above the boundary between warm tropical air and the underlying cold polar air, known as the **polar front** (Figure 2.17). The jet stream tends to develop large undulations, typically three to six in number, which eventually become so extreme that cells of tropical air become isolated at relatively high latitudes, and cells of polar air become isolated at relatively low latitudes, as shown in the third part of Figure 2.16. This mechanism results in the horizontal transport of enormous amounts of warm air polewards and cold air equatorwards. Meanwhile, the vertical air movements that of necessity accompany these large-scale horizontal wave motions (known, incidentally, as *Rossby waves*), lead to the development of cyclonic flow and low pressure centres, and anticyclonic flow and high pressure centres, near the Earth's surface.

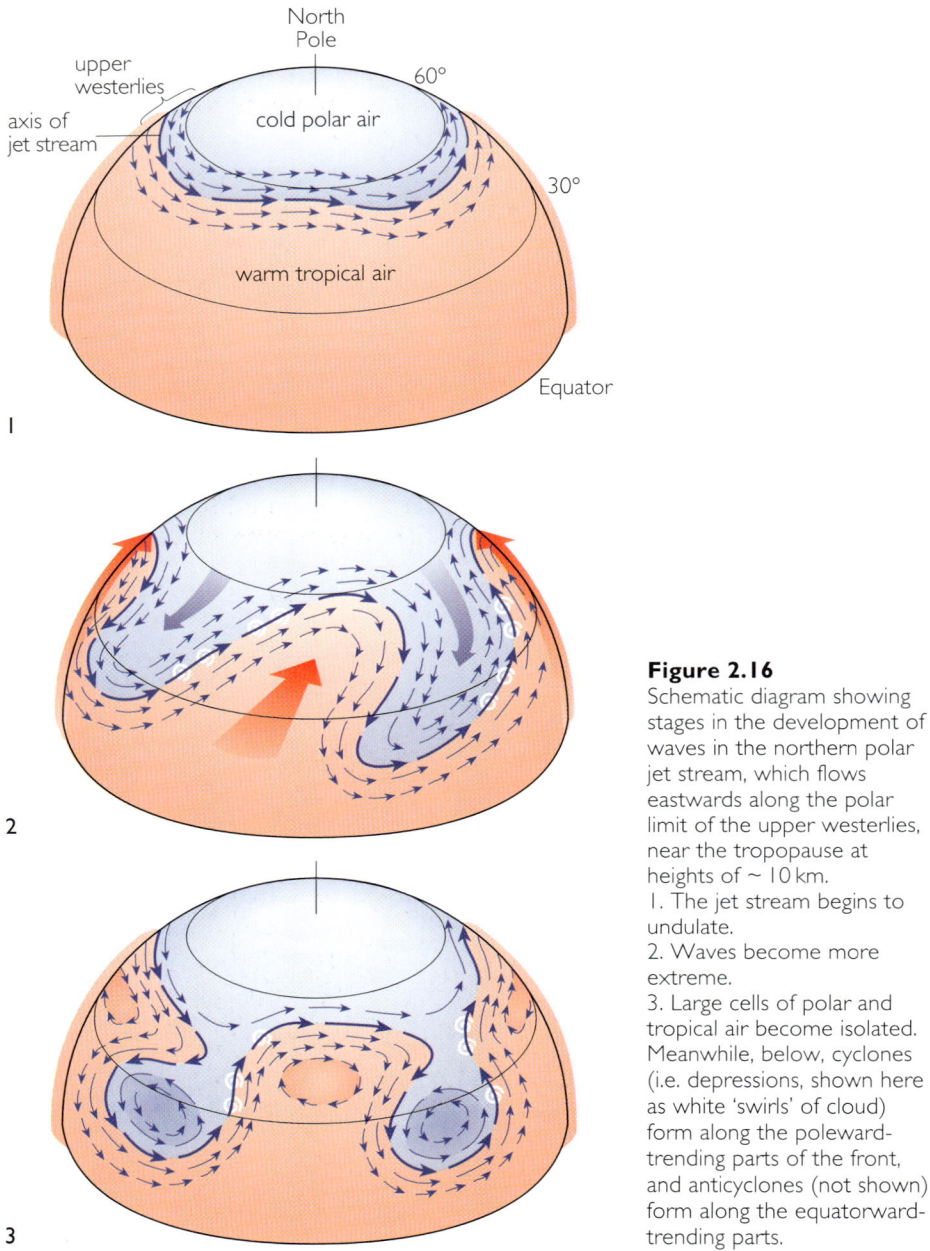

Figure 2.16
Schematic diagram showing stages in the development of waves in the northern polar jet stream, which flows eastwards along the polar limit of the upper westerlies, near the tropopause at heights of ~ 10 km.
1. The jet stream begins to undulate.
2. Waves become more extreme.
3. Large cells of polar and tropical air become isolated. Meanwhile, below, cyclones (i.e. depressions, shown here as white 'swirls' of cloud) form along the poleward-trending parts of the front, and anticyclones (not shown) form along the equatorward-trending parts.

- What would be the effect of the undulations in the jet stream becoming more pronounced?

- Cold polar air will flow to lower latitudes than normal, while warm tropical air will flow to higher latitudes than normal.

The areas of the globe that would be affected by pronounced undulations in the jet stream would depend on the configuration of the jet stream at the time; the undulations themselves travel westward around the Earth, relative to the air flow. In recent years, unusual behaviour of the jet stream (perhaps related to global warming) has resulted in droughts and crop failures in the United States. The effect

Keeping the Earth habitable

of droughts on agriculture, whether in the United States or the African Sahel, reminds us that whether a particular type of vegetation flourishes or dies is determined by the prevailing climatic conditions. Natural ecosystems are more robust than modern agricultural monocultures, but they are attuned to the prevailing climate, and if that changes, then so will they.

The polar jet streams are not the only high-level atmospheric jets. Figure 2.17a is a schematic cross-section through the atmosphere showing the approximate average positions of these intense high-level winds. Figure 2.17b shows the approximate northerly and southerly limits of meanders in the northern polar jet stream, and the mean winter position of the northern subtropical jet stream. The paths of such high-level winds may have been different in the past, with far-reaching effects on global climate.

Now look at Figure 2.18 (overleaf), which shows the prevailing winds at the Earth's surface. Note the zones of westerly winds, between the subpolar low pressure regions and the subtropical highs. These are the parts of the globe that – depending on the configuration of the polar jet stream (Figure 2.16) – may be affected by cold polar air or warm tropical air. These regions are often referred to as *temperate*.

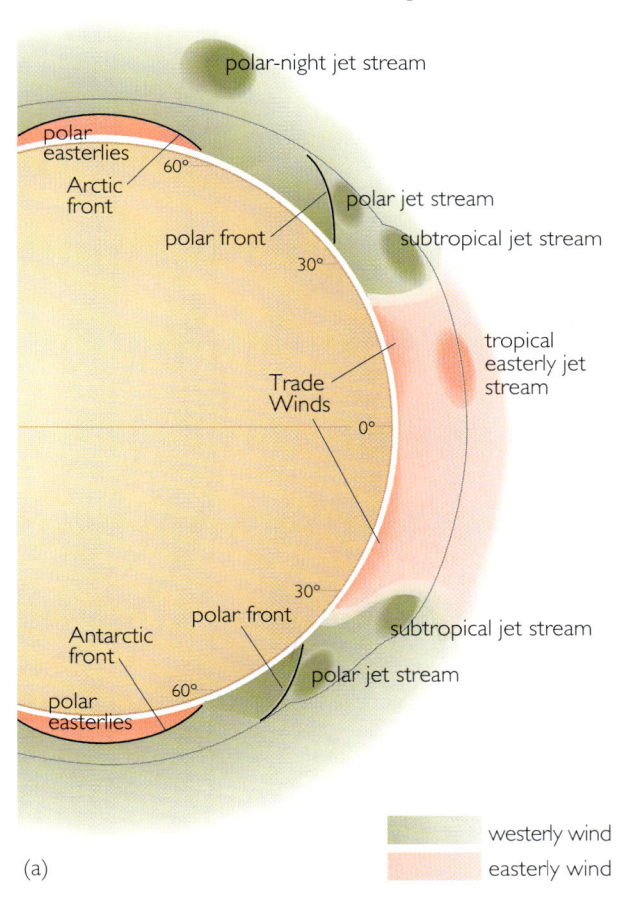

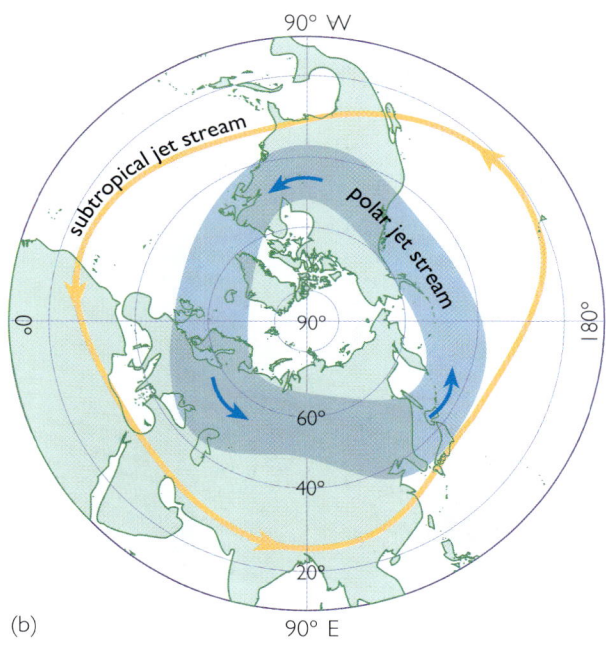

Figure 2.17
(a) Schematic cross-section showing direction and intensity of zonal winds. Red corresponds to easterlies (i.e. flow 'out of the page'), green to westerlies (flow 'into the page'). The narrow grey line is the tropopause; the full black lines are major fronts between warmer and colder air masses. Regions of deep colour at high altitude are jet streams, the positions and intensities of which vary according to time of year. The jet streams that seem to have most influence at the Earth's surface are the polar jet streams and subtropical jet streams; see (b).
(b) The belt of winter activity of the northern polar jet stream (blue tone), and the mean winter position of the axis of the northern subtropical jet stream (yellow tone).

(a) JULY

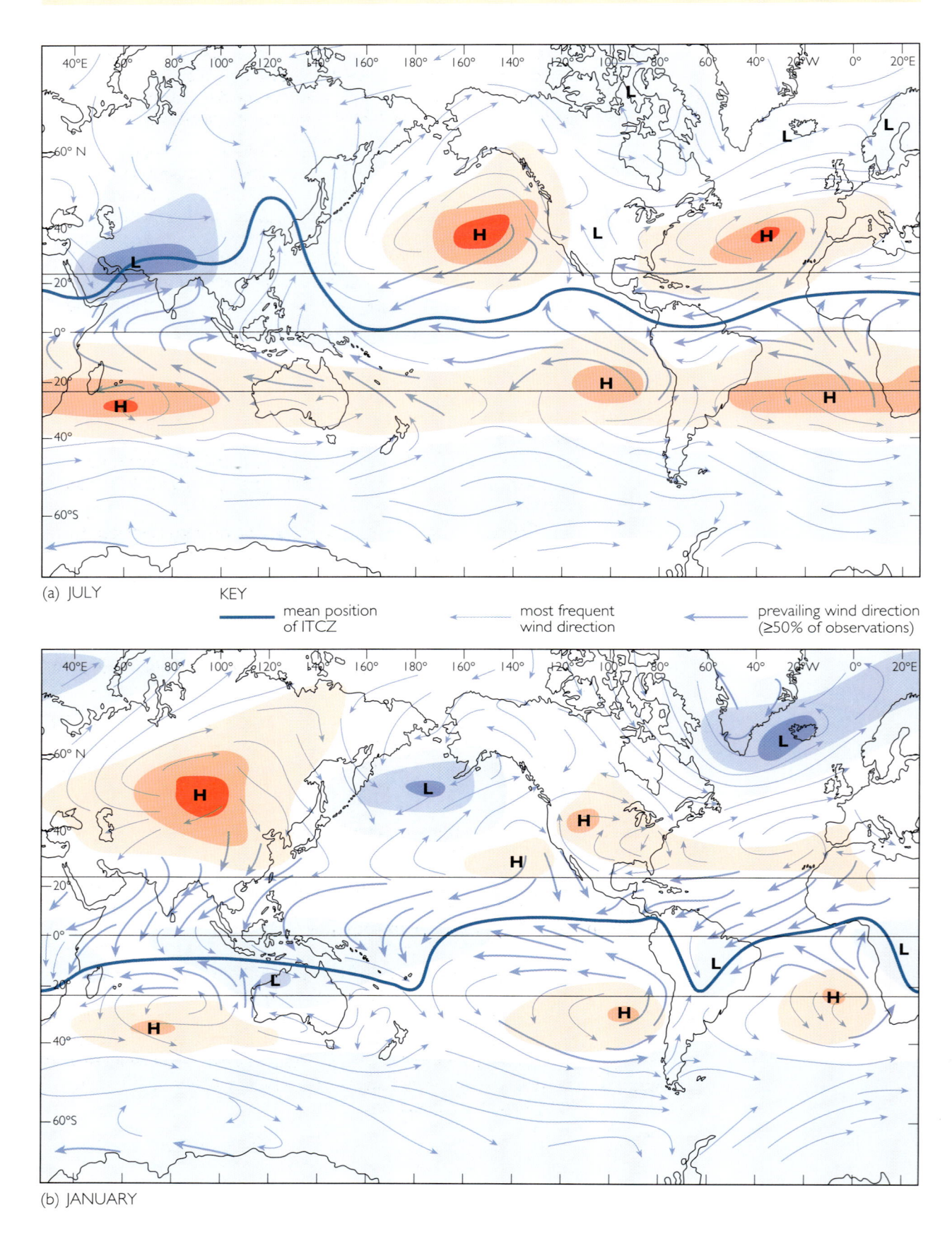

KEY: mean position of ITCZ — most frequent wind direction — prevailing wind direction (≥50% of observations)

(b) JANUARY

Keeping the Earth habitable

Figure 2.18 (opposite) The prevailing winds at the Earth's surface, and the position of the Intertropical Convergence Zone (ITCZ) where the wind systems of the two hemispheres meet, in (a) July (northern summer/southern winter) and (b) January (southern summer/northern winter). Also shown are the positions of the main regions of high and low surface pressure (red and pink tones, respectively) for these seasons of the year.

■ Compare the wind patterns in Figure 2.18 with those shown in Figure 2.7. Over which parts of the globe are the subtropical highs and subpolar lows most evident? Can you therefore say what features of the real world cause the actual wind pattern to differ from the hypothetical one shown in Figure 2.7?

□ Subpolar lows and subtropical highs are most evident over the oceans; this is not surprising given that Figure 2.7 shows the wind pattern appropriate to an Earth completely covered in water. It is the presence of the continental landmasses that makes the real wind pattern so much more complicated.

So, it is the presence of the continents that 'distorts' the global wind system from the hypothetical east–west zones of polar high/subpolar low/subtropical high/equatorial low. To discover more about the effects of the distribution of land and sea, try Question 2.2.

Question 2.2

The ITCZ, where the wind systems of the two hemispheres meet (Figure 2.18), generally follows the zone of highest temperature at the Earth's surface.

(a) In general terms, what are the main differences between the position and/or shape of the ITCZ in July (Figure 2.18a) and January (Figure 2.18b)? How does Figure 2.1 and the related discussion help to explain these differences?

(b) The ITCZ moves seasonally northwards and southwards between the extreme positions shown in Figure 2.18. What implication does the changing position of the ITCZ have for the prevailing wind direction at the Earth's surface? To answer this, describe how the wind direction changes over tropical west Africa (Mauritania, Senegal, Guinea) between July and January.

(c) Over what region is the north–south seasonal shift in the position of the ITCZ greatest?

Many low latitude regions experience such seasonal changes in wind direction. Air masses that travel over the oceans pick up moisture, and those that move over arid regions are dry, so seasonally reversing winds often bring with them contrasting climatic conditions (Figure 2.19, overleaf).

■ Bearing this in mind, how might you expect conditions in the south of the Eurasian continent to differ between July and January?

□ Conditions will be wet in July, when the prevailing winds are south-westerlies blowing off the Indian Ocean, and dry in January, when the prevailing winds are north-easterlies blowing off the Eurasian continent.

The rains that fall over southern Asia in the summer are often referred to as the 'monsoon' rains. However, as the word 'monsoon' derives from the Arabic for 'winds that change seasonally' there are really two Asian monsoons – the warm, moist South-West Monsoon, and the relatively cool, dry North-East Monsoon. It has been worth considering monsoonal reversals in some detail because, as you will see later, the interaction between winds and continents may also have a more subtle, and longer term, influence on global climate.

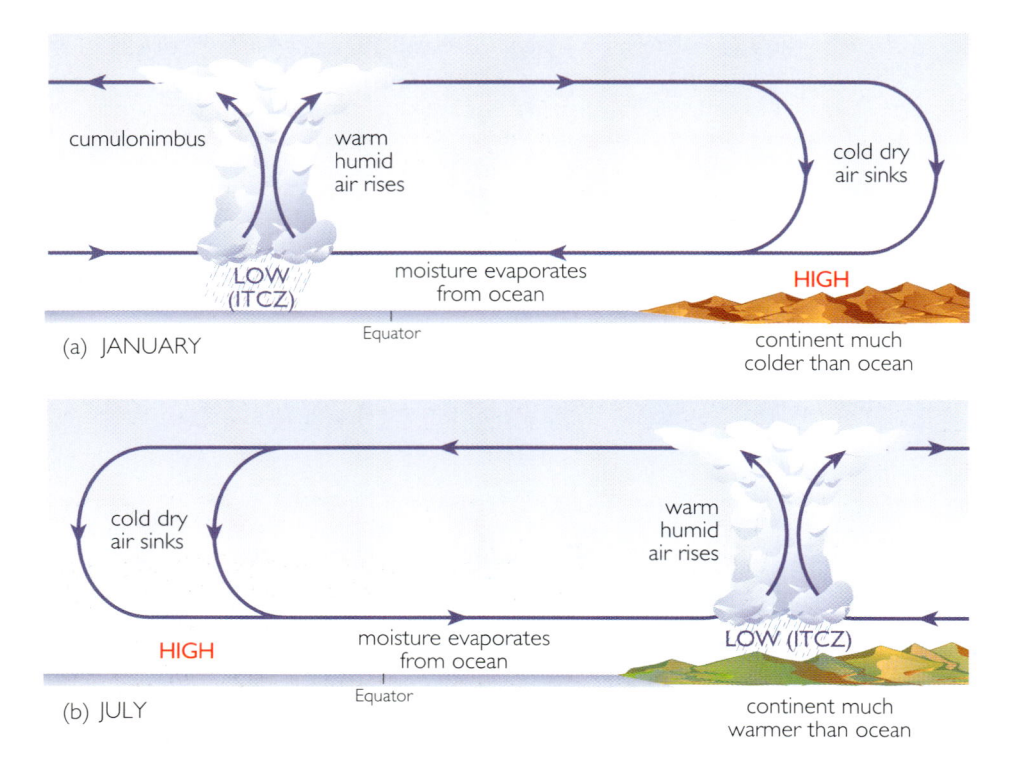

Figure 2.19
Schematic diagrams to illustrate the changing conditions over continent and ocean in a monsoon climate, drawn assuming that the continent is in the Northern Hemisphere.
(a) Northern winter: dry, cooled air subsides over the continent which is a region of high surface pressure; winds blow off the continent picking up moisture from the much warmer ocean, and eventually moist air rises at the Intertropical Convergence Zone, causing abundant rain.
(b) Northern summer: land is now much warmer than the ocean, and so the region of low surface pressure corresponding to the ITCZ, and its zone of rain, has moved northwards towards the interior of the continent.
(*Note*: the latitudes shown are notional, as the northernmost and southernmost positions of the ITCZ vary from place to place, as you can verify from Figure 2.18, which should be studied in conjunction with this figure.)

In fact, the large-scale reversals in atmospheric circulation over southern Asia and the Indian Ocean are particularly marked because the continental landmass involved is not only large but in places very high. We will come back to this specific case later, but in general it is important to remember that climatic conditions are affected not only by the distribution of land and sea, but by the shape of the landmasses in the vertical dimension – their topography. Air that is forced to rise over high ground may be triggered to convect, leading to rainstorms. Furthermore, moisture-laden air that is forced to rise by the presence of a topographic high will cool adiabatically, and being cooler, it will no longer be able to hold as much moisture. If cooling is sufficient, condensation will occur, clouds will form and rain or snow will result (Figure 2.20a). By contrast, air subsiding over the lee slope will warm adiabatically and having no moisture source may become very dry, with the result that there is a *rain shadow*. Rain and snowfall triggered by mountains (known as *orographic* precipitation) occurs at all scales – on small volcanic islands in mid-ocean and over mountain chains. In the south-western USA, the western slopes of the Coast Range and the Sierra Nevada (capped by snow in Figure 1.3) receive ample precipitation from the moisture-laden north-westerlies blowing off

Keeping the Earth habitable

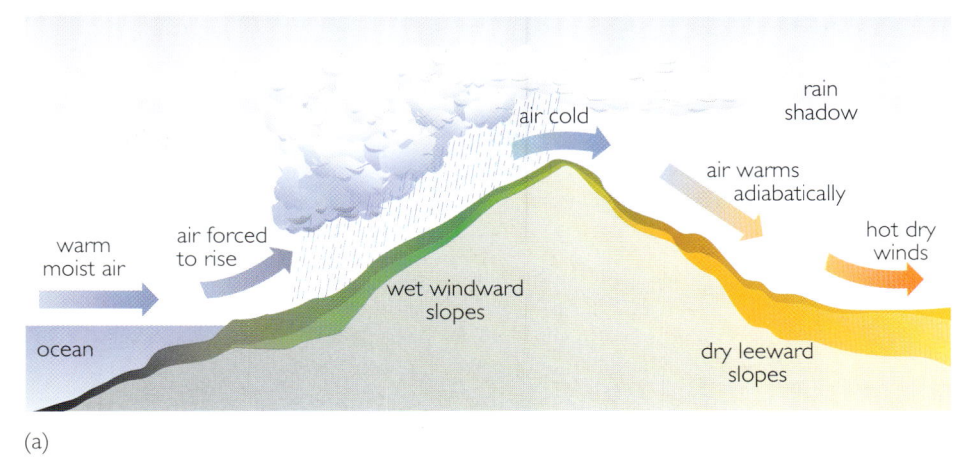

(a)

(ii)

Figure 2.20
(a) Orographic precipitation: mountains cause moisture-laden air to rise, with the result that the windward slopes are wet and well vegetated, while to leeward is an area of rain shadow. (b) Two locations on the Hawaiian Island of Kauai: (i) the verdant eastern side of the island, thought to be one of the wettest places on Earth; (ii) Waimea Canyon on the western side, which has a much drier climate and desert-like vegetation.

(b)(i)

the Pacific (Figure 2.18); to the south-east of these mountains, in their lee, are the desert regions of Nevada and eastern California (also visible on Figure 1.3). Figure 2.20b shows the dramatic effect of orographic precipitation and rain shadow on the vegetation of the mountainous island of Kauai.

Before moving on to look more closely at the role of the ocean in the climate system, we should remember the influence that living terrestrial organisms have on their environment. This is most dramatically illustrated by the effects of deforestation, so often shown on our television screens. Removal of large areas of rainforest means that rainwater is no longer trapped, but runs away, carrying with it the topsoil (itself largely a product of the forest). As a result, productive ecosystems which trapped and recycled water, and provided moisture and heat to the overlying atmosphere, are all too often converted to arid wastes.

2.2.2 Heat transport by the ocean

As shown by the summary diagram in Figure 2.2, the ocean has a surface current system, and a deep convection system driven primarily by cooling at the surface in high latitudes. The surface current system is wind-driven, and is mostly confined to the uppermost wind-mixed layer of the ocean, which is separated from deeper colder water by a zone of marked decrease in temperature known as the **thermocline**.

If you compare Figure 2.21, which shows the average surface current pattern, with the winds on Figure 2.18 (particularly (b)), you can see a fairly close correspondence between the two, although current patterns are of necessity modified by the presence of landmasses, which cause flow within ocean basins to form more-or-less closed circulatory systems, or **gyres**. In the Atlantic, anticyclonic wind systems blowing around the high pressure regions in mid-latitudes give rise to anticyclonic gyres, often referred to as 'subtropical gyres', in both hemispheres; similar subtropical gyres can be seen in the Pacific.

■ To what extent can cyclonic gyres be identified, corresponding to the cyclonic winds associated with subpolar low pressure regions (cf. Figure 2.7)?

■ Cyclonic gyres can be seen in the subpolar regions of both the North Atlantic and the North Pacific, but *not* in similar latitudes in the Southern Hemisphere. Here, in the Southern Ocean, where there are no land barriers, the westerly winds drive the Antarctic Circumpolar Current (once known as the West Wind Drift) eastwards around the globe.

To see how surface currents influence the distribution of heat, compare Figure 2.21 with Figure 2.1a and b, showing surface temperature. Although it is quite hard to see from Figure 2.1, in low to mid-latitudes, sea temperatures on the western sides of the Atlantic and the Pacific are generally higher than on the eastern sides.

■ Why is this?

■ Water warmed at low latitudes is carried polewards in the western 'limbs' of the subtropical gyres. Some water is carried to high latitudes, but some must circulate in the subtropical gyres for a considerable time. However, because flow is anticyclonic, the western sides of the gyres have always been warmed more recently than the eastern sides.

One of the clearest manifestations of surface current flow in the temperature distributions in Figure 2.1 is the shape of the isotherms in the North Atlantic, following the north-eastwards flow of the Gulf Stream (cf. Figure 2.21). By contrast, in the South Pacific, the 'finger' of low temperatures extending northwards up the west coast of South America (particularly clear in Figure 2.1b) may be correlated with flow in the cold Peru or Humboldt Current, carrying water cooled at high latitudes equatorwards; a similar though less marked influence on the temperature distribution may be seen in the South Atlantic, associated with the Benguela Current (Figure 2.21).

The flow of warm water north-eastwards across the North Atlantic in the Gulf Stream moderates the climate of Britain and north-west Europe, making the region much warmer than it would otherwise be (more of this later). The Gulf Stream is the narrow, fast-flowing western side of the North Atlantic subtropical gyre. It is known as a 'western boundary current', and its intensity is a consequence of the

Keeping the Earth habitable

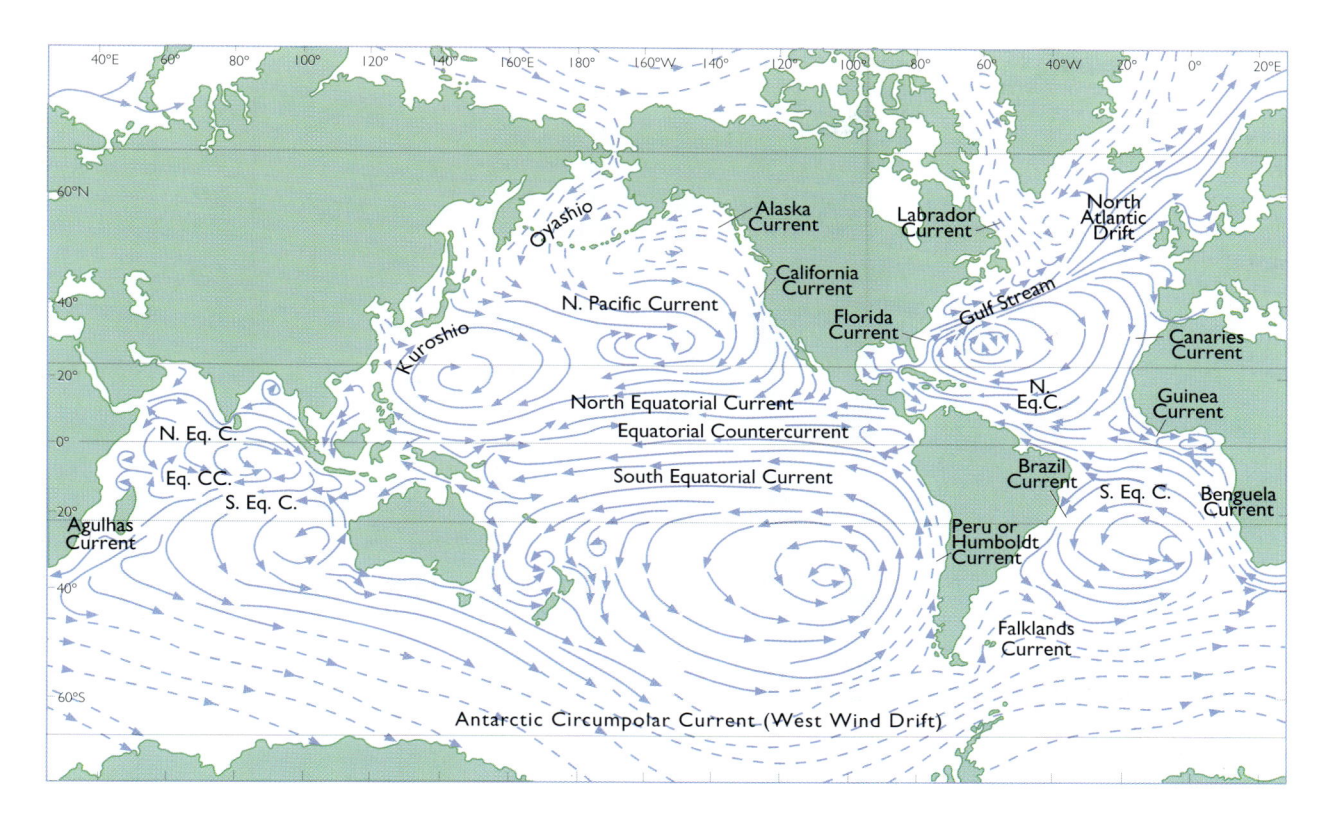

Figure 2.21
The global surface current system in the northern winter; this is the long-term average pattern – at any one time, the pattern will differ in detail. There are local differences in the northern summer, particularly in regions affected by monsoonal reversals; see also Figure 2.2 and Section 2.2.1. Cold currents are dashed. Note that even in strong currents, such as the Gulf Stream, current speeds rarely exceed a few m s^{-1}.

fact that the Coriolis effect increases with latitude. Were it not for the speed at which the Gulf Stream flows poleward, much more of its heat would be lost to the atmosphere and adjacent ocean *en route*, and so be unavailable to warm higher latitudes.

The gyres and the Gulf Stream, and other current patterns on maps like Figure 2.21, are examples of large-scale, long-term features of the oceanic circulation. They represent the *average* situation in the ocean (at any one time the current in a particular place could be flowing in the opposite direction to that shown on the current map), and are often thought of as the ocean's 'climate'. The ocean also has 'weather' – short-term, smaller-scale features. Most significant in this context are eddies about 50–250 km across (i.e. about a quarter of the size of mid-latitude cyclones and anticyclones); because they are of intermediate size, they are known as **mesoscale eddies**.

Such eddies are frequently generated by meanders in strong currents like the Gulf Stream or the Antarctic Circumpolar Current. Because these currents flow along boundaries between bodies of water with contrasting temperatures they are known as 'frontal' currents; the formation of eddies from meanders in frontal currents is similar to the formation of weather systems along atmospheric fronts (Figure 2.22, overleaf). Mesoscale eddies clearly play an important role in transporting heat around the oceans, particularly across oceanic frontal regions, but their importance to the climate system as a whole is still unclear. This is partly because, until recently, computer simulations of the ocean have not been of sufficiently fine resolution to be able to include the eddies. Indeed, computer modelling of oceanic phenomena generally needs more 'computing power' than is needed to model atmospheric phenomena, as the high density and viscosity ('stickiness') of water compared with those of air mean that the length-scales of current patterns are generally shorter, and the time-scales longer, than is the case for winds.

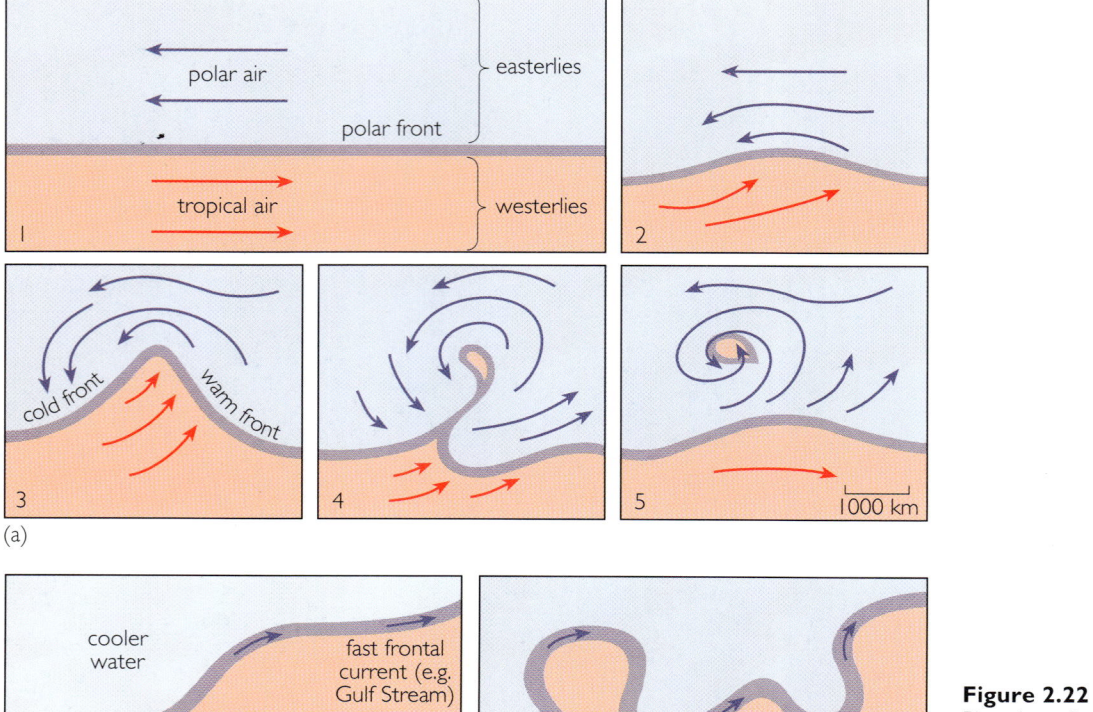

Figure 2.22
Plan view diagrams showing weather in the atmosphere and in the ocean. (a) Stages in the development of a mid-latitude cyclone (depression) in the Northern Hemisphere, showing how it contributes to the poleward transport of heat. (b) The formation of mesoscale eddies at a strong frontal current like the Gulf Stream, showing how they act to transport heat from one side of an oceanic front to the other.

2.2.3 Atmosphere–ocean coupling

It is well known that what happens in the lower atmosphere can affect the underlying ocean – the most obvious example is that winds drive surface currents. Less well known is the fact that what happens in the ocean can affect the overlying atmosphere. In this context, let's look briefly at a tropical cyclone – a powerful atmospheric phenomenon which, for reasons that will become obvious, can *only* be initiated over the ocean.

Tropical cyclones – also known as hurricanes or typhoons – are quite different from mid-latitude 'lows' or 'depressions' (Figure 2.23). They are fast-moving regions of vertical convection combined with strong cyclonic winds. The energy that drives a tropical cyclone comes from the release of latent heat as the water vapour in the rising air condenses to form clouds and rain; the resultant warming of the air

Keeping the Earth habitable

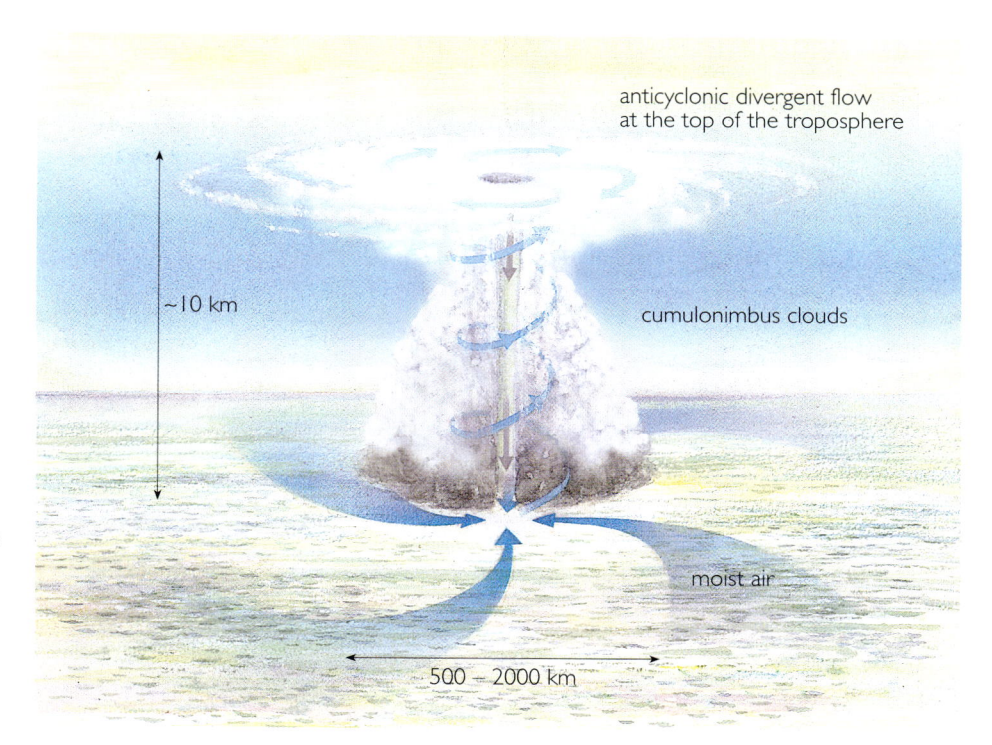

Figure 2.23
Schematic diagram of a tropical cyclone showing air movements and areas of cloud formation and heavy precipitation. The cumulonimbus clouds are arranged in bands which form a spiral pattern around the core region – the 'eye' of the storm. Subsidence of air and adiabatic warming occur within this core region, which is a region of light winds and little cloud.

around the central region of the cyclone causes it to become less dense and to rise yet more, and in a **positive feedback** loop, the rising air draws in more air, which takes up yet more moisture from the sea-surface, and so on. Only when they drift over land do tropical cyclones begin to die out. They tend to move westwards and polewards, and transport enormous amounts of heat away from the tropical ocean.

Question 2.3
(a) Why do tropical cyclones only form over the ocean?
(b) In the event of continued global warming, would you expect tropical cyclones to become more or less frequent?

Given their source of energy, it is not surprising that tropical cyclones only occur over relatively large areas of ocean where the surface temperature is high. In fact, it seems that cyclones are only generated above a sea-surface at 28 °C or more. This link between sea-surface temperature and cyclone generation is an example of the strong interaction or **coupling** that exists between the atmosphere and ocean, particularly in tropical areas. Despite the strength of the interactions between the atmosphere and the ocean, researchers investigating the climate system have only recently begun to build such interactions into computer models.

Perhaps the best known example of large-scale atmosphere–ocean coupling is the climatic perturbation known as El Niño, centred on the tropical Pacific. For a short discussion of this famous, or rather infamous, natural phenomenon, see Box 2.4.

Box 2.4 El Niño

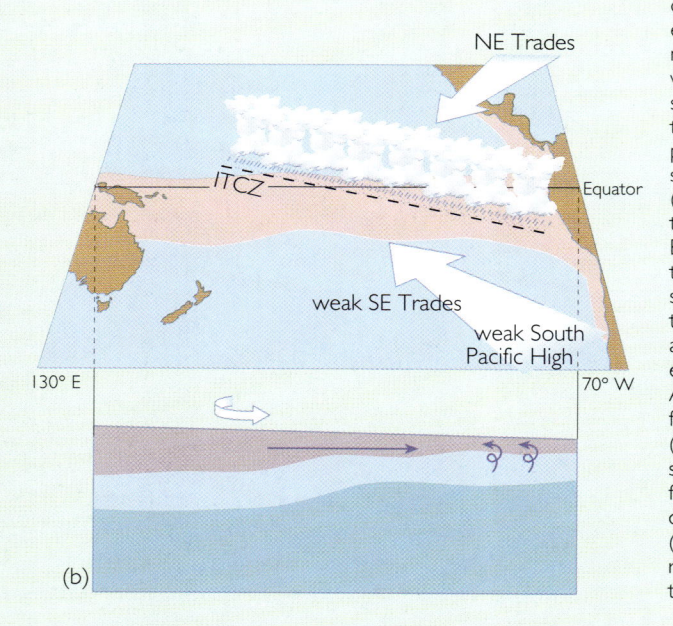

El Niño means the Christ Child (literally, the Boy Child), and was originally the name given by the fishermen of the Peruvian port of Paita to a warm nutrient-rich current that flows southwards from the Gulf of Guayaquil around Christmas time. Being associated with good fish catches, this current was regarded as beneficient. However, the name El Niño has now come to be associated with change in the Pacific equatorial current system, first noticeable around the turn of the year, that brings mainly disaster.

El Niño events occur every 4–10 years or so, and generally last about a year. The South-East Trades are weaker than usual and bursts of westerly winds occur in the western equatorial Pacific. The collapse of the South-East Trades allows a pool of warm water usually in the western equatorial Pacific to spread into the central and eastern Pacific, and then polewards along the coasts of North and South America. Regions of vigorous convection and heavy rain, normally over Indonesia, move into the central Pacific as the ITCZ moves southwards (Figure 2.24b).

El Niño's notoriety is due to the loss of human life and livestock, and the stress on natural ecosystems, that severe events can bring. Cyclones occur much further east than usual; regions that are usually dry experience torrential rain (Figure 2.25) which in turn may cause mudslides and encourage epidemics; other areas, usually with plentiful rainfall, may be stricken with drought.

One important consequence of El Niño is that nutrient-rich water is no longer brought to the surface, or 'upwelled', along the coast of Peru and Chile, or in the vicinity of the Equator in the eastern Pacific. Normally, these **upwelling** regions sustain high primary productivity (see Figure 1.4), and hence productive fisheries. During El Niño events, the fisheries crash.

El Niño events are sometimes referred to as El Niño–Southern Oscillation (ENSO) events. They occur when the pressure difference between the high pressure region centred in the south-eastern Pacific and the low pressure region usually over Indonesia (see Figures 2.18 and 2.26a) is at a minimum. This pressure difference continually rises and falls on time-scales of several years

Figure 2.24
(a) Climatic conditions in the Pacific during a normal year. In the upper plan-view diagram, the pink area has the highest sea-surface temperatures; the green tone represents high biological productivity due to upwelling. As shown in the cross-section along the Equator, the South-East Trades drive surface waters westward, with the result that the sea-surface slopes up towards the west and the thermocline slopes downwards. The shallowness of the thermocline in the eastern tropical Pacific allows nutrient-rich subthermocline water to be mixed up to the surface, with the result that there is high biological productivity here. The vertical scale is greatly exaggerated.
(b) During an El Niño event, the weakness of the South-East Trades, and westerlies in the western Pacific, cause the slopes of the sea-surface and thermocline to collapse, allowing warm water to flow eastwards along the Equator. Any upwelling now occurs from within the surface layer. (All El Niño events are slightly different, but the features shown here seem to occur in most events.)
(*Note*: You do not have to remember all the details in this figure.)

Keeping the Earth habitable

(a)   (b)

Figure 2.25
Christmas Island (a) under normal conditions, (b) during the severe El Niño event of 1982–83. Christmas Island is situated at 2°N and just west of the International Date Line (180°) (cf. Figure 2.24). Between August 1982 and January 1983, it received more than double its annual rainfall, with the result that vegetation grew luxuriantly. The unusual conditions meant that the 17 million bird population all but disappeared.

(a phenomenon known as the Southern Oscillation; see Figure 2.26b), so it could be argued that El Niño events are not 'perturbations' of the climate at all, but a natural though relatively uncommon state of the atmosphere–ocean system. Indeed, many climatologists now see the climate system of the tropical Pacific as oscillating between extreme states, and have named the periods of particularly large pressure differences and very strong South-East Trade Winds 'La Niña' ('the Girl Child').

Since 1990, however, there has been no clear oscillation in the pressure difference between the Indonesian Low and the South Pacific High, and the Southern Oscillation Index has remained below the long-term average almost continuously (cf. Figure 2.26b). As a result, the Pacific has tended to be in El Niño rather than La Niña mode. The underlying reasons for this are not understood, but may be related to global warming.

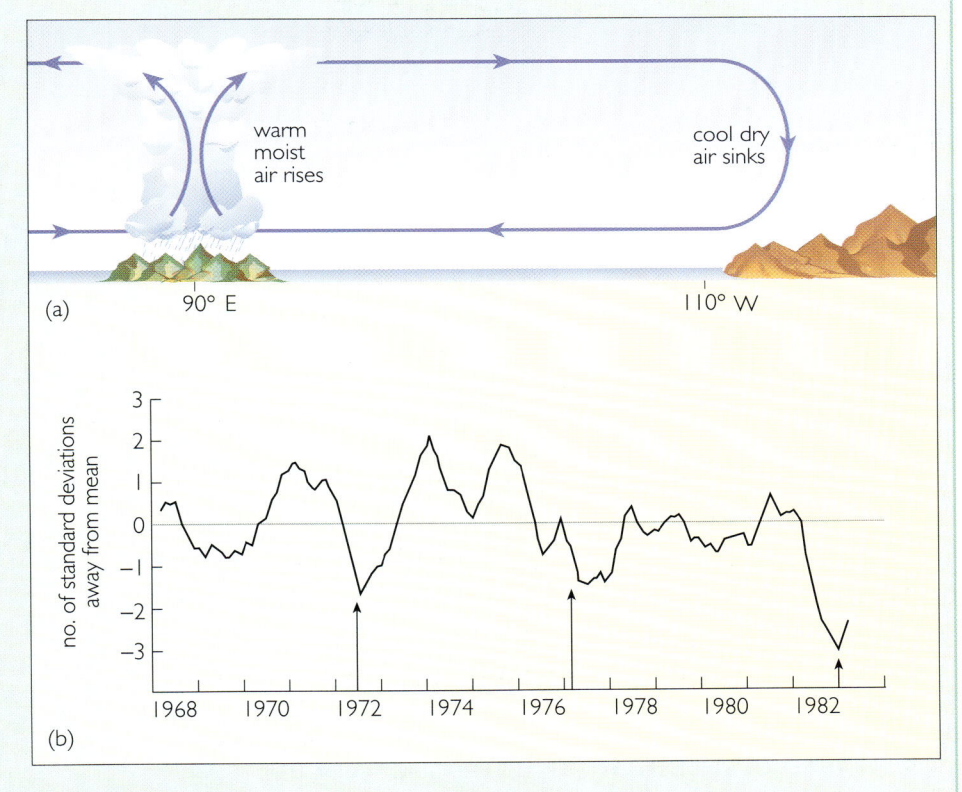

Figure 2.26
(a) Schematic diagram to show the atmospheric circulation between the Indonesian Low and South Pacific High (cf. Figure 2.18). The greater the surface pressure difference between the two, the stronger the South-East Trades. (b) Variation of the Southern Oscillation Index between 1968 and 1983. The two meteorological stations used here are Tahiti for the South Pacific High and Darwin, Australia, for the Indonesian Low, and the Index is expressed in terms of the departure from the normal difference in sea-level atmospheric pressure between the two stations divided by the standard deviation for the appropriate month. Negative values mean pressure differences less than usual. The arrows indicate the start of El Niño events, and the record stops towards the end of the most extreme El Niño in recent times.

For those interested in the workings of the Earth's climate, El Niño events are particularly fascinating because they are clear examples of the atmosphere and ocean acting as one system: it is not even possible to say whether an El Niño event is initiated in the ocean or in the atmosphere, so closely are the two coupled together. The vigorous convection in the atmosphere in the central and eastern Pacific – the cyclones and the heavy precipitation – are all a consequence of eastwards movement of the 'pool' of warm water usually in the western Pacific (see Figures 2.1 and 2.24), but the eastwards flow of warm water is itself a result of changes occurring in the atmosphere. In other words, El Niño events develop through positive feedback loops between atmosphere and ocean; for this reason, the Pacific atmosphere–ocean system is inherently *unstable* during such an event, and must eventually revert to its 'normal' state.

The effects of El Niño events are felt around the globe. They spread to higher latitudes in the Pacific through the action of wave-like fluctuations in the upper ocean and thermocline (including Rossby waves like those in the polar jet stream), and they affect other oceans because changes in the pressure field over the Pacific cannot occur independently of other pressure systems (Figures 2.18 and 2.26a). Even the strength of the Asian monsoons is affected by the state of the Southern Oscillation in the Pacific.

Similar (though generally less dramatic) oscillations between pressure systems are found in the tropical Indian Ocean and the tropical Atlantic, and are also found at higher latitudes. The pressure difference between the low pressure normally over Iceland and the high pressure normally centred over the Azores in mid-Atlantic oscillates like that between the Indonesian Low and the South Pacific High. It has been found that when the pressure difference is large, westerly winds over the Atlantic are stronger and flow in the Gulf Stream (and hence northwards transport of heat in the Atlantic) is greater than usual; and when it is low, eastward flow in winds and currents is less than usual.

The realization that widely separated phenomena may be linked because they are manifestations of the same climatic event has lead to the invention of the term **teleconnections**. Such linked phenomena may include effects on living organisms, which are much more marked than changes in individual aspects of climate, such as rainfall or air temperature. Organisms are affected by a number of environmental variables, so their life-histories are much subtler indicators of climatic conditions than trends in any one measured variable. Recently, for example, it has been found that the maize yield in Zimbabwe in southern Africa can be more reliably predicted from the state of the Southern Oscillation in the Pacific than from local weather forecasts.

Of course, El Niño events have dramatic effects on ecosystems in the tropical Pacific itself. By temporarily rearranging climatic zones over the course of weeks or months (so that regions that are usually wet become arid, etc.) they lead to mass mortalities of plants and animals, on land and in the ocean. Perhaps most fascinating of all, they may leave lasting marks on the compositions of ecosystems, particularly those on mid-ocean islands such as the Galápagos. During El Niño events, plants and animals, especially seeds, eggs or larvae, may be carried in winds and currents *eastwards* from the Indo-Pacific region, enabling them to colonize regions usually out of reach when the South-East Trades are strong (compare Figure 2.24a and b). It is also possible that by pushing certain species to the edge of extinction, leaving only isolated 'outposts', extreme conditions such as occur during El Niño events may set the stage for the evolution of new species.

Figure 2.27 (opposite) (a) Equatorward winds along the coast of Peru and Chile lead to offshore current flow (i.e. to the left of the wind, as this is in the Southern Hemisphere), causing divergence of water from the coast and upwelling of nutrient-rich water from below the thermocline.
(b) Schematic diagrams to show other types of wind fields that lead to upwelling:
(i) Trade Winds crossing the Equator (where the Coriolis force is zero) lead to a zone of upwelling along the Equatorial Divergence.
(ii) Cyclonic winds lead to divergence of surface water and mid-ocean upwelling (here shown for the Northern Hemisphere); by contrast, anticyclonic winds lead to convergence of surface water and downwelling.

Keeping the Earth habitable

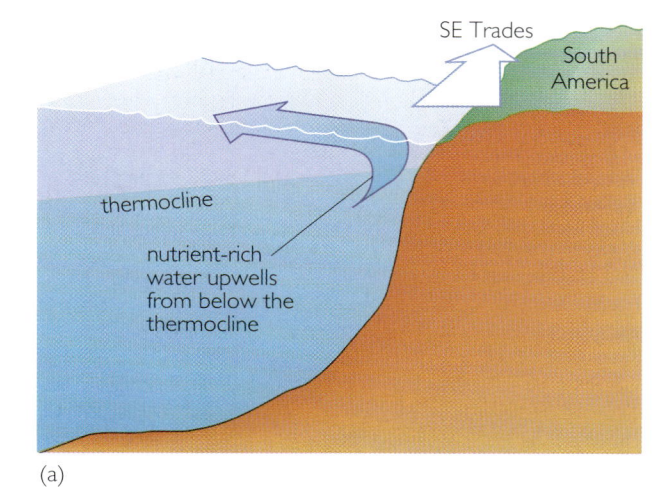

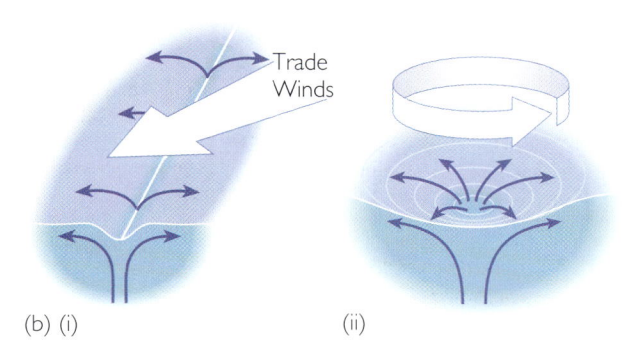

Upwelling and downwelling

Before moving on, we should look briefly at *why* a slackening of the South-East Trades causes a reduction in upwelling in the eastern Pacific. Water upwells from below if the pattern of wind acting on surface waters (often referred to as the *wind field*) causes them to move apart, or diverge. Conversely, water sinks, or downwells, if the wind acts so as to drive surface waters together, i.e. causes them to converge. Because of the Coriolis force, water acted on by the wind moves at an angle to it, so it is not immediately obvious which wind fields will give rise to upwelling or downwelling. As discussed in Section 2.2.1, the Coriolis force tends to turn flows to the right in the Northern Hemisphere and to the left in the Southern Hemisphere. As a result, the South-East Trades blowing *along* the western side of South America cause surface water to diverge from the coast, so that subsurface water rises to take its place. You do not need to worry about the details of exactly *how* coastal upwelling occurs, but instead look at Figure 2.27 which shows various wind patterns that lead to upwelling, (a) along coasts and (b) in the open ocean. As you will see in later chapters, patterns of upwelling and downwelling have important consequences for life in the oceans, and indirectly affect the climate.

Figure 2.28, which shows the upwelling and downwelling flow in July, was computed using the known wind field. You can clearly see the Equatorial Divergence, the band of upwelling that occurs where the South-East Trades cross the Equator (cf. Figure 2.27b(i)), and the areas of coastal upwelling along the western coasts of North and South America, and of Africa, caused by the Trade Winds (cf. Figure 2.27a).

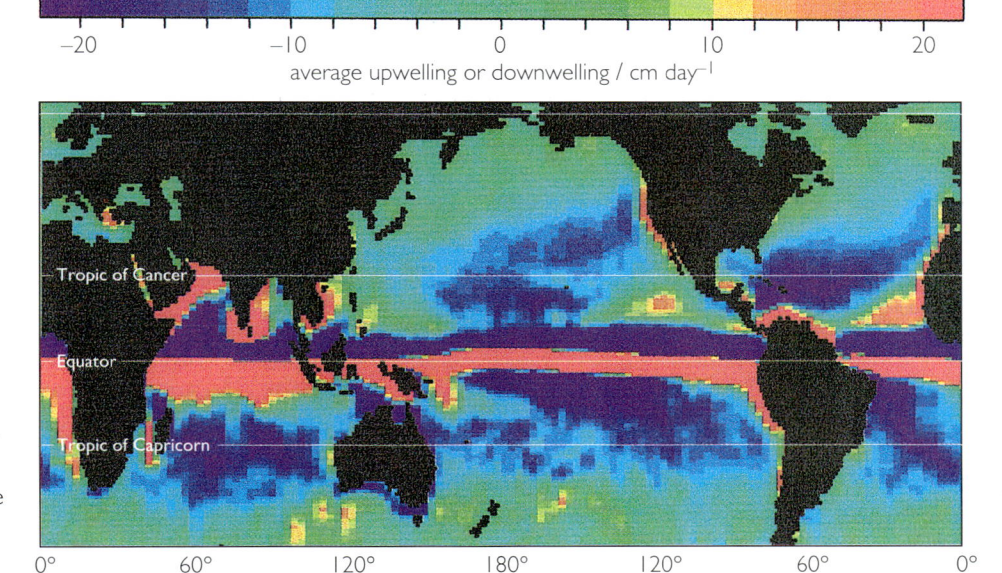

Figure 2.28 Upwelling and downwelling (vertical current flow in cm day^{-1}) for July, computed from the wind field. Areas of strongest upwelling (large positive values in the key) are pink; areas of strongest downwelling (large negative values) are blue.

- By reference to Figures 2.7 and 2.18, can you suggest why there is also some open ocean upwelling at latitudes above about 45°?

- Divergence and upwelling will occur in the subpolar gyres, which are driven by the cyclonic winds of the subpolar low pressure systems (cf. Figure 2.27b(ii)).

By contrast, the centres of the subtropical gyres, beneath the anticyclonic winds of the subtropical highs, are sites of convergence of surface water and sinking.

2.2.4 The deep circulation

Now let's turn to another type of atmosphere–ocean interaction – one that occurs mainly at high latitudes but is of profound importance for the global climate. As discussed earlier, the ocean's convection system is driven by cooling at the surface. At high latitudes in winter, in certain locations, cold winds cool surface water to such an extent that it becomes denser than the water beneath it. This results in instability, and the denser water sinks, displacing the water beneath, which rises to the surface to be cooled in turn. Convection cells are set up and the deep mixing leads to the formation of homogeneous bodies of water known as **water masses**. In particular regions, the water may become well mixed all the way to the sea-bed. The dense water masses that result from deep mixing circulate at great depths in the ocean and are known as *deep* (or *bottom*) *water masses*.

Figure 2.29 shows the main features of the convective or deep circulation in the Atlantic. The two northward-flowing water masses shown are known as Antarctic Bottom Water and Antarctic Intermediate Water; both of these water masses extend northwards in all three ocean basins, Antarctic Bottom Water flowing along the sea-bed, and Antarctic Intermediate Water at a depth of 1000 m or so.

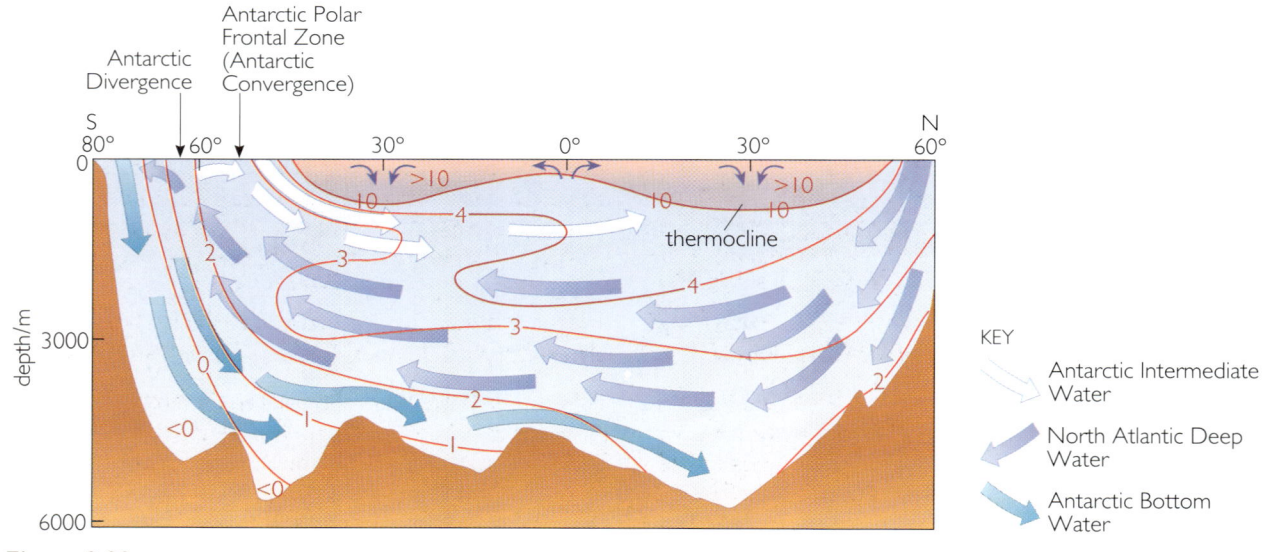

Figure 2.29
Generalized north–south cross-section of the Atlantic between 80° S and 60° N, showing flow of main water masses and resulting temperature distribution; values on the isotherms are in °C. At the Antarctic Polar Frontal Zone (or Antarctic Convergence), Antarctic surface waters converge with warmer subtropical waters and sink beneath them: at the Antarctic Divergence, water that sank in high northern latitudes eventually reaches the surface again. Small downward arrows near the surface at about 30° of latitude represent convergence in the subtropical gyres, and small upward arrows near the surface at the Equator represent upwelling at the Equatorial Divergence (cf. Figures 2.27b(i) and 2.28). *Note*: vertical scale is greatly exaggerated.

Keeping the Earth habitable

If you look at the temperature contours between about 40° S and the Equator, you will see that Antarctic Intermediate Water may be distinguished as a 'tongue' of water that is cooler than both the overlying surface water and the underlying deep water mass – North Atlantic Deep Water.

- Why, at first sight, would this seem to be an unlikely situation?
- Because cooler water overlying warmer water would seem to be an unstable situation, unlikely to persist for any length of time.

The solution to this conundrum is that Antarctic Intermediate Water is relatively fresh (for seawater), both because it forms in the subpolar regions where precipitation is high (Figure 2.7), and because it contains ice meltwater. North Atlantic Deep Water, by contrast, is relatively saline; that is to say, its concentration of dissolved salts – its **salinity** – is relatively high. So, despite being cooler than North Atlantic Deep Water, Antarctic Intermediate Water is less dense and so flows above it. Because the density of seawater is not simply determined by its temperature, but also by its salinity, the ocean's deep convective system is known as the **thermohaline** circulation, where 'thermo-' refers to its dependence on temperature, and '-haline' to its dependence on salinity.

Salinity of seawater may be altered in two ways: (1) addition of freshwater by precipitation (rain or snow) or its removal by evaporation; and (2) by addition of meltwater from ice, or removal of freshwater by freezing. When sea-ice forms (Figure 2.30), the ice itself is freshwater, so droplets of brine-rich water collect beneath it. This *brine-rejection* in cold polar waters may produce water sufficiently dense to sink to the sea-bed. It occurs extensively around Antarctica, contributing to formation of Antarctic Bottom Water; it also occurs around the Arctic Ocean, where ice forms over the continental shelf, and here it leads to formation of a cold, saline water mass which sinks to fill all the deep basins of the Arctic. Eventually, it escapes into the North Atlantic, either by flowing through the Canadian Arctic islands to the west of Greenland and out into the Labrador Sea, or by flowing out into the deep basins to the east of Greenland, and thence into the north-eastern Atlantic.

Figure 2.30 Newly forming sea-ice (also known as 'grease ice' or 'frazil ice') in the Greenland Sea. Note that this ice, forming from seawater, should not be confused with the polar and Greenland ice-caps, which form from snowfall.

Either way, this cold, saline Arctic water flows below regions in the northernmost Atlantic where surface water may be cooled by winter winds to such an extent that deep convection is triggered. Surface water may sink down to mix with the Arctic water at several thousand metres depth, and the resulting mixture is North Atlantic Deep Water. As the water sinking from the surface has been carried north in the Gulf Stream, it is initially fairly warm, at about 12–15 °C. It has been estimated that in winter about 1300 km^3 of water sinks from the surface of the northern North Atlantic each day, having been cooled by winds from 12–15 °C down to 1–4 °C.

Question 2.4
Assuming for convenience that the warm surface water cools by 11 °C, how much heat is given up each day to the atmosphere over the northern North Atlantic? (Use information from Box 2.1 as appropriate, and remember that 1 m^3 of water has a mass of 10^3 kg and there are 10^9 m^3 in 1 km^3.)

This is an enormous amount of heat – four orders of magnitude more than that supplied by solar radiation at ~ 55° N in winter (cf. Figure 1.9).

Having sunk, the waters flow equatorwards as North Atlantic Deep Water – still carrying heat, but much less than they carried polewards. Meanwhile, much of the heat given up to the atmosphere is carried eastward in the prevailing winds, so formation of North Atlantic Deep Water has important implications for the climate of Europe.

Furthermore, it is thought that as water sinking from the surface draws in yet more water in to fill its place, North Atlantic Deep Water formation effectively increases flow in the Gulf Stream. This idea, combined with what is known about the paths of water through the ocean, has led to the concept of the 'thermohaline conveyor belt', illustrated in Figure 2.31. Overall, heat is carried northward in both the North and the South Atlantic, and this is compensated for by a net transfer of heat into the South Atlantic from the Pacific and Indian Oceans. Of course, the flow pattern shown is a generalization – a particular parcel of water would be very unlikely to follow this particular path. However, to give you some idea of the time-scales involved, a parcel of water sinking in the northern North Atlantic would take of the order of a thousand years to come to the surface again in the northern Indian Ocean or the Northern Pacific.

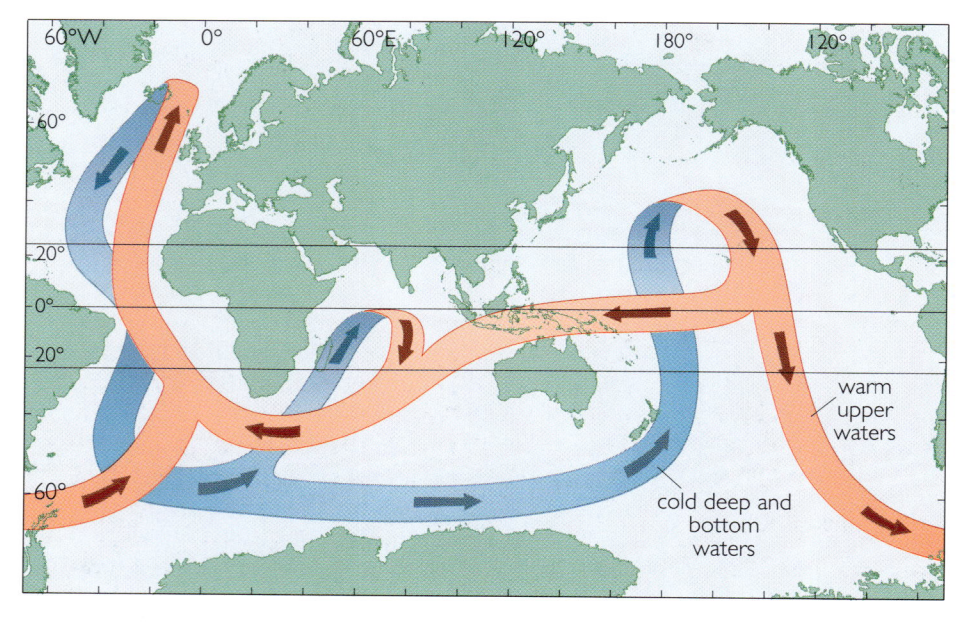

Figure 2.31
Schematic diagram of the 'global thermohaline conveyor', driven by the sinking of cold water at high latitudes. Warm water in the upper 1000 m or so of the ocean generally follows a pathway towards the northern North Atlantic; after sinking, cold water generally follows a path towards the northern Pacific and the northern Indian Ocean. You may see slightly different versions of this elsewhere, as the current flows in the Pacific and Indian Oceans (and the seas between) are not as well known as those in the Atlantic.

Interestingly, the Pacific has no water mass analogous to North Atlantic Deep Water. The reason for this seems to be that because of high precipitation, the salinity of surface seawater in the northern Pacific is so low that cooling and evaporation can never increase its density sufficiently for sinking to occur. So does this mean that formation of North Atlantic Deep Water can be 'turned off' by a decrease in the salinity of Atlantic surface waters? Perhaps so. Indeed, many scientists believe that NADW formation was turned off by a 'lid' of low-salinity meltwater between 11 000 and 10 000 years ago, so causing the period of cold climate known as the Younger Dryas.

As they flow through the oceans, water masses gradually mix with adjacent waters, but they may nevertheless still be identified many thousands of kilometres from their sites of formation. Because they acquire their characteristic temperatures and salinities when they are in contact with the atmosphere at the surface, and maintain these characteristics over long periods of time, water masses – or at least 'imprints' of them – are invaluable for studying past climate, as you will see in Chapter 5.

2.3 A closer look at climate

So far, we have been looking at the atmosphere, ocean and solid Earth mainly from a physical point of view, as if their composition was of no consequence. This is a convenient starting point but when we look closer and consider the chemistry of these 'spheres' of the Earth, and how they interact with the biosphere and with one another, the story becomes even more fascinating.

2.3.1 The atmosphere – bringing in chemistry

Table 2.1 lists the gases in the atmosphere in order of abundance. By far the most abundant gases are nitrogen and oxygen, which together make up ~99% of the total. Apart from nitrogen and argon, all of these gases affect the Earth's climate through their interaction with incoming (short-wave and long-wave) radiation, outgoing long-wave radiation, or both. As shown in Figure 2.32b, each gas absorbs particular wavelengths, often in more than one part of the spectrum.

Table 2.1 Gases naturally present in the atmosphere, and current concentration of each.

Gas	Approximate concentration by volume*
nitrogen, N_2	78%
oxygen, O_2	21%
argon, Ar (and other noble gases)	1%
water vapour, H_2O	~3000 p.p.m. (0.3%), but very variable
carbon dioxide, CO_2	~360 p.p.m. (0.036%)[†]
ozone, O_3	~0.01–0.1 p.p.m.
methane, CH_4	1.7 p.p.m.[§]
nitrous oxide, N_2O	0.3 p.p.m.

* Measured as the proportion of the total number of molecules in the atmosphere contributed by each component, equivalent to its proportion by volume in the atmosphere; p.p.m. = parts per million. Note that these values are averages.

[†] Increasing by ~1–2 p.p.m. annually.

[§] Increasing by ~0.01 p.p.m. annually.

Figure 2.32a shows, in effect, the spectrum of wavelengths emitted by the Sun, and the spectrum of wavelengths re-radiated by the Earth. The curve in (b) shows the percentage of absorption of radiation passing through the atmosphere, as a function of wavelength. This is a composite curve with the effects of the different atmospheric gases added together – some peaks are entirely attributable to a particular gas, others result from the combined effect of two or more gases.

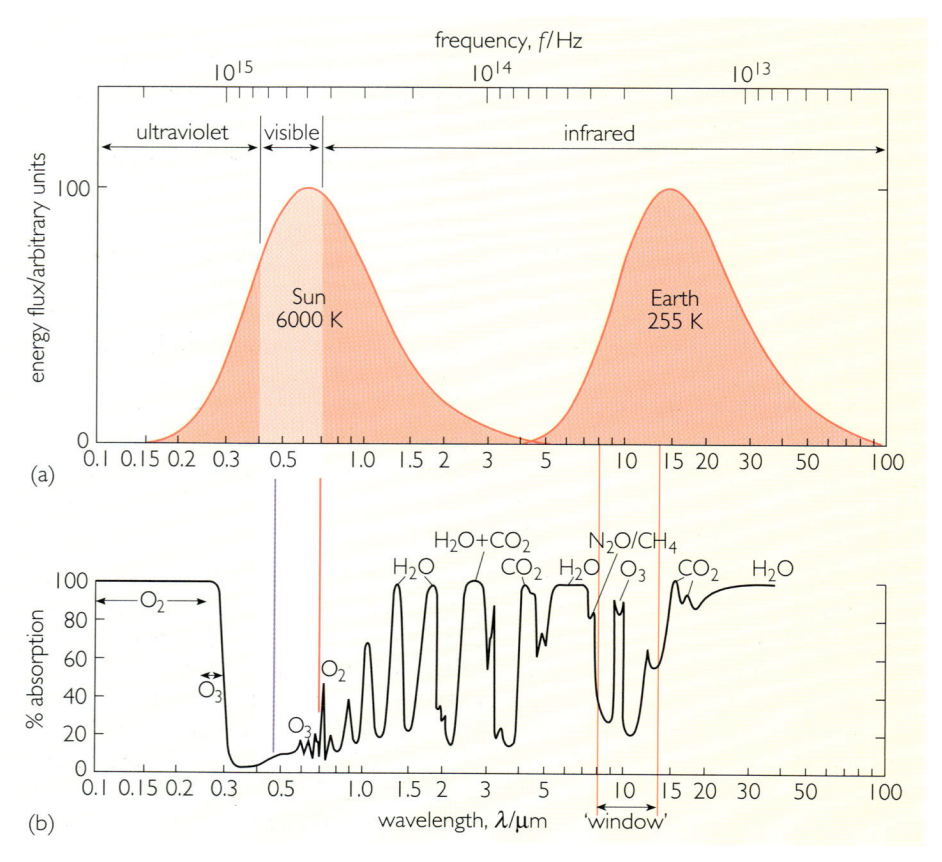

Figure 2.32
(a) The spectrum of electromagnetic radiation emitted by the Sun (assuming that it is a black body at 6000 K) and the Earth (assuming that it is a black body at 255 K – see the caption to Figure 1.16). Note that the Sun does emit radiation in wavelengths outside the range shown, but these contain such a tiny proportion of the total energy that they do not show up on this scale. (b) The 'absorption spectrum' of the Earth's atmosphere, showing the total percentage of radiation absorbed on passing through the atmosphere as a function of wavelength. This is a composite curve with the effects of the different atmospheric gases added together; CH_4 is methane and N_2O is nitrous oxide. Also shown within the visible band are the wavelengths of the radiation most used by photosynthesizing organisms (~ 0.45 and ~ 0.65–0.7 μm); note that these are the wavelengths *absorbed*, not the ones reflected (it is the latter which determine the colour of chlorophyll pigment). Both horizontal scales are logarithmic.

As far as oxygen and ozone are concerned, most absorption of incoming solar radiation occurs in the stratosphere. This absorption of energy results in that layer heating up directly, and it is *this*, not absorption of long-wave radiation from below, that causes the temperature inversion which defines the stratosphere (Figure 2.6). The radiation absorbed by oxygen and ozone in the stratosphere is in the short-wavelength 'tail' of the Sun's spectrum – the ultraviolet. Its absorption has important implications for life on Earth, because proteins and nucleic acids are damaged by radiation with wavelengths less than about 0.29 μm. As Figure 2.32b shows, for most ultraviolet radiation, absorption by oxygen provides an effective filter, but for frequencies approaching 0.29 μm, only ozone has any impact.

Keeping the Earth habitable

Figure 2.33
Schematic diagram to show the overall energy budget of the Earth–ocean–atmosphere system. Values for the outgoing radiation have been measured by satellite-borne radiometers; whereas the *re-radiated* radiation ('back radiation') has a longer wavelength than the incoming radiation, *reflected* radiation has the same wavelength after reflection as before. Other values are derived from model calculations or measurements, and you may find slightly different values given elsewhere. Because of the effect of greenhouse gases and clouds, the energy that the Earth's surface radiates (115 units) is greater than that originally absorbed as solar radition. This is why the Earth's average surface temperature (~15 °C) is greater than its effective planetary temperature (~−18 °C) (Figure 1.16).

Question 2.5

(a) To what extent is (i) visible radiation, and (ii) incoming infrared radiation, absorbed by atmospheric gases? Which gases are mainly responsible in each case?

(b) Which atmospheric gases are mainly responsible for the absorption of outgoing long-wave radiation?

So, the gases that absorb most energy from incoming solar radiation are oxygen and ozone, water vapour and carbon dioxide. For outgoing long-wave radiation, ozone, water vapour and carbon dioxide are again important, along with nitrous oxide (N_2O) and methane (CH_4). Remember that gases that absorb radiation, re-radiate energy at longer wavelengths. The absorption of outgoing long-wave radiation and the subsequent re-radiation, much of it back towards the Earth's surface, is what is referred to as the **greenhouse effect**. The range of wavelengths of 'terrestrial' radiation *not* subject to absorption by the atmosphere, and which can therefore be lost to space, is mainly in the range 8–13 µm; this is known as the **atmospheric window**.

Figure 2.33 summarizes the overall energy budget for the Earth and its atmosphere, including all the factors considered so far. The 100 'units' of incoming solar radiation represent the effective solar flux at the top of the atmosphere of ~343 W m^{-2} (Section 1.2). Note that (1) *at the top of the atmosphere*, incoming and outgoing radiation are in balance (100 units = 31 units reflected + 69 units re-radiated); while (2) *within the atmosphere*, total energy absorbed, from both the Sun and back radiation (23 + 106 + 24 + 7 = 160 units), balances the total emitted, both as back radiation and from the top of the atmosphere (100 + 40 + 20 = 160 units). The percentage of incoming solar energy reflected by the Earth (i.e. 6 + 17 + 8 = 31%) is close to our earlier estimate for the Earth's albedo (Section 1.2).

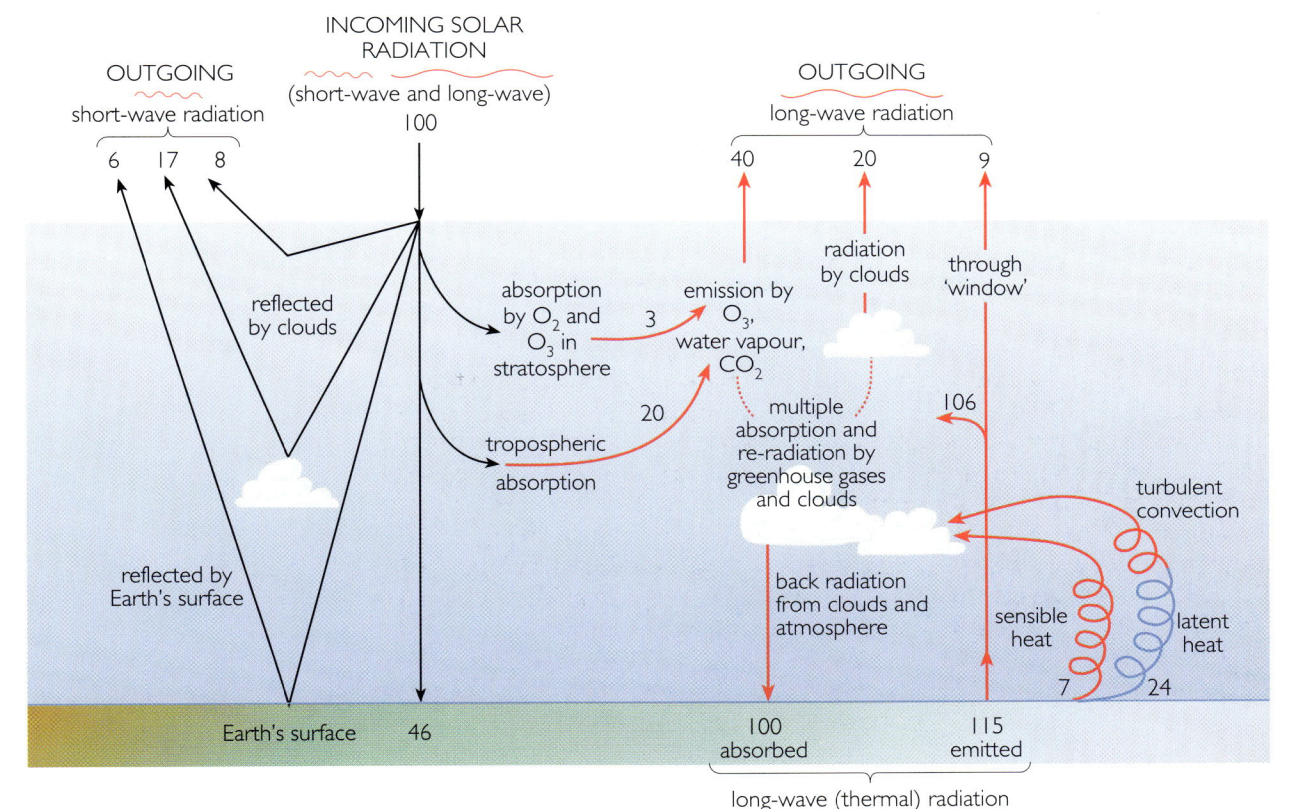

Note also that clouds absorb outgoing long-wave radiation (trapping warmth) as well as reflecting incoming solar radiation (cf. Box 2.2). Variations in cloud cover are thought to be the greatest modulator of the Earth's radiation balance, but determining the balance between reflection and absorption for different types of clouds is no mean feat.

2.3.2 Atmosphere–ocean–Earth: the support system for life

In Section 2.2, we considered the Earth's heating and air-conditioning system in terms of two convecting fluid envelopes – the atmosphere and ocean – meeting and interacting at the sea-surface. But atmosphere, ocean, and indeed the solid Earth and living organisms, are interlinked by the movement between them of actual material. Though the time-scale of movement can be months, millennia or even millions of years, depending on the circumstances and the elements involved, this continual redistribution may be described in terms of *biogeochemical cycles* – a concept which will be fundamental to our discussion of carbon in Chapter 3.

We've already considered the transport of water (and hence heat) by means of evaporation (particularly from the sea-surface at low latitudes), followed by condensation in cooler environments. This process is just one part – albeit a very important part – of the **hydrological cycle**, illustrated in Figure 2.34.

As mentioned at the start of this book, the Earth is unique in the Solar System in that it has large oceans of liquid water.

> **Question 2.6**
> (a) According to Figure 2.34, what percentage of the Earth's water inventory is in the oceans at the present time? And what percentage is in ice-caps and glaciers?
> (b) There is currently much concern about rise in sea-level as a result of global warming. If, for the sake of argument, one-quarter of the ice-caps eventually melted, what percentage increase in the volume of the oceans would result?

So while the ice-caps and glaciers contain most of the water that is on the land surface, they contain only a few percent of the global water inventory; all but 3% of the Earth's water is in the oceans.

The hydrological cycle is just one of a number of cycles that you will come across in this book, so it is worth using Figure 2.34 to define the concept of **residence time**, which in this case is the length of time an individual water molecule stays, on average, in any particular stage, or **reservoir**, of the cycle. It is calculated by dividing the amount of water in that particular reservoir by the amount entering (and leaving) it in unit time.

- How much water (in 10^{15} kg) moves through the atmosphere annually? And what is the average residence time of water in the atmosphere?
- According to Figure 2.34, the amount of water entering the atmosphere yearly is $336 + 64 = 400 \times 10^{15}$ kg (or the amount leaving it per year is $100 + 300 = 400 \times 10^{15}$ kg); in other words, 400×10^{15} kg of water moves through the atmosphere annually. On average, 13×10^{15} kg of water resides in the atmosphere at any one time, so the residence time of water in the atmosphere is: 13×10^{15} kg $/ 400 \times 10^{15}$ kg yr^{-1} = 0.033 yr or about 12 days.

Keeping the Earth habitable

![Figure 2.34: The hydrological cycle diagram]

low latitudes — **mid-latitudes** — **high latitudes**

total in atmosphere 13

36 blows over land
100 falls on land
300 falls on oceans
64 in evaporation and transpiration
336 evaporates from ocean
36 runs off from land
29 300 in ice-caps + glaciers
2 in biomass
229 in lakes + inland seas
70 in soil moisture
8400 in groundwater

(All quantities are in 10^{15} kg)

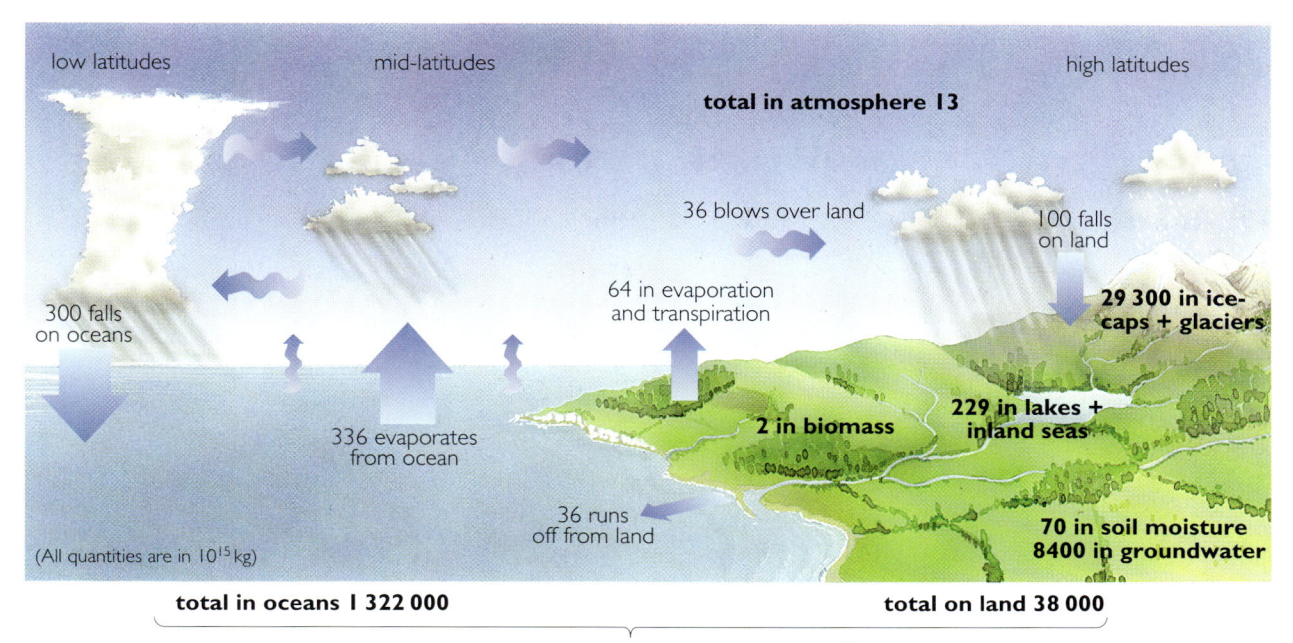

total in oceans 1 322 000 **total on land 38 000**

total world water supply = 1 360 000 × 10^{15} kg

Figure 2.34
The hydrological cycle, showing annual movements of water through the cycle and amounts of water stored in different parts of the cycle (often referred to as 'reservoirs' and here shown in bold type). Note the role played by vegetation in storing and cycling water. All quantities are × 10^{15} kg; note that 10^{15} kg liquid water occupies 10^3 km^3. The values given are estimates, and you may see different values elsewhere. For example, some authorities consider that groundwater storage accounts for a much larger proportion of the total water on land than shown here. The labels 'low latitudes', 'mid-latitudes' and 'high latitudes' are intended to remind you of the high precipitation at the ITCZ and in subpolar regions, and the high net evaporation in mid-latitudes; they are not supposed to imply that the processes shown are unique to those areas.

Figure 2.34, and the calculations you have just done, are appropriate to the Earth as it is at the present time. In the past, the Earth's climate has been very different (and, presumably, it will be so again): the hydrological cycle and processes driven by it will also have been different, operating at different rates or in different ways.

This is a convenient point to stop and consider the implications of what you have read so far, for life on Earth. In Chapter 1, you saw how the tilt of the Earth's axis and the shape of its orbital path affect the distribution of incoming solar energy over the surface of the globe, and in this chapter you have seen how the resulting pattern of surface temperature is modified by winds and currents. Intimately linked to the transport of heat over the globe is the transport of water.

▪ So, to what extent is the following statement true?

'Together, energy from the Sun and water-supply determine the distribution of primary production (mainly of plants) over the surface of the globe (cf. Figure 1.4).'

If you considered the statement carefully, you will probably have realized that it is only partly true. As any gardener or farmer knows, some areas of land are more fertile – i.e. will support more plant growth – than others. A fertile soil is one that contains in abundance the elements necessary for plant growth – i.e. **nutrients** – in a form that can be used by plants. To a large extent, fertility is determined by rainfall – except at oases, deserts are generally devoid of vegetation. For some time after an eruption, the slopes of a volcano are generally barren – only after rainfall has begun to break down the ashes and lavas does colonization by plants begin. Once the plants and associated organisms have begun to contribute to the formation of a layer of soil, the slopes are well on their way to being as verdant as those of older volcanoes (cf. Figure 2.20b(i)).

Some landscapes will never be very green, however much rain falls on them, because the minerals in the bedrock cannot be made *available* for plant growth (i.e. as dissolved ions) in sufficient amounts to support large stands of vegetation. Examples of such barren regions are steep rocky surfaces where no soil can accumulate, sand dunes and limestone pavement (Figure 2.35). In general, however, the distribution of land plants is limited by the availability of light for photosynthesis, temperature, and the availability of water. In the ocean, of course, water is never in short supply; furthermore, temperature is never too low to inhibit plant growth (see Figure 1.2). Nutrients, on the other hand, are not so plentiful. How do they get into the ocean in the first place?

The movement of water – a powerful solvent – through the hydrological cycle contributes to the cycling of other constituents through the Earth–atmosphere–ocean system. Let's start, for convenience, in the atmosphere, where gases, including CO_2 and SO_2, dissolve in rainwater. The result is a weak acid which, when it falls onto land, is neutralized by reaction with minerals in soil and rocks. **Chemical weathering** of these minerals by rainwater and soil water releases their constituent ions into solution.

Figure 2.35 Limestone pavement at the Burren in southern Eire. Plants can only grow in the cracks where some soil has built up and water from the abundant rainfall can be retained.

Looking at Figure 2.34, it would be tempting to conclude that because rivers flow into the ocean, seawater is simply a concentrated form of river water. But is it? Figure 2.36 shows the average chemical composition of rainwater, river water and seawater. Study these histograms carefully, taking note of the different vertical axes (and for the moment ignoring the fact that some columns of the river water histogram consist of two parts).

- ▪ Just comparing the histograms by eye (and ignoring the differences in concentration), would you say that seawater is closer in composition to rainwater or to river water?
- ▪ To rainwater: the rainwater and seawater histograms bear a striking similarity to one another.

You probably realized that one of the main differences between rainwater and seawater, on the one hand, and river water on the other, is the relatively high concentrations of dissolved Ca^{2+}, HCO_3^- and SiO_2 in the latter. These are supplied to rivers as a result of chemical weathering of carbonate and silicate minerals, both common constituents of the rocks making up the Earth's crust. What happens to remove these constituents from solution in seawater will become clear in Chapter 3.

Another marked difference is the relatively large amount of chloride in rainwater and seawater and the fairly small amount in river water. Only a tiny proportion of the chloride in the oceans comes from weathering. It originated in HCl gas released in volcanic eruptions (especially early in Earth's history when volcanism was more widespread than it is now) and is continually recycled via oceanic aerosols. These originate as droplets of seawater ejected into the atmosphere when conditions are rough, along with any dissolved constituents, gases and particulate matter that

Keeping the Earth habitable

Figure 2.36
The average chemical composition of (a) rainwater, (b) river water and (c) seawater, shown in terms of the eight most abundant dissolved constituents, some at concentrations too low to appear. TDS = total dissolved salts. In (b), the darker lower parts of the columns are the contributions from weathering; the lighter upper parts correspond to salts supplied via aerosols (see text). Note the changes in scale of ×15 from (a) to (b), and ×400 from (b) to (c). Concentrations increase partly because of the *addition* of new dissolved constituents during weathering, etc., and partly because of *loss of water* through evaporation and transpiration. Comparison of the top and bottom histograms shows that seawater is 5000 times $(34.4/(7.1 \times 10^{-3}))$ more concentrated than rainwater.

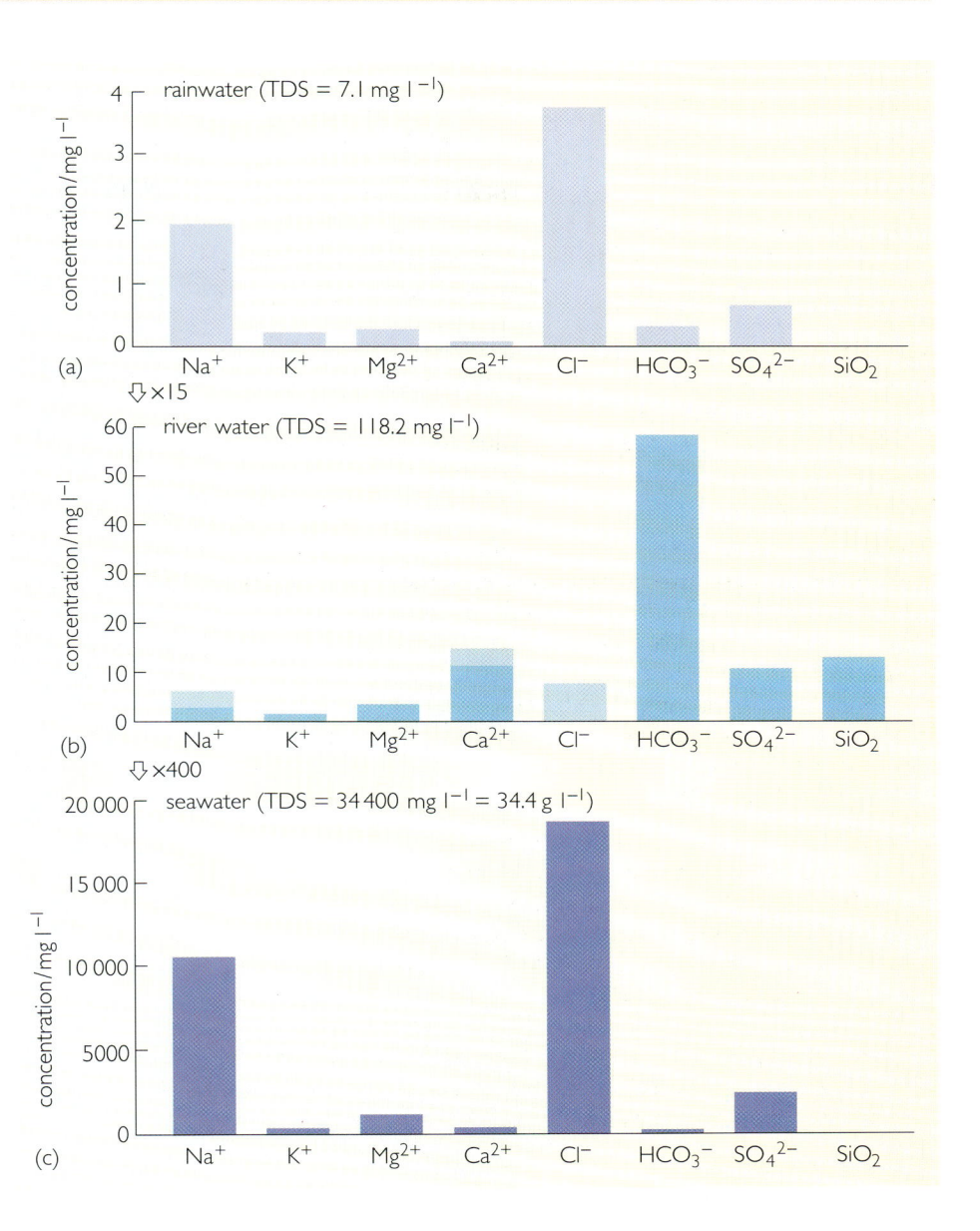

they may contain. Particles of sea salt that end up in the atmosphere in this way are amongst those acting as nuclei for cloud condensation (cf. Box 2.2), and their constituents are eventually rained out. As shown in Figure 2.36, while most dissolved ions in river water result from weathering (darker parts of histogram columns), a significant proportion of both sodium (Na^+) and calcium (Ca^{2+}), in addition to virtually all the chloride (Cl^-), comes from the ocean via rainwater (lighter parts of columns).

All of the 92 naturally occurring elements have been detected in seawater, in widely varying concentrations. Figure 2.37 (overleaf) summarizes the various ways in which dissolved constituents enter and leave the oceans. Note that for some elements, hydrothermal vents at spreading axes are an important source. Don't worry if you can't understand all the details shown; those aspects that we haven't yet touched on will be covered later.

The Dynamic Earth

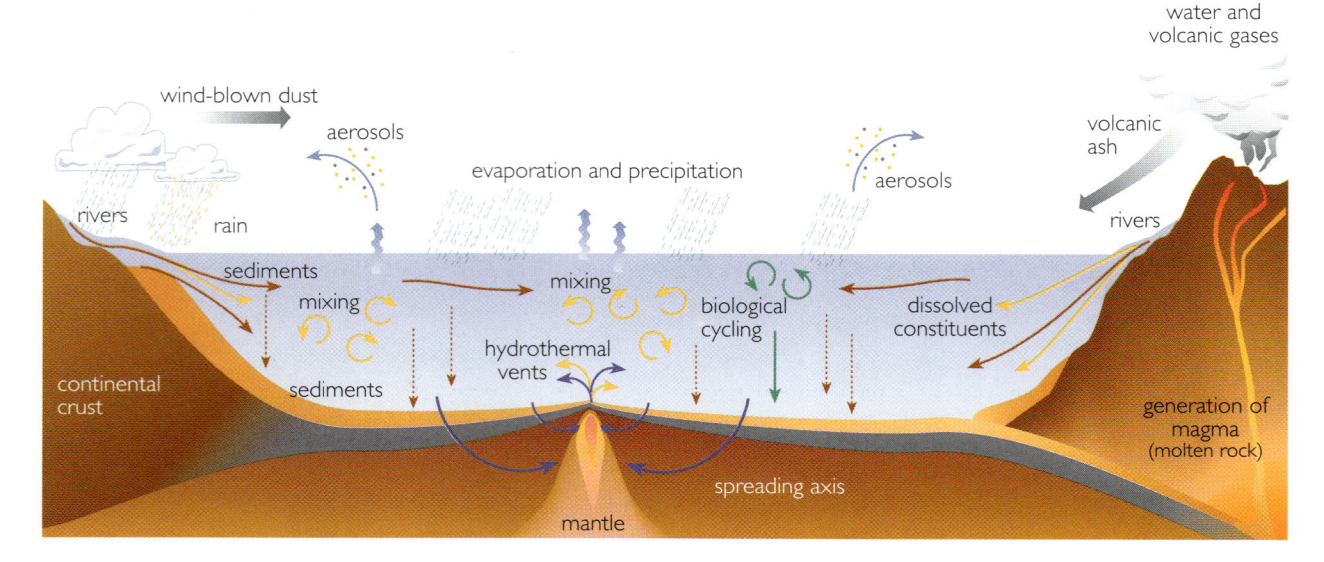

Figure 2.37
Diagrammatic cross-section illustrating the important role of the ocean in the global cycling of elements. Any unfamiliar terms will be explained later.

Gases are found in the ocean, as well as particulate and dissolved material. Figure 2.38 (below) shows the concentration in seawater of the four most abundant gases in the atmosphere. The data in (b) are given for a particular temperature and pressure because the solubility of gases generally *decreases* with increasing temperature and salinity, and *increases* with increasing pressure.

Figure 2.38
(a) Proportions by volume of the four most abundant gases in the atmosphere (together totalling 99.9%).
(b) Concentrations by volume of these gases dissolved in seawater at 24 °C as controlled by the solubilities of the gases at their atmospheric concentrations. Note the different logarithmic scales on the vertical axes. (b) has been plotted assuming that there is equilibrium (see Chapter 3) between atmosphere and ocean across the air–sea interface; for the four gases shown, this is probably valid, but only to a first approximation, for reasons that will become clear.

Keeping the Earth habitable

■ Which of the gases shown in Figure 2.38 is the most soluble in seawater? And which is essential for respiration of marine organisms?

■ Carbon dioxide is the most soluble, by a long way (the reason for this will become clear in the next chapter). Oxygen is essential for respiration.

Atmospheric gases dissolve into or escape from the surface ocean, depending on the relative concentrations either side of the air–sea interface (this is discussed further in Chapter 3). If it is very windy, and the sea-surface is rough, air and water mix together and the rate of gas transfer between them is increased. Once a gas is dissolved in the upper ocean, its transport throughout the ocean is the result of current flow and turbulent mixing.

■ Given what you know about ocean currents (Figures 2.2 and 2.29), would you expect to find living organisms at great depths in the ocean?

■ Yes, because dissolved oxygen is carried down into the deep ocean in the cold water masses sinking in polar regions.

So the ocean does not just contain dissolved constituents, particulate material and dissolved gases – it is the largest 'living space' on Earth, a soup of living organisms, large and small. If asked to name a marine organism, many people might say 'fish' or 'whale', but the most important organisms are too small to be seen by the naked eye. They are the primary producers, the minute floating algae, sometimes called 'the grass of the sea' – the **phytoplankton** (Figure 2.39a) – and the decomposers, **bacteria**. Feeding on these smaller organisms and on each other are minute animals, the **zooplankton** (Figure 2.39b).

Figure 2.39
(a) Living phytoplankton: the needle-shaped, cylindrical and chain-like organisms are diatoms, which are encased in silica. The irregular ones with 'horns' or spines are dinoflagellates. The field of view is about 1.75 mm across.
(b) Living zooplankton, including copepods (planktonic crustaceans) and the planktonic larvae of various animals. The field of view is about 1.75 cm across. (These photographs show samples collected by special plankton nets which concentrate the organisms; they are not found so closely packed in the open sea.)

Not surprisingly, biological activity plays an important role in changing the concentrations of oxygen and carbon dioxide below the sea-surface. Respiration by marine organisms, including bacteria, takes place throughout the ocean, at all depths.

■ Is this also true of photosynthesis?

■ No. Photosynthesis uses light, and can therefore only occur in near-surface waters.

The oceans are pitch dark below about 1000 m, and light is too weak for photosynthesis below ~ 250 m, even in the clearest water. Because they are plants, and need to photosynthesize to grow (Equations 1.1 and 1.2), phytoplankton (and the zooplankton that feed on them) live in the sunlit surface waters, in the **photic zone**. As far as phytoplankton are concerned, the most important of the non-gaseous constituents dissolved in seawater are nitrate (NO_3^-), phosphate (PO_4^-) and silica (SiO_2) – the nutrients. Mysteriously, N and P occur in seawater in the same molar ratio (15 : 1) that they occur in the soft tissues of marine organisms. Nobody knows whether marine organisms adapted to this seawater nutrient ratio, or whether they themselves established it over time.

(a)

(b)

Other constituents are also used by living organisms – for example, calcium, like silica, is used to build shells and skeletons, and sodium is one of a number of elements used in soft tissues. However, calcium and sodium are available in great abundance in seawater and so their precise concentrations are not important for living organisms. Nitrate, phosphate and silica, on the other hand, may be completely used up by phytoplankton in surface waters and so their concentration may be the limiting factor determining whether or not phytoplankton can grow. For this reason, these three constituents are described as **biolimiting**.

Recently, there has been much speculation as to whether, in certain circumstances and certain locations (e.g. the Southern Ocean around Antarctica), phytoplankton growth might be limited by lack of dissolved iron. Iron is found in seawater and in organisms in concentrations several orders of magnitude less than those of N, P and Si. It is therefore sometimes referred to as a *micronutrient*.

Most of the organic matter forming the soft tissue of phytoplankton is 'recycled' in near-surface waters, through consumption by zooplankton (and other pelagic* animals) and bacterial breakdown of detritus and excretion products. However, a small amount of debris escapes from surface waters by sinking down through the thermocline. As a result of decomposition of organic remains and dissolution of shells and skeletons, their constituents, including the nutrient elements, are gradually returned to seawater below the thermocline.

- How might they be returned to surface waters, where they can again be used by planktonic organisms?

- By upwelling of subsurface water, in response to divergence of surface water (cf. Figure 2.27).

Question 2.7
Figure 2.40 shows the concentration of chlorophyll pigment in the surface waters of the North Atlantic in May and December. Bearing in mind Figures 1.9 and 2.18:
(a) Explain why there is such a difference in the primary productivity in the northern North Atlantic at the two seasons of the year.
(b) Explain why, even in May, there is very little chlorophyll pigment in the central North Atlantic.

So, as mentioned earlier, vertical current flows in the ocean have important implications for phytoplankton populations.

Incidentally, it is important to bear in mind that the only reason we can so easily observe changes in marine primary productivity using satellite images is that phytoplankton increase their populations very fast but only live a short time, i.e. they have a very fast 'turnover rate'. It would not be so easy to observe the changes in productivity of a forest of long-lived trees.

As touched on above, it is possible that in some regions of the ocean, primary productivity may be limited not by the availability of the usual nutrients – nitrate, phosphate and silicate – but by the availability of iron. The clear blue waters of the tropical oceans are often described as 'barren' because of their small phytoplankton populations. To test the hypothesis that tropical waters are iron-limited, in 1995 a thousand square kilometres of the tropical Pacific was 'fertilized' by the addition of dissolved iron. The area concerned was to the south-east of the Galápagos Islands, at about 4° S, 104° W (cf. Figure 1.4) and the experiment was named IRONEX II.

* Pelagic means 'of the water column'; the equivalent term relating to the sea-bed is *benthic*.

Keeping the Earth habitable

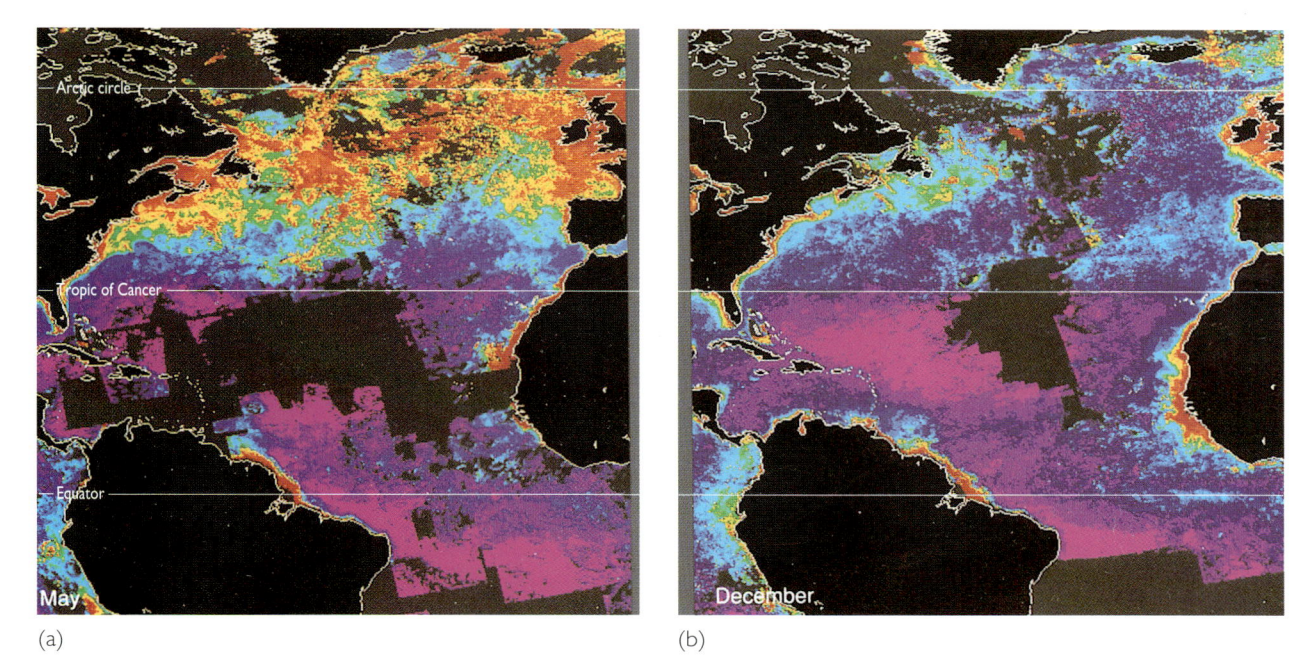

(a) (b)

Figure 2.40
Seasonal changes in primary productivity in the North Atlantic, as indicated by concentration of chlorophyll pigment in (a) May and (b) December, measured by the satellite-borne Coastal Zone Color Scanner. Pink corresponds to lowest chlorophyll concentrations, bright red to highest chlorophyll concentrations. Note that the areas of bright colour around the UK may be due partly to large amounts of suspended sediment and partly to real high primary productivity in response to nutrients (including agricultural fertilizer) contained in river runoff.

The large phytoplankton bloom that resulted from the addition of iron showed conclusively that primary production in tropical waters *is* iron-limited, at least in that region of the Pacific. As you will see from the next Chapter, such controls on the amount of marine primary production may have important implications for climate.

It seems likely that the more we learn about our planet, the more links we will find between the biosphere and the other components of the climate system – the atmosphere, hydrosphere and lithosphere. In the late 1980s, a link was proposed between phytoplankton in the ocean and cloud formation. Many phytoplankton have population explosions or 'blooms' during spring and summer (cf. Figure 2.40), and blooms of certain species produce the volatile sulfur compound dimethyl sulfide (DMS). Once in the atmosphere, DMS undergoes a series of photochemical oxidation reactions, so producing sulfur dioxide, SO_2, and then sulfuric acid aerosols, which – as you know from Box 2.2 – act as cloud condensation nuclei. This discovery excited many researchers, including James Lovelock, the originator of the Gaia hypothesis, because increased concentrations of 'greenhouse' CO_2 in the atmosphere might also lead to increased marine primary productivity – i.e. more phytoplankton and bigger plankton blooms.

Question 2.8
How and why might such increased primary productivity affect the Earth's albedo?

So, by producing bigger blooms, and hence more clouds, phytoplankton might counteract, or at least moderate, the effect of increased amounts of CO_2 in the atmosphere – an example of *negative* feedback, stabilizing the climate system. This idea was favourably received when it was first proposed in the 1980s, but was criticized by some scientists on the grounds that the large amounts of SO_2 added to the atmosphere by industry and fossil-fuel burning had not led to measurable increases in cloud cover.

We should remember that primary production *does* occur in the dark of the deep sea, by bacteria using chemical energy rather than the energy of sunlight. Like their counterparts photosynthesizing in surface waters, these primary producers form the bases of whole ecosystems, of which the best-known are the communities of animals found at hydrothermal vents. Unlike phytoplankton, however, chemosynthetic bacteria do not interact with the climate system in any significant way – at least as far as we know.

2.4 Sunspots and climatic change: the importance of keeping an open mind

Many people believe that there is a connection between the number of sunspots and the weather on Earth, but until recently nobody could see how this could be. We now know that sunspots are more numerous when the Sun is more active (Section 1.2), but it is still hard to see how weather patterns can be affected by the tiny fluctuations that occur in the Sun's brightness – about 0.2% over the ~ 11-year cycle. However, most of the change occurs in the ultraviolet which, as you know (Figure 2.32), is absorbed by oxygen and ozone in the stratosphere, causing warming. Furthermore, ultraviolet radiation breaks down oxygen to produce ozone, and so the more incoming ultraviolet radiation, the more ozone there is relative to oxygen. It seems that when the Sun is very active (large numbers of sunspots), stratospheric ozone levels can rise by as much as 2%, and the stratospheric warming that results is enough to cause poleward shifts in winter storms of ~ 700 km – which is consistent with observed shifts in weather patterns.

But what about 'average weather'? Can solar variability on the time-scale of the sunspot cycle affect the Earth's climate? Before moving on to consider this apparently unlikely possibility, try Activity 2.1.

Activity 2.1

Note: This is an important Activity: do try it, and study the answer carefully if you have any problems.

It has probably struck you that the factors that control conditions on the Earth's surface, or which drive climatic change (i.e. the *forcing factors* discussed in Chapter 1), act over a wide range of time- and space-scales. The longest-acting factor (though one it's easy to forget about) is the continual increase in solar luminosity since the formation of the Solar System. The most transient factors, which affect only relatively small areas of the Earth at any one time, are the phenomena we think of as weather.

Figure 2.41 is a partially completed diagram in which phenomena that affect conditions on Earth are grouped in 'bubbles' according to their time-scales and spatial scales. All the groups are more or less incomplete. The aim of this Activity, which we will come back to at several points during the book, is to progressively complete the diagram. For now, we ask you to plot onto it the phenomena listed below.

You do not need to plot the items very accurately. Don't worry if you are not sure of the spatial scale over which a phenomenon has an effect – in general you can assume that the longer the periodicity or time-scale, the more of the Earth will be affected, and vice versa.

1 Tropical cyclones. Meanders in the polar and subtropical jet streams.
2 El Niño–Southern Oscillation.

Please look at the answer to this Activity.

Keeping the Earth habitable

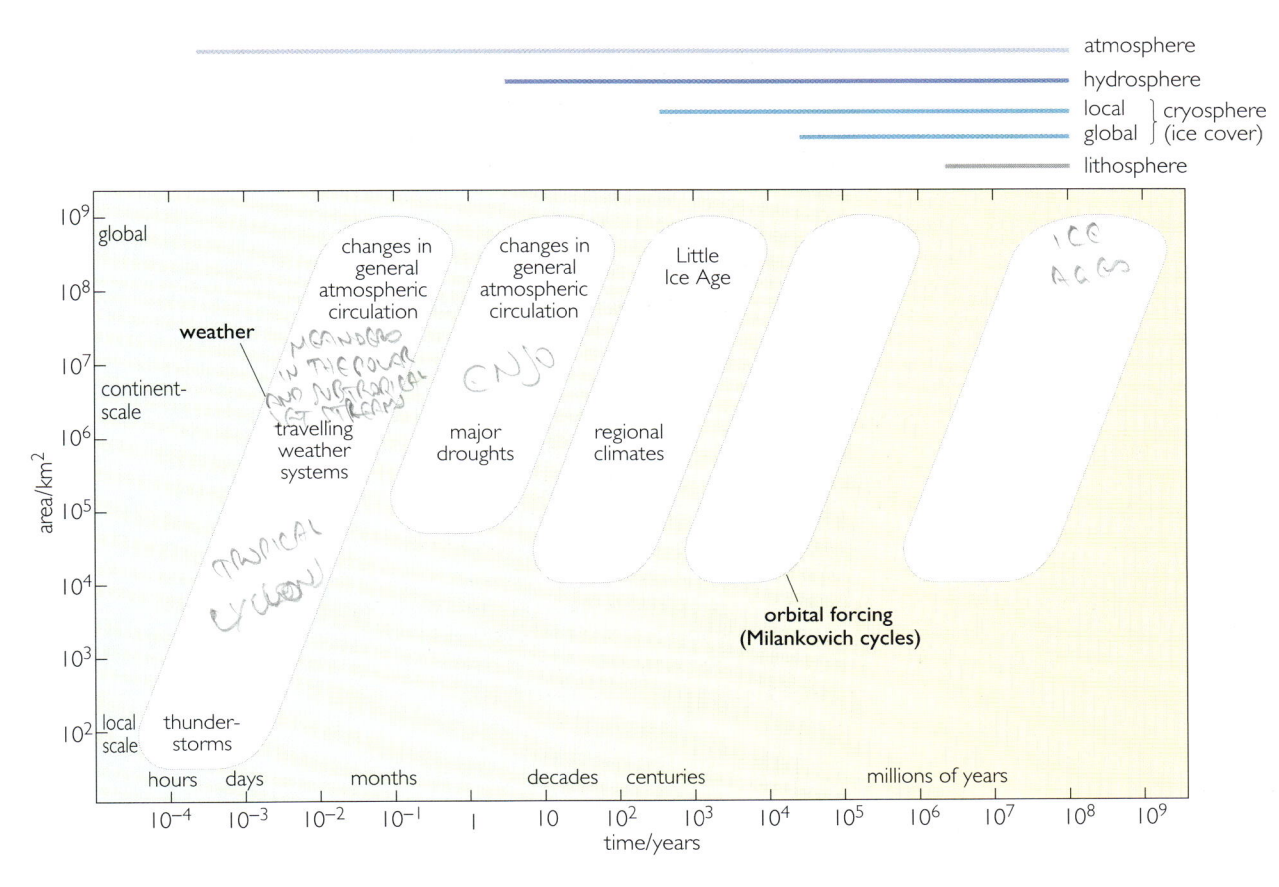

Figure 2.41
(For Activity 2.1.) Schematic diagram showing the time- and space-scales of various phenomena affecting conditions on Earth. (The continual increase in solar luminosity since the formation of the Solar System is not included.) The bars along the top are to indicate the general principle that while short-term variations are mostly limited to the atmosphere, longer-term variations involve progressively more components of the Earth system (as you will see in later chapters).

Figure 2.41 can be used to illustrate an important lesson: if you are looking for a mechanism to explain climatic change on a particular time-scale, you should look for forcing factor(s) *acting on the same time-scale*. For example, climatic variations occurring on time-scales of thousands of years (such as variations in the strength of the monsoons) are very likely to be related to changes in the Earth's orbit and/or tilt, i.e. to the Milankovich cycles.

For decades, some scientists have been claiming that waxing and waning of the Sun might drive climatic change, but the idea has never really been taken seriously by climatologists generally. Until the 1980s, there was no indication that the Sun's brightness was other than constant (at least on time-scales longer than centuries). Even after the ~ 11-year sunspot cycle was shown to be related to changes in the solar flux (Figure 1.18), there seemed to be no mechanism whereby changes in solar luminosity of as little as 0.2% acting over such short time-scales could be responsible for detectable changes in the climate. As discussed in Chapter 1, feedback effects in the Earth system generally take a while to come into play.

Recently, however, correlations have been established that make it difficult to ignore possible connections between solar activity and climate. Sunspot cycles are not all of the same length – cycles with high activity peaks are shorter than those with lower activity peaks (Figure 1.18a) – so the length of a solar cycle is effectively also a measure of solar activity.

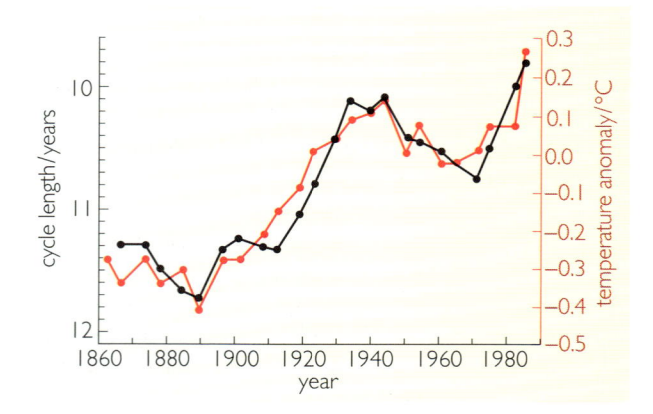

Figure 2.42
Black plot: variation of the length of the sunspot cycle (plotted at the middle of each cycle). Red plot: variation in the average air temperature over land, expressed as departures from the average for 1951 to 1980. (Plots of sea-surface temperature and ice cover instead of land air temperatures show similar correlations with the sunspot cycle.) (Adapted from Friis-Christensen, E. and Lassen, K. (1991) 'Length of solar cycle...', *Science*, **254**, 1 Nov., copyright © 1991 American Association for the Advancement of Science.)

- Bearing this in mind, what does Figure 2.42 suggest about the relationship between solar activity and the Earth's temperature, over time-scales of decades?

- It indicates that there is a strong correlation between solar activity and the Earth's temperature.

But how can this strong correlation be explained, given the short time-scale available for feedback loops to come into play? As hinted above, the time-scale may itself provide a clue.

- Which climatic phenomenon occurs at intervals of the same order as the periodicity of the sunspot cycle?

- El Niño, which generally occurs every 4–10 years.

This suggests that the types of mechanisms that operate to cause the periodic occurrence of El Niño events – notably strong coupling between the tropical atmosphere and ocean, with positive feedback between the two – might also act to magnify the effect of changes in incoming solar radiation acting over decadal time-scales. Thus, even changes in incoming solar radiation of as little as 0.2% could have disproportionately large effects on the Earth's climate.

We have concluded this chapter on the Earth's climate by looking again at the role of the Sun for two reasons. First, we should not forget that it is solar energy that drives the global climate system. Secondly, this new insight into how solar activity might drive climatic change on decadal time-scales provides an object lesson in scientific thinking: never dismiss an idea just because at first sight it looks unlikely.

2.5 Summary of Chapter 2

1. The continents have a much lower thermal capacity than the oceans, and so heat up and cool down much faster. This strongly affects the Earth's surface temperature distribution, particularly in the land-dominated Northern Hemisphere; in the oceanic areas of the Southern Hemisphere, temperature decreases more smoothly from Equator to poles, the temperature zones simply shifting northwards and southwards with the seasons.

2. Heat is redistributed over the surface of the Earth by winds, surface currents and the thermohaline circulation. Convection in the atmosphere is driven by

warming from below; that in the deep ocean is driven by cooling (plus evaporation and/or ice formation) at the surface. The evaporation, transport and condensation of freshwater is intimately linked to the redistribution of heat. Rising air generally has a high moisture content (as at the Intertropical Convergence Zone); sinking air (as at the subtropical highs) is usually dry.

3. The Coriolis force acts to deflect flows to the right in the Northern Hemisphere and to the left in the Southern Hemisphere. The Coriolis effect increase with increasing latitude, with the result that atmospheric flow in low latitudes is largely vertical, and takes the form of Hadley cells, while that at higher latitudes is more slantwise or horizontal, forming anticyclonic and cyclonic weather systems. The formation of mid-latitude weather systems is determined by the behaviour of the polar jet stream, whose undulations at the top of the troposphere lead to the transport of enormous amounts of warm air polewards and cold air equatorwards. Extreme undulations of the polar jet stream may lead to severe droughts and unusual weather patterns. The ocean also has 'weather' in the form of mesoscale eddies, which cause problems for those modelling climate because they are an order of magnitude smaller than atmospheric cyclones and anticyclones.

4. Tropical regions where the ITCZ moves seasonally over land are subject to seasonally reversing winds (monsoons) and hence dry seasons alternating with seasons of heavy rainfall.

5. The atmosphere and ocean are tightly coupled by positive feedback loops, especially in tropical areas. Tropical cyclones are a small-scale manifestation of this coupling; El Niño–Southern Oscillation (ENSO) events are Pacific-wide and have effects in the other oceans and at higher latitudes. Connections between widely separated phenomena, linked by the same climatic event, are known as teleconnections. Positive feedback loops cause instability in systems: negative feedback loops are stabilizing.

6. While light, suitable temperatures and availability of water are the factors limiting primary production (and hence most other life) on land, in the oceans the limiting factors are light and the availability of nutrients. Nutrients depleted from the sunlit surface layer may be returned to it via upwelling. Upwelling occurs along the Equator, along western coastlines under the Trade Winds, and as a result of cyclonic winds.

7. The formation of North Atlantic Deep Water results in the release to the atmosphere of large amounts of heat, responsible for warming north-western Europe.

8. Gases in the atmosphere, notably carbon dioxide and water vapour, absorb outgoing long-wave radiation and re-radiate energy, much of it back to the Earth's surface. This trapping of radiant energy (which also involves clouds) is known as the greenhouse effect.

9. In the hydrological cycle, water moves through various reservoirs, of which the largest by far is the ocean. Residence time in a given reservoir is defined as the mass in the reservoir at any one time divided by the rate of transport into (or out of) it.

10. Rainwater is weakly acidic and interacts with minerals in rocks and soil causing chemical weathering. This results in constituents of the minerals (notably Ca^{2+}, HCO_3^- and SiO_2) going into solution, and eventually entering the ocean. The ocean plays a major role in the cycling of elements through the Earth system.

11   While phytoplankton live in near-surface waters, in the photic zone, animals and bacteria live throughout the ocean, at all depths. This is possible because the cold water masses sinking at high latitudes carry dissolved oxygen down into the deep ocean.

12   Clouds both reflect incoming solar radiation and absorb long-wave radiation. Although they are a very important component of the climate system, and potentially the main driving mechanism for changes in the Earth's radiation budget, their role is by no means fully understood. Cloud condensation nuclei include sulfate aerosols, produced by volcanoes and industry, as well as forming from DMS, a waste product of phytoplankton populations.

13   Climatic change on a particular time-scale is likely to be driven by forcing factors acting on a similar time-scale. For example, the 11-year sunspot cycle could drive decadal climatic change through the types of mechanisms that come into play during El Niño events, involving strong coupling between atmosphere and ocean in the tropics.

Figure 2.43 summarizes much of what has been discussed about conditions at the surface of the Earth at the present time. As you will see, in the past the Earth has been very different.

Figure 2.43
Highly schematic map, showing general climatic features for land and sea, and the general position of the polar fronts in the atmosphere. HIGH and LOW refer to atmospheric pressure at the Earth's surface. Regions of deep and bottom water-mass formation are shown by blue hatching. AABW formation, in particular, is very localized, occurring at specific places within the region shown. Regions of upwelling are shown in green. This map is not intended to be comprehensive, but to remind you of some of the main points described in this chapter.

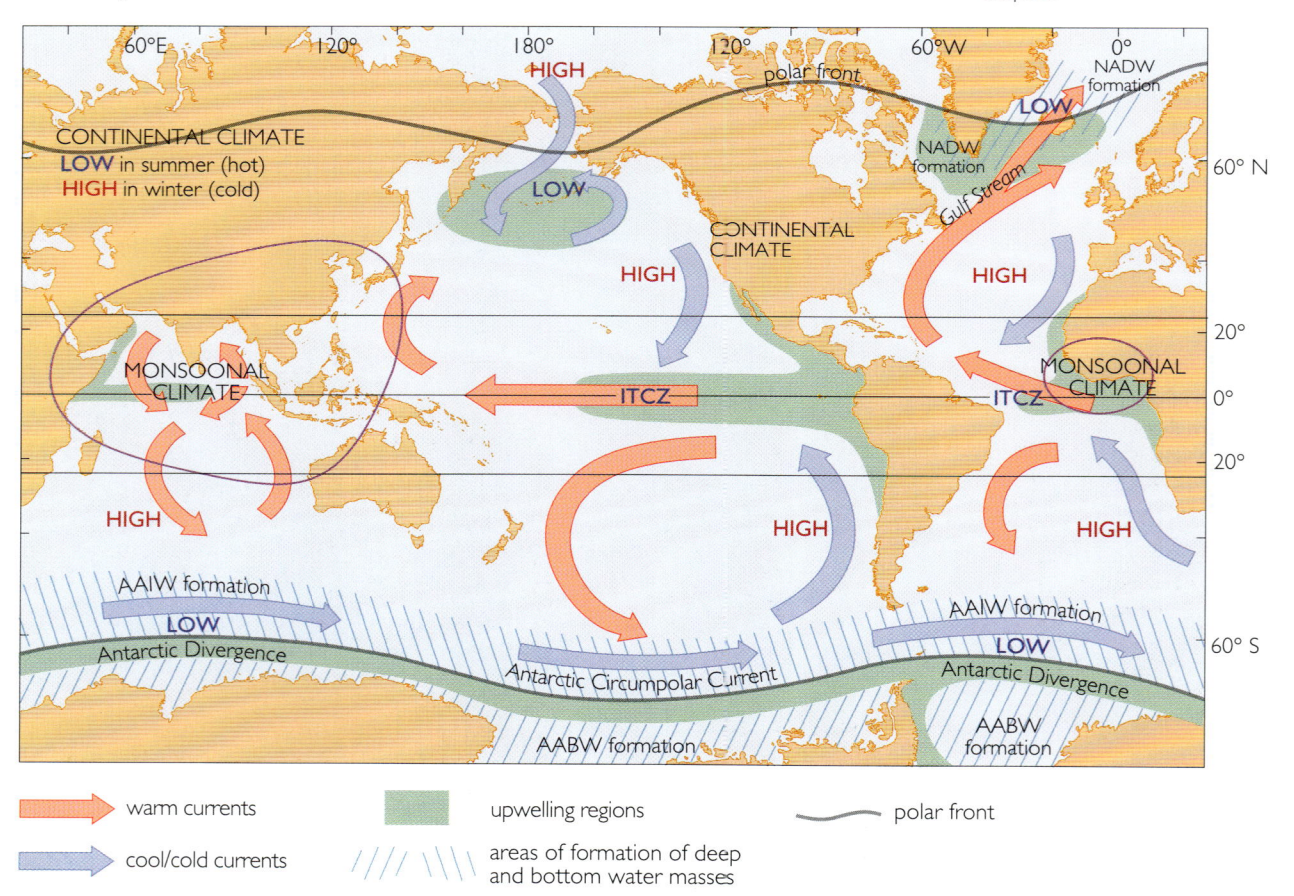

Keeping the Earth habitable

Now try the following questions to consolidate your understanding of this chapter.

Question 2.9
Suggest at least two ways in which vegetated regions (e.g. rainforest) play a different role in the climate system from arid desert regions.

Question 2.10
Figure 2.44 shows variation with latitude (averaged over the course of a year) of the incoming solar radiation at the top of the atmosphere, the radiation reflected by clouds and the radiation reflected by the Earth's surface.

(a)   (i) Very approximately, how much more radiation is reflected by clouds than by the Earth's surface? (To answer this question, you need to compare the areas under these two curves.)

      (ii) Is it possible to estimate the Earth's albedo from Figure 2.44? If so, is the value estimated for the albedo similar to that given in the text?

(b)   Why is there a maximum in the radiation reflected by clouds in the vicinity of the Equator?

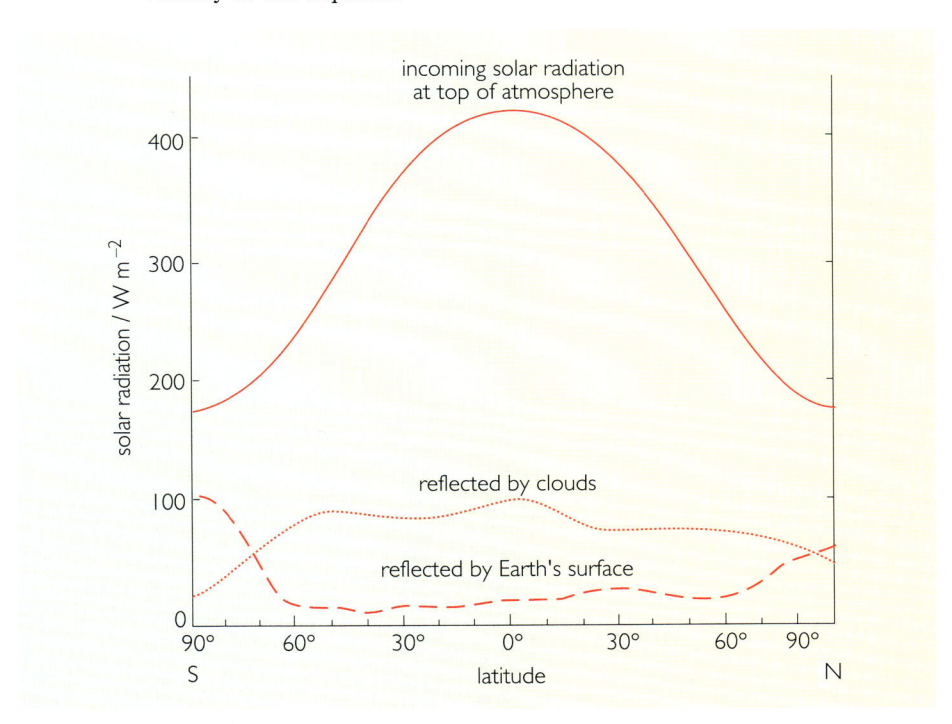

Figure 2.44 The variation with latitude (averaged over a year) for the incoming solar radiation at the top of the atmosphere (solid line), the radiation reflected by clouds (dotted line) and the radiation reflected by the Earth's surface (long-dash line).

Question 2.11
Figure 2.45 (overleaf) shows two images of the distribution of phytoplankton pigment around the Galápagos Islands, which lie on the Equator in the eastern Pacific. One shows the distribution when the region was under the influence of easterly winds, and the other the distribution when the region was under the influence of westerly winds.

(a)   Given this information, and remembering Figure 2.21 and the circumstances that lead to upwelling of subsurface nutrient-rich water, which of images (a) and (b) corresponds to the normal situation in the Pacific and which is more characteristic of the region during an El Niño event (see Figure 2.24b)?

(a) (b)

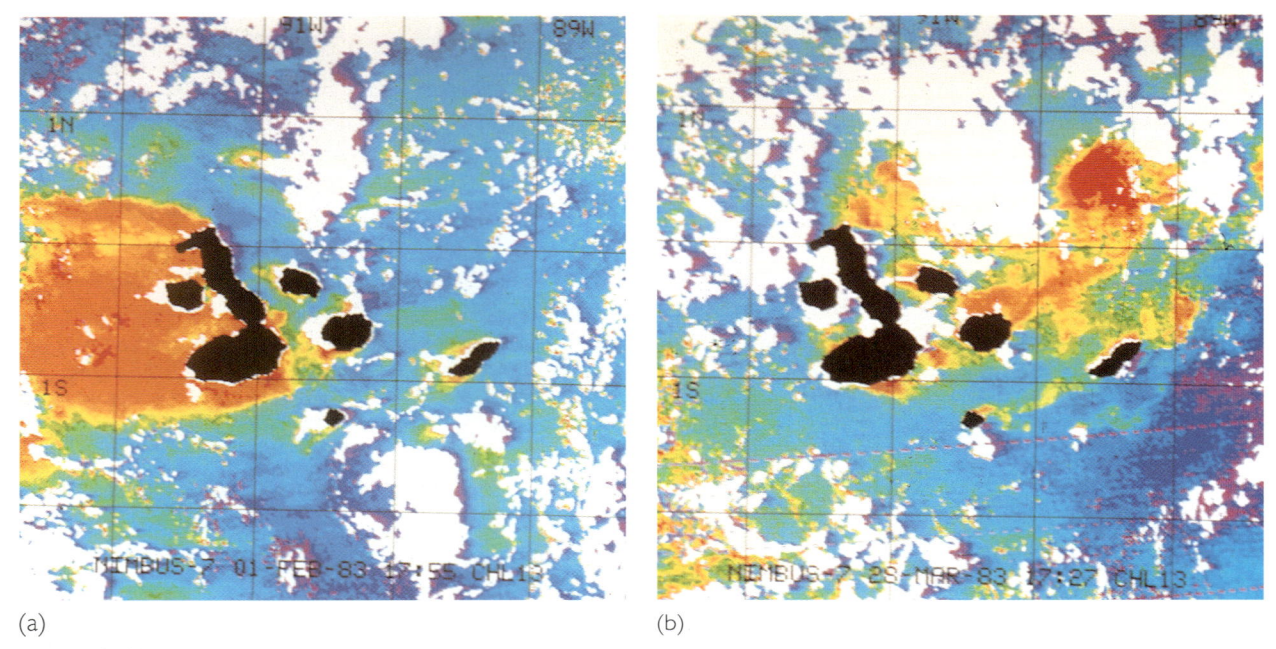

Figure 2.45
The distribution of phytoplankton pigment around the Galápagos Islands at two different times, based on data obtained by the satellite-borne Coastal Zone Color Scanner. Red corresponds to high productivity, blue to low productivity (the normal state of most tropical waters). (Feldman, G., Clark, D. and Halpern, D. (1984) 'Satellite color observations …', *Science*, **226**, 30 Nov., copyright © 1984 American Association for the Advancement of Science.)

(b) It has been suggested that the pattern of primary productivity shown in Figure 2.45 could be at least partly explained by 'iron-fertilization' of the surface waters. Where would the iron have come from?

Question 2.12
To what extent is the following statement true?

'The distribution of primary production over the surface of the globe is largely determined by the ways in freshwater and heat are redistributed by the circulatory patterns in the atmosphere.'

Chapter 3
The carbon cycle

All life, from bread mould to beetles, and from streptococci to students, is composed of similar combinations of carbon-containing molecules and water. In fact, life on this planet is based on the chemistry of carbon, which is why carbon-based compounds are often referred to as 'organic' (this can be confusing, as not all compounds so described have in fact been synthesized by organisms). Carbon can form literally millions of different chemical compounds in an enormous variety of forms: atmospheric gases, rocks, pond ooze, your breath warming your hands on a cold winter's day, and the skin that is being warmed – there is no end to the list.

3.1 Carbon and life

There are good reasons for carbon's unique role in the living world. Here are the main ones:

1 **Carbon can form compounds that readily dissolve in water as well as insoluble compounds.** Life is thought to have originated in water, and water is essential to life. Elements fundamental to the development of living organisms must be able to interact freely with one another, and that occurs most readily in the presence of water. Also, as discussed in Section 2.3.2, carbon dioxide in the atmosphere readily dissolves in water, as do many organic molecules that make up living organisms.

2 **Carbon can combine with other elements to make more compounds than any other element, and these compounds may be solid, liquid or gaseous.** It is continually cycled through living organisms, and between living, dead and inorganic components of the Earth.

3 **Carbon permits the storing and passing on of information.** The chemical properties of carbon allow the construction of large, complex three-dimensional molecules (RNA and DNA) which are unique in the extent to which they can store, replicate and convey large amounts of information.

4 **Carbon can store and release energy.** Carbon exists in both oxidized and reduced compounds. The energy released when carbon is transformed from a reduced to an oxidized state can be used for biochemical reactions in cells.

The underlying reason for all of these characteristics is the unique chemistry of the carbon atom. If you feel you need to be reminded about this and its consequences for carbon compounds, look at Box 3.1 (overleaf).

Chemical reactions can be of two types: **endothermic** reactions, which *require* energy, or **exothermic** reactions, which *release* energy. Reactions involving carbon are no exception. For instance, burning coal is an *exothermic* reaction in which the carbon in coal is combined with the oxygen in air, releasing energy in the form of heat and light.

From the perspective of life, the most important *endothermic* reaction of carbon is photosynthesis, whereby plants make organic compounds and oxygen.

■ The outside energy source is light from the Sun, but what are the raw materials used?

Box 3.1 Carbon compounds

The carbon atom has four unpaired electrons in its outer shell and forms covalent bonds by sharing these four electrons with other atoms. As a result, carbon atoms may form up to four separate bonds with other atoms, and so build complex three-dimensional structures. For example, carbon can share four electrons with four hydrogen atoms to form the simplest organic carbon compound, methane, CH_4, or share two pairs of electrons with two oxygen atoms to form carbon dioxide, CO_2 (Figure 3.1b,c). Moreover, carbon readily forms similar shared-electron bonds with oxygen, nitrogen, phosphorus, and sulfur – all key elements in living organisms. Most important of all, carbon can share electrons with other carbon atoms. This enables it to form large molecules such as sugars (e.g. glucose, Figure 3.1d), chain-like molecules such as **carbohydrates** (which are sugars strung together), and cyclic molecules. Figure 3.1e shows the structure of glycine, an amino acid – one of the building blocks of the large and complex protein molecules.

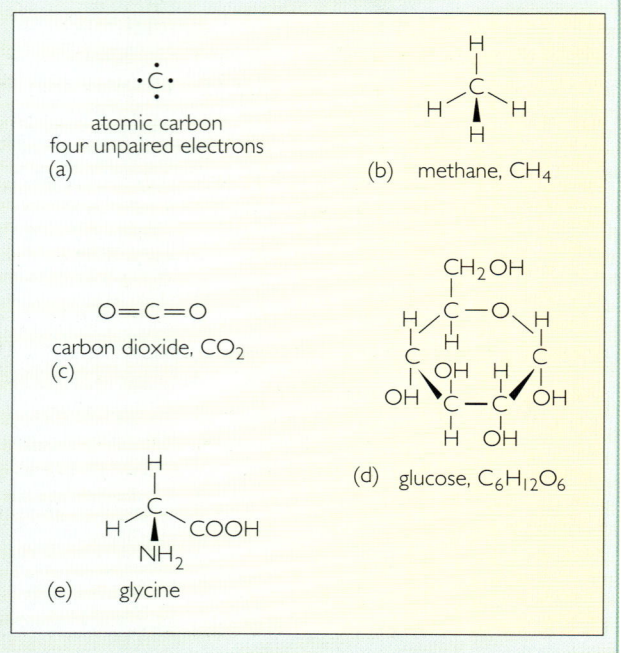

Figure 3.1
The carbon atom and some common carbon compounds.
(a) Atomic carbon, with its four unpaired electrons, which enable each carbon atom to form up to four bonds; (b) methane, a tetrahedral molecule; (c) carbon dioxide; (d) glucose, a sugar; (e) glycine, an amino acid.

■ They are carbon dioxide from the air, and hydrogen in water.

The carbon dioxide diffuses into the plant through *stomata* – special cells (mainly on the underside of leaves), which can open and close to the air and, along with the roots, regulate the gas and water balance of the plant.

Photosynthesis is a complex process involving many chemical reactions, but as mentioned in Section 1.1, it may be expressed as follows:

$$\underbrace{6CO_2}_{\text{carbon dioxide from atmosphere}} + 6H_2O \xrightarrow{\text{energy}} \underbrace{C_6H_{12}O_6}_{\text{organic matter}} + \underbrace{6O_2}_{\text{oxygen}} \qquad \text{(Equation 1.2)}$$

where organic matter is represented by the carbohydrate molecule $C_6H_{12}O_6$ (see Figure 3.1). The *total* amount of carbon fixed in this way by plants is called **gross primary production (GPP)**. Primary production (and, by extension, organic matter) is usually expressed in terms of grams of carbon (written gC) or kilograms of carbon (kgC).

Some of the CO₂ taken up by plants is returned to the atmosphere through the plants' **respiration**, which is essentially the reverse reaction:

$$C_6H_{12}O_6 + 6O_2 \longrightarrow 6CO_2 + 6H_2O + \text{energy} \qquad \text{(Equation 3.1)}$$

Respiration releases the energy stored in organic matter and provides energy for the plant to sustain itself – synthesizing tissues, reproducing and so on. The carbon that is *not* released back into the atmosphere via plant respiration is called the **net primary production (NPP)**. In other words, net primary production is the carbon that becomes incorporated into plant tissue. (Note that the term *respiration* encompasses what we commonly refer to as 'breathing' *and* 'eating'. Eating supplies organic matter to be oxidized with oxygen, releasing energy and CO₂ as a by-product. Breathing brings in the oxygen and removes the carbon dioxide. Both processes are combined in Equation 3.1.)

Because plants can manufacture their own food, they are referred to as **autotrophic** (from auto meaning 'self-', and trophic meaning '-feeding'). Plant tissue may be consumed by organisms that cannot photosynthesize and depend ultimately upon primary producers for energy; these include carnivores that consume animals that consume plants. Such organisms – be they fungi, cows or humans – are referred to as **heterotrophic** (hetero- means 'other-'). Like plant respiration, aerobic (oxygen-using) heterotrophic respiration uses oxygen to convert organic carbon back into CO₂ and water, releasing stored energy in the process (Equation 3.1). So plants cannot live without sunlight and we cannot live without plants.

> **Question 3.1**
> How would you classify the chemical reactions involved in (a) respiration (Equation 3.1) and (b) photosynthesis (Equation 1.2) in terms of oxidation or reduction (cf. point 4 at the beginning of this section)?

It is important to appreciate that the oxygen liberated in photosynthesis comes from breakdown of the *water* molecules. The hydrogen so produced reduces the CO₂ to form the carbohydrate molecules. In addition to being a reducing reaction, the conversion of CO₂ into organic matter during photosynthesis is an endothermic reaction (as we saw above), taking in energy in the form of sunlight. Taking in energy is a primary characteristic of living things; on death (and during respiration), the energy is released to their surroundings.

Incidentally, you may wonder why, if it releases energy, Reaction 3.1 does not simply occur spontaneously, so that we carbon-based life-forms would tend to spontaneously combust in the oxygen-rich atmosphere – which does occasionally happen if stories in the press are to be believed! The answer is that the reaction does occur, but its rate is very slow. Both plants and animals secrete enzymes (which are a type of catalyst) which allow them to regulate the rate of the reaction.

3.2 Carbon and climate

As discussed in Chapter 2, CO_2 and CH_4 are greenhouse gases that absorb long-wave (infrared) radiation emitted by the Earth's surface and re-radiate energy at longer wavelengths, much of it back towards the Earth (Figure 2.33). The warming caused by these and the other natural greenhouse gases (water vapour, ozone and nitrous oxide) means that rather than being about −18 °C, the average surface temperature of our planet is about +15 °C, favourable for the maintenance of liquid water and life. Indeed, without the 'blanket' of greenhouse gases, the Earth's surface would be completely covered in ice.

The current interest in atmospheric CO_2 in the context of global warming is heightened because we have good evidence that its concentration has varied over the history of the Earth, and that these variations are at least partially linked to changes in the Earth's climate. We are therefore very keen to pin down the factors that control or affect atmospheric CO_2. For reasons that will become clear later, the best way to begin to attack this problem is to study the fluxes of carbon through the natural carbon cycle.

3.3 The natural carbon cycle: a question of time-scale

Largely because of its involvement in living processes, carbon continually cycles through the different components of the Earth – the biosphere itself, the atmosphere, the ocean and even the outer part of the solid Earth, the lithosphere. This global cycling of an element used by organisms – for example, carbon, oxygen, nitrogen, phosphorus or sulfur – in and out of living, dead and inorganic reservoirs (cf. Figure 2.37), is a **biogeochemical cycle** (a term you have already met in Section 2.3.2). Such elements periodically move between reservoirs and rearrange themselves into different compounds; they may be thought of as the currency exchanged between Earth and life.

For convenience, we can begin by identifying six main carbon reservoirs, as shown in Figure 3.2.

- What are these reservoirs, in order of decreasing size?

- They are:
  1. the deep ocean;
  2. deep-sea sediments;
  3. soil, including organic debris;
  4. the upper ocean (including its biota);
  5. the atmosphere;
  6. terrestrial organisms (the land biota).

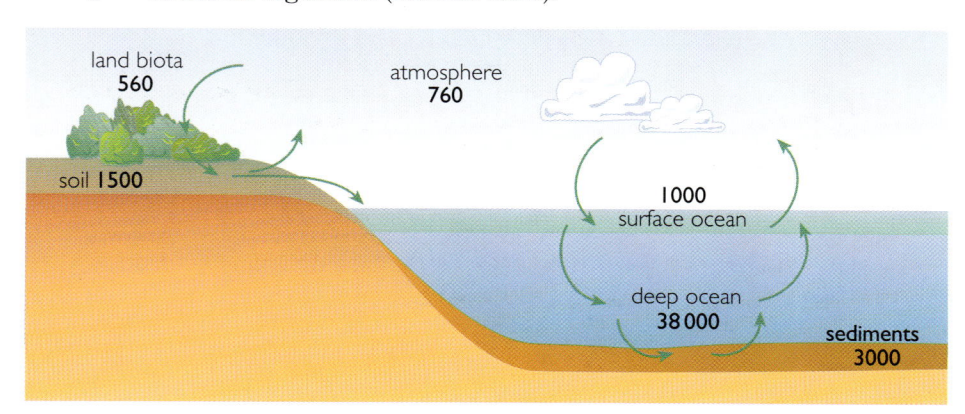

Figure 3.2
A schematic representation of the natural carbon cycle, with six main reservoirs as described in the text. 'Biota' means living organisms, in this context mainly plants. Bold numbers give the sizes of the reservoirs in 10^{12} kg). In the rest of this chapter, we will be putting estimated values onto the fluxes indicated by the arrows.

The carbon cycle

The different amounts of time that carbon atoms spend in each of these reservoirs, on average, have enormous implications for their potential to affect climate.

The overall cycle in Figure 3.2 is actually a number of cycles that link together and, as you will see, the cycling of carbon occurs on a number of different time-scales: a short biological time-scale of months/years to decades, an intermediate time-scale of up to hundreds of thousands of years involving chemical, biological and physical components, and – not shown on Figure 3.2 – a geological time-scale of up to hundreds of millions of years, involving rocks and sediments.

As we have to begin somewhere, let's start by looking at the carbon cycle driven by land-based organisms, often referred to as the *terrestrial carbon cycle*.

3.3.1 Short time-scales: the terrestrial carbon cycle

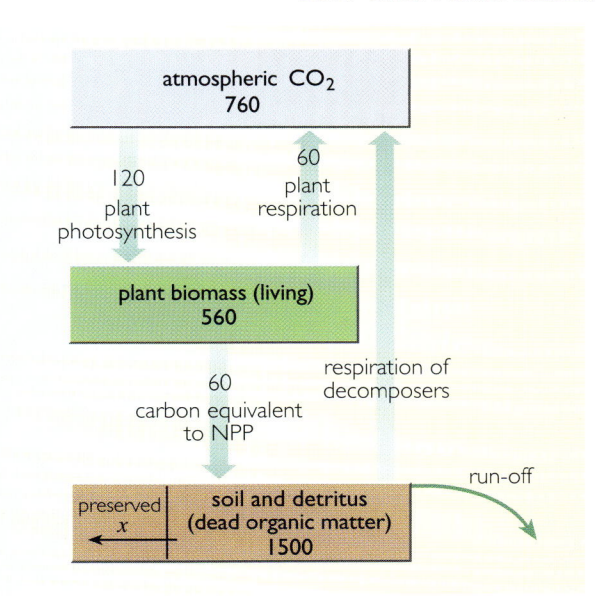

The terrestrial carbon cycle is dominated by the processes of photosynthesis and respiration and involves three major reservoirs: the atmosphere; plant biomass; and soil organic matter (including detritus) (Figure 3.3). The reservoir sizes of plant biomass and the atmosphere are similar at approximately 560×10^{12} kg and 760×10^{12} kg, respectively; soils store about three times as much carbon as plants (about 1500×10^{12} kg).

- Why have we not shown a box for terrestrial animal biomass?

- Animals are an insignificant carbon reservoir when compared with plant biomass (in fact, only about 0.01% of it).

The land surface of the Earth is about 150×10^6 km^2, of which a large proportion is covered by plants (Figure 1.4).

Figure 3.3 The biospheric carbon cycle on land. The bold numbers are the sizes of the reservoirs and the numbers on the arrows are the sizes of the annual fluxes (both in 10^{12} kgC).

- If terrestrial biomass were spread evenly over the land surface, how much (in kilograms of carbon, kgC) would there be per square metre?

- Dividing 560×10^{12} kg by 150×10^6 km^2 gives ~3.7×10^6 kgC per km^2 land surface or 3.7 kgC per m^2.

Clearly, this average conceals huge variations – think of the amount of plant biomass in 1 m^2 of lawn compared with the average in 1 m^2 of forest.

Plants convert about 120×10^{12} kg of carbon from atmospheric CO$_2$ into organic carbon each year. This is the annual terrestrial gross primary production, or GPP. About half of this is returned to the atmosphere and soil as CO$_2$ through plant respiration, so that *net* primary production (NPP) – the annual *net* fixation of carbon from the atmosphere into terrestrial biomass – is about 60×10^{12} kg yr^{-1}. Most of this carbon goes into soil organic matter when vegetation dies or trees shed leaves or needles, so the amount of carbon deposited in the soil is also about 60×10^{12} kg yr^{-1}. This is balanced globally by an approximately equal amount of carbon released back to the atmosphere when plant debris decomposes or vegetation burns in natural fires (at this point, we shall ignore the effect on this flux of human activities, such as deforestation). (As mentioned above, the amount of carbon that goes into animal biomass is trivial compared with plant biomass; respiration fluxes from animals – particularly the less numerous 'higher' animals – are also small.) The return of CO_2 to the atmosphere from 'recycled' plant material is therefore almost entirely accounted for by the respiration of decomposers (bacteria, etc.) in the soil.

As discussed in connection with the hydrological cycle (Figure 2.34), if we know the size of a reservoir and the flux of a substance into and out of it, it is a simple matter to calculate the **residence time** of the substance in the reservoir.

- How?
- Residence time = $\dfrac{\text{mass of substance in reservoir}}{\text{flux into (or out of) it}}$.

In doing such a calculation, we are assuming that an equilibrium has been reached, i.e. that the fluxes in and out are equal. Imagine a plant that has been uprooted. Within a short time it wilts: the flux of water molecules into the roots has been cut off, but the flux out of the leaves has not. The fluxes of water are no longer in balance and so the size of the reservoir (the amount of water in the plant) decreases.

Question 3.2

(a) What is the average residence time of carbon in plant biomass? What is it in soil organic matter?

(b) In global terms, 560×10^{12} kg is a relatively small reservoir of carbon (cf. Figure 3.2), and 60×10^{12} kg yr^{-1} is a relatively large flux. Would you expect large fluxes through a small reservoir to result in long or short residence times?

The terrestrial biospheric carbon cycle is characterized by relatively large global fluxes through small reservoirs of biota (living or dead), with relatively short residence times. Although the average residence time of organic carbon in soils is about 25 years, it varies greatly depending on the composition of the organic matter and the location: most carbon in fresh litter is decomposed in a year or two, but highly resistant organic matter in the same soil (such as the carbon in plant structural material, **lignin**) may take much longer to decompose. In fact, carbon in **anoxic** (oxygen-poor) soils may never fully decompose to release CO_2 back to the atmosphere, and may be stored as residues in peat bogs, swamps or similar environments, perhaps eventually becoming coal. However, the amount of carbon preserved in this way is a very small proportion of the total plant debris, and very difficult to quantify; it is indicated by x in the 'soil and detritus' reservoir in Figure 3.3.

- So what is the size of the 'respiration of decomposers' flux in Figure 3.3?
- It must be $(60 - x) \times 10^{12}$ kg yr^{-1}.

In other words, Figure 3.3 shows that the carbon in the CO_2 produced by respiration of organisms decomposing plant debris in the soil over the course of a year, must equal the carbon added in plant debris *minus* the amount that has been preserved and effectively removed from the cycle *plus* that carried away in run-off. Both of these are very small, so the carbon added to the soil in plant debris and the carbon removed from the soil in the CO_2 produced by respiration of decomposers are therefore more or less equal (both about 60×10^{12} kg yr^{-1}), and together equal the amount of carbon fixed as (gross) primary production (about 120×10^{12} kg yr^{-1}).

Although the amount of organic carbon removed from soils in run-off and exported by rivers to the sea each year is small, removal rates are locally higher if there is soil erosion, because of exceptionally high rainfall and floods, or as a consequence of human activities such as farming, mining or forestry. This organic carbon may be either **particulate organic carbon (POC)** (fragments of soil or organic debris) or **dissolved organic carbon (DOC)**. This, together with inorganic carbon removed in the weathering of rocks (dissolved inorganic carbon, DIC – see later, in Box 3.2), gives a run-off flux of about 0.4×10^{12} kgC yr^{-1}.

Difficulties in estimating the *sizes* of the different reservoirs – the amount of carbon stored globally in each of the various categories we have identified – impose significant limitations on the accuracy of any calculations of residence time. At its simplest, making a global estimate of a carbon reservoir involves multiplying two numbers: the global area (or volume) of the reservoir and the average amount of carbon in a representative area (or volume) of the reservoir. In practice, arriving at these two numbers is difficult and requires making assumptions and approximations. For example, to estimate the amount of carbon stored globally as plant biomass, one could begin by categorizing the vegetation of the Earth into a number of distinct ecosystem types, or **biomes**, which characteristically store different amounts of carbon. Figure 3.4 shows a global distribution of terrestrial biomes based on climatological and geographical considerations; in practice, satellite images are often used to determine such distributions (cf. the land areas on Figure 1.4). For each biome, one could:

1. Lay out sampling grids (e.g. 10 m^2, 1 km^2, etc. depending on the type of biome) in numerous different parts of the biome covering all the known variability.
2. Measure the amount of carbon contained in each of the grid squares.
3. Convert that to an average for the entire biome.
4. Multiply that average amount by the global area of the biome.

At best, the final number can only represent an approximate biomass for an extremely varied reservoir. Total global biomass would be obtained by repeating the process for every biome defined, and adding the individual totals together.

The fourth column of Table 3.1 (overleaf) shows the results of one such calculation for the total mass of carbon in terrestrial biomass. Here, the land surface is divided into 14 different ecosystem types (column 1), the area of each ecosystem type is calculated (column 2), and then for each the mean plant biomass (in terms of carbon) per square metre is estimated (column 3). Columns 2 and 3 are then multiplied together to give an estimate of the total mass of carbon in this ecosystem type at any one time. All such estimates may then be summed to provide a total global estimate. The last two columns show the net primary productivity (NPP) for each of the ecosystem types, i.e. the global net fixation of carbon into plant material per year.

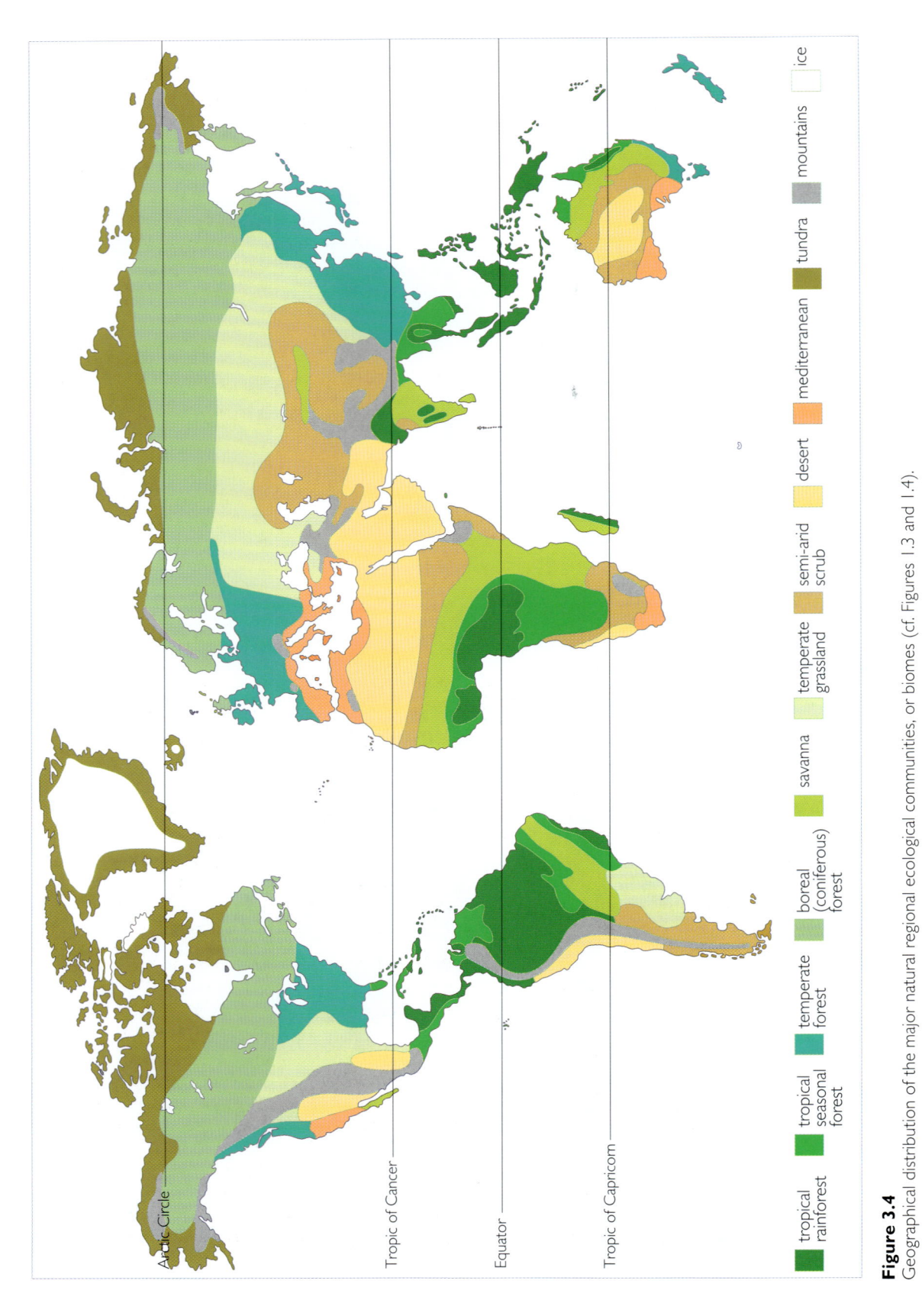

Figure 3.4
Geographical distribution of the major natural regional ecological communities, or biomes (cf. Figures 1.3 and 1.4).

The carbon cycle

Table 3.1 Estimates of terrestrial primary production (standing stock) and net primary productivity (i.e. rates of primary production). You are not expected to remember the details of this table. It is included so that you can see the sizes of numbers involved, and the magnitude of the problem of making global estimates of production and productivity. (The last column in the last row has been left empty for completion in Question 3.3.)

Ecosystem type	Area/ 10^6 km^2	Mean plant biomass per unit area/ kgC m^{-2}	Total plant biomass/ 10^{12} kgC	Mean NPP per unit area/ kgC m^{-2} yr^{-1}	Total NPP/ 10^{12} kgC yr^{-1}
swamp, marsh	2	6.8	13.6	1.125	2.2
lake, stream	2.5	0.01	0.02	0.225	0.6
cultivated land	14	0.5	7.0	0.290	4.1
rock, ice, sand	24	0.01	0.2	0.002	0.04
desert scrub	18	0.3	5.4	0.032	0.6
tundra, alpine meadow	8	0.3	2.4	0.065	0.5
temperate grassland	9	0.7	6.3	0.225	2.0
woodland, shrubland	8	2.7	22	0.270	2.2
savanna	15	1.8	27	0.315	4.7
boreal forest	12	9.0	108	0.360	4.3
temperate deciduous forest	7	13.5	95	0.540	3.8
temperate broadleaf evergreen forest	5	16	80	0.585	2.9
tropical seasonal forest (deciduous)	7.5	16	120	0.675	5.1
tropical rainforest (evergreen)	17	20	340	0.900	15.3
total (or global mean)	149	(5.55)	827	(0.324)	48.3

According to Table 3.1, the global total for carbon in terrestrial vegetation reservoirs is estimated as 827×10^{12} kgC, rather different from the 560×10^{12} kgC used in Figure 3.3. Variations between estimated totals should not be surprising, as there are serious difficulties in making the individual estimates in columns three and four of Table 3.1. The value of 827×10^{12} kg is in fact at the upper end of the range of estimates for the total global carbon in terrestrial plants; some estimates are as low as $\sim 420 \times 10^{12}$ kg. However, all the estimates are of the same order of magnitude, and the biggest is only about twice the smallest – such agreement is actually quite impressive for a quantity that is so hard to assess. The differences among these estimates is almost entirely due to variation in values for tropical forests, which have not yet been studied as much as ecosystems in mid-latitudes.

Even identifying the biomes is not straightforward, as an apparently homogeneous ecosystem may in fact be made up of several subcategories, and errors may be introduced by 'smoothing' the borders between biomes. Other problems include mapping or locating the biomes, measuring the amount of carbon in each grid, choosing representative grid samples and measuring sufficient samples. In addition, there may be errors in laboratory analysis and inconsistencies arising because different researchers are using different techniques.

Other reservoirs are more uniform and easier to estimate. For example, the global reservoir of atmospheric carbon is known to high accuracy because the air in the troposphere is well mixed and therefore fairly homogeneous (in contrast to the very 'patchy' and heterogeneous distribution of vegetation), its volume is known and we have many direct and precise measurements of atmospheric carbon concentrations (in CO_2 and CH_4) around the Earth. By contrast, the total amount of carbon in soils and detritus, or in marine sediments, may be significantly larger or smaller than the current best estimate.

Estimating fluxes in and out of a carbon reservoir is even more tricky than estimating the reservoir size, because it is often very difficult to quantify the *rates* of the processes (such as photosynthesis, respiration, decomposition) which contribute to these fluxes. In practice, many of the processes are measured indirectly. For example, the rate at which various types of organic material decompose may be estimated in the field by the loss in mass over a given time period of a bag containing a known quantity of litter; the bag is made of a mesh through which organisms can move freely, decomposing the litter to CO_2. Similarly, GPP may be measured by monitoring the CO_2 decrease or O_2 increase in a sealed box containing a plant or leaf (Figure 3.5), and NPP may be estimated by the change in the weight of plant biomass over a given area at the beginning and the peak of the growing period.

Figure 3.5
Estimating gross primary productivity by monitoring the decrease in CO_2 concentration in a sealed chamber (supported on a tower) enclosing a branch of Norway spruce (*Picea abies*).

Before moving on, let's look more closely at the information in Table 3.1, and see how it relates to the geographical distribution of ecosystems/biomes shown in Figure 3.4.

Question 3.3

(a) In terms of net primary productivity (NPP) m^{-2}, tropical rainforests (darkest green in Figure 3.4) are by far the most productive ecosystems.
(i) What is the annual flux of carbon into, and out of, tropical rainforests?
(ii) According to the data in the last row and last column of Table 3.1, what percentage of the total global NPP is contributed by these forests?

The carbon cycle

(b) (i) Area for area, how much more productive are tropical forests than boreal forests (i.e. forests that grow at high northern latitudes; see Figure 3.4)?
(ii) Bearing in mind what you have read earlier, particularly relating to Figures 1.9 and 2.7, briefly discuss the various possible reasons for this difference in productivity.
(iii) How does the residence time of carbon in living plants in swamps and marshes compare with that in boreal forests?

Of course, in reality we cannot assume that areas of the globe covered by particular ecosystems remain constant. For example, tropical rainforests are known to be decreasing – adding to the widespread concern about global warming. This apart, our assumption that reservoirs and fluxes are constant does not in practice exclude small-scale fluctuations that average out over relatively short periods of time. The amount of carbon stored in living Northern Hemisphere forests, for instance, may greatly fluctuate from summer to winter, but over a number of years the average amount stays relatively constant. Likewise, as you will see, the concentration of atmospheric CO_2 fluctuates seasonally.

- If, on average, the input fluxes currently equal the output fluxes, how do you think these various carbon reservoirs built up in the first place?

- At various time(s) in the Earth's history the fluxes have *not* been in balance, and inputs have been larger than outputs.

Thus, in the Carboniferous (354–290 million years ago) the terrestrial reservoir of carbon grew dramatically as plants spread over the land.

Now let us move from the terrestrial biological cycle, which acts on a time-scale of months/years to decades, to a cycle that acts on a time-scale of hundreds to thousands of years – the marine carbon cycle.

3.3.2 Intermediate time-scales: the marine carbon cycle

Although the only 'one-way' flux of carbon to the ocean is in rivers and other run-off (Figure 3.2), by far the greatest carbon fluxes into and out of the ocean are those to and from the atmosphere, across the air–sea interface. For this reason, it's convenient to start our discussion of the marine carbon cycle in the atmosphere. To make progress, therefore, we first need to understand how gaseous CO_2 interacts with liquid water.

Carbon dioxide in water

Carbon dioxide is the most soluble of the major atmospheric gases (Figure 2.38b). The reason for this is that some of the CO_2 molecules that diffuse into water then react with the water to produce a variety of dissolved ions (sometimes referred to as ionic 'species'). Box 3.2 sets out the details of the reactions that occur between gaseous carbon dioxide and its various aqueous forms – known collectively (for reasons that should become clear) as the **carbonate system**. You do not need to remember all the details of this box, but you should take away these main points:

1. Very little of the CO_2 dissolved in, say, rainwater or the ocean is in the form of dissolved gas; most is in the form of bicarbonate and carbonate ions (HCO_3^- and CO_3^{2-}, respectively) or as molecules of carbonic acid, H_2CO_3.

2. Given time, the carbonate system (see Box 3.2) will always tend to adjust to an equilibrium situation.

Box 3.2 Carbon dioxide and water: the carbonate system

Very little of the carbon dioxide dissolved in natural waters is in the form of dissolved gas as such. When molecules of CO_2 gas diffuse into water, some reacts with water to produce the weak acid, H_2CO_3 (known as carbonic acid), but most occurs as hydrated CO_2 (written as $CO_2(aq)$), where each CO_2 molecule is surrounded by water molecules. Because it is difficult to distinguish analytically between $CO_2(aq)$ and $H_2CO_3(aq)$, in practice dissolved carbon dioxide is normally referred to simply as carbonic acid. If we follow this convenient shorthand, we can write the chemical equation for the solution of CO_2 gas in water as:

$$CO_2(g) + H_2O \longrightarrow H_2CO_3(aq) \qquad \text{(Equation 3.2a)}$$

At any particular temperature, the amount of CO_2 that diffuses into the water depends upon the concentration of CO_2 in the atmosphere on the one hand, and the concentration of carbonic acid in the water on the other.

When enough H_2CO_3 builds up in the water, some of the dissolved carbon is released as CO_2 back to the atmosphere in the reverse reaction:

$$H_2CO_3(aq) \longrightarrow CO_2(g) + H_2O \qquad \text{(Equation 3.2b)}$$

Eventually, the forward reaction (Equation 3.2a) and the reverse (Equation 3.2b) reactions are occurring at equal rates, and a state of 'dynamic balance' or **chemical equilibrium** has been established. At equilibrium, *reactants* (on the left-hand side of the equation) are forming *products* (on the right-hand side of the equation) at the same rate as products are decomposing back into their constituent reactants. Although the concentrations of product and reactant do not change, at the molecular level there is a constant and equal exchange between them. To represent this, we write equilibrium systems with two half-headed arrows pointing in opposite directions, as in Equation 3.2c:

$$CO_2(g) + H_2O \rightleftharpoons H_2CO_3(aq) \qquad \text{(Equation 3.2c)}$$

The system is, however, more complicated than this. Carbonic acid, like all acids, tends to *dissociate*, i.e. lose a hydrogen ion. A second equilibrium is established between the carbonic acid and its dissociation products, a hydrogen ion (H^+) and a **bicarbonate ion** (HCO_3^-):

$$H_2CO_3(aq) \rightleftharpoons H^+(aq) + HCO_3^-(aq) \qquad \text{(Equation 3.3)}$$

Furthermore, bicarbonate itself dissociates to form a **carbonate ion** (CO_3^{2-}) and a hydrogen ion. This third equilibrium is written:

$$HCO_3^-(aq) \rightleftharpoons H^+(aq) + CO_3^{2-}(aq) \qquad \text{(Equation 3.4)}$$

So carbon dissolved in water achieves a state of dynamic chemical equilibrium between the CO_2 in the air and that partitioned among three dissolved carbon compounds (shown in bold):

$$\underset{\text{(Eqn 3.2)}}{\mathbf{CO_2}(g) + H_2O \rightleftharpoons \mathbf{H_2CO_3}(aq)} \underset{\text{(Eqn 3.3)}}{\rightleftharpoons H^+(aq) + \mathbf{HCO_3^-}(aq)}$$

$$\Updownarrow$$

$$H^+(aq) + \mathbf{CO_3^{2-}}(aq) \quad \text{(Eqn 3.4)}$$

$$\text{(Equation 3.5)}$$

This dynamic equilibrium among dissolved carbon compounds is called the carbonate equilibrium system, and the dissolved carbon compounds H_2CO_3, HCO_3^- and CO_3^{2-} are collectively termed **dissolved inorganic carbon**, or **DIC**. At equilibrium, the *concentrations* of the different carbon compounds are constant (but not necessarily equal) and for each pair of compounds the amounts of carbon *exchanged* are equal. Therefore, the relative *proportions* of the reactant and product remain constant. You may visualize this by imagining four boxes, each containing a different number of marbles but with a constant exchange of two marbles back and forth between them (Figure 3.6). In the same way (Figure 3.7), at equilibrium, the

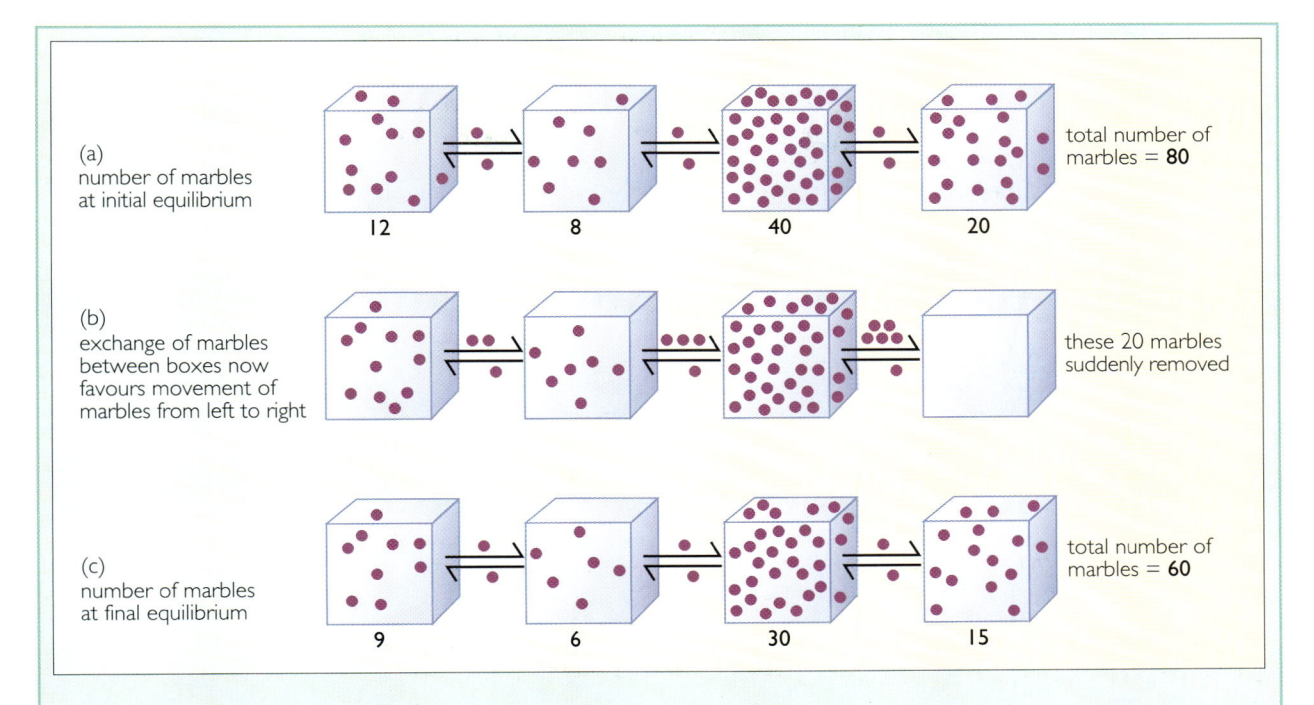

The carbon cycle

Figure 3.6
(a) Equilibrium among four reservoirs in a closed system; as far as the carbonate system is concerned, the four reservoirs are loosely analogous to atmospheric CO_2, H_2CO_3, HCO_3^- and CO_3^{2-}.
(b) If one of the reservoirs is depleted, more reactants will form products in a direction tending to re-establish equilibrium.
(c) A new equilibrium is established.

amount of CO_2 leaving the surface of the water is the same as the amount going into the water (Equation 3.2), and the amount of carbonate being formed is the same as that being recombined to form bicarbonate, HCO_3^- (Equation 3.4). Note that the concentration of the components is determined by two equilibrium reactions, Equations 3.3 and 3.4.

When a chemical system in equilibrium is subjected to an external constraint, the system responds in a way that tends to lessen the effect of the constraint; this expression of a fundamental observation is known as Le Chatelier's principle. In a reaction, any factor which removes some or all of the product(s) will cause the reactant(s) to form more product(s), and re-establish the equilibrium.

Returning to our marble analogy, according to Le Chatelier's principle, if the box at the right-hand end is depleted (Figure 3.6b), the exchange of marbles between boxes will begin to favour the movement of marbles towards the right-hand end so that eventually, equilibrium – a different equilibrium from the first – is established (Figure 3.6c).

■ As far as the carbonate system is concerned, what would be the effect on the rate of diffusion of CO_2 into the water of any process that removes H_2CO_3 from solution?

■ Anything that removes carbonic acid from solution will increase the rate of diffusion of CO_2 *into* the water so that it is greater than the rate of diffusion *out of* the water; hence the rate of dissolution of atmospheric CO_2 will increase.

Conversely, if the concentration of atmospheric CO_2 increased, more carbon dioxide would dissolve in water, forming more H_2CO_3 and ultimately partitioning itself among the various dissolved inorganic carbon compounds (Figure 3.7).

$$CO_2(g)$$
$$\updownarrow$$
$$CO_2(g) + H_2O \rightleftharpoons H_2CO_3(aq) \rightleftharpoons H^+(aq) + HCO_3^-(aq)$$
$$\updownarrow$$
$$H^+(aq) + CO_3^{2-}(aq)$$

Figure 3.7 Pictorial representation of the carbonate system expressed by Equation 3.5. (In reality, H_2CO_3 is mostly $CO_2(aq)$.)

As mentioned at the end of Box 3.2, if the concentration of atmospheric CO_2 increased, more would dissolve in natural waters. Figure 3.8 is a schematic representation of this situation applied to a body of water such as the ocean. The flux into the ocean would remain greater than the flux out (Figure 3.8b) only until a new equilibrium had been established. Conversely, if the concentration of atmospheric CO_2 decreased, more would come out of solution in the ocean than would dissolve, until a new equilibrium had been established.

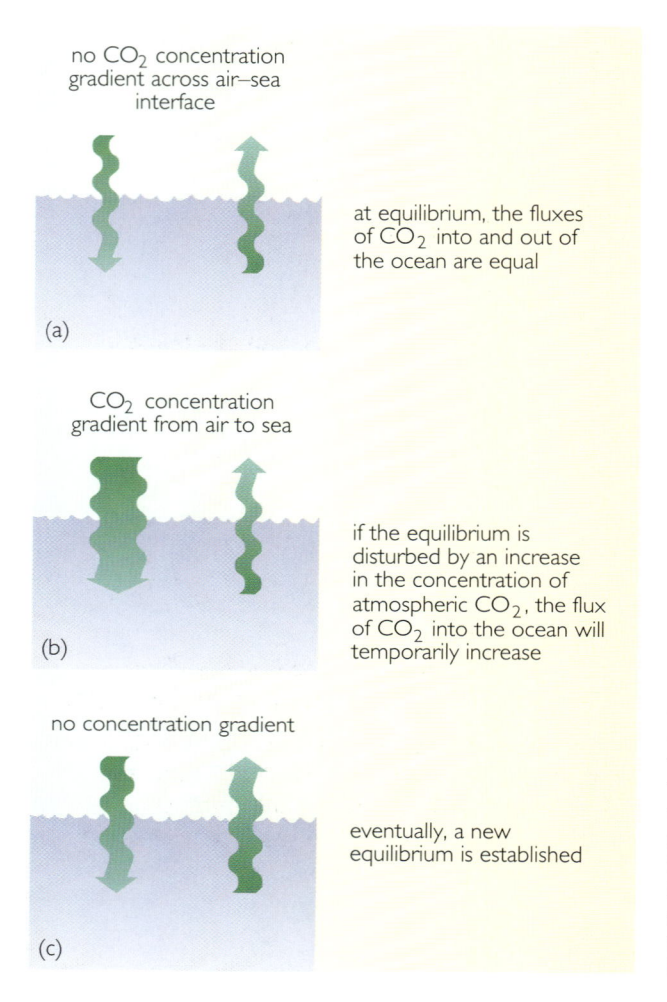

Figure 3.8
Schematic illustration of the change from one equilibrium situation to another. (a) At equilibrium, the fluxes of CO_2 into and out of the ocean are equal. (b) If the equilibrium is disturbed by an increase in the concentration of CO_2 in the atmosphere, the flux of CO_2 into the ocean (i.e. the rate at which CO_2 goes into the ocean) will temporarily increase. (c) Eventually, a new equilibrium is established, in which more CO_2 is dissolved in the ocean.

■ Bearing in mind what you read in Section 2.3.2 about the solubility of gases, can you suggest another circumstance that would lead to an increased flux of CO_2 from the ocean to the atmosphere?

■ As all gases are more soluble in cold than warm water, warming of surface waters would cause more CO_2 to come out of solution, increasing the flux of CO_2 from ocean to atmosphere.

In other words, the equilibrium situation for warmer water is a lower concentration of dissolved CO_2. This new equilibrium would take a while to establish itself, and until this happened, there would be a net flux from sea to air.

A component of the carbonate equilibrium system that we have not yet discussed is the hydrogen ion (H^+). This has an important role, as its concentration in solution

The carbon cycle

determines the relative proportions of the different DIC compounds (H_2CO_3, HCO_3^- and CO_3^{2-}). Hydrogen ion concentration in water – often written using square brackets: $[H^+]$ – is generally expressed as **pH**, where a low pH corresponds to a high hydrogen ion concentration and vice versa. For example, a pH of 5 represents a hydrogen ion concentration, $[H^+]$, of $10^{-5}\,\text{mol}\,l^{-1}$, while a pH of 10 represents a $[H^+]$ of $10^{-10}\,\text{mol}\,l^{-1}$. Low pH (below about 5) favours the formation of carbonic acid, H_2CO_3, whilst higher pH (7 to 8) favours the formation of bicarbonate, HCO_3^-; still higher values (above 9) favour the formation of carbonate ion, CO_3^{2-}. These relationships are shown schematically in Figure 3.9, along with the chemical equilibria that are affected (i.e. Equations 3.2–3.5).

$$CO_2(g) + H_2O \rightleftharpoons H_2CO_3(aq) \rightleftharpoons H^+(aq) + HCO_3^-(aq) \rightleftharpoons H^+(aq) + CO_3^{2-}$$

Figure 3.9
Generalized diagram showing approximately how the relative proportions of the three principal components of DIC in the aqueous carbonate system vary with pH in natural waters. The pH of rainwater is 5.6 or less; the pH of seawater averages about 7.7 and can range from about 7.2 to 8.2. The positions of the curved lines can vary with temperature and, in the ocean, also with salinity and pressure.

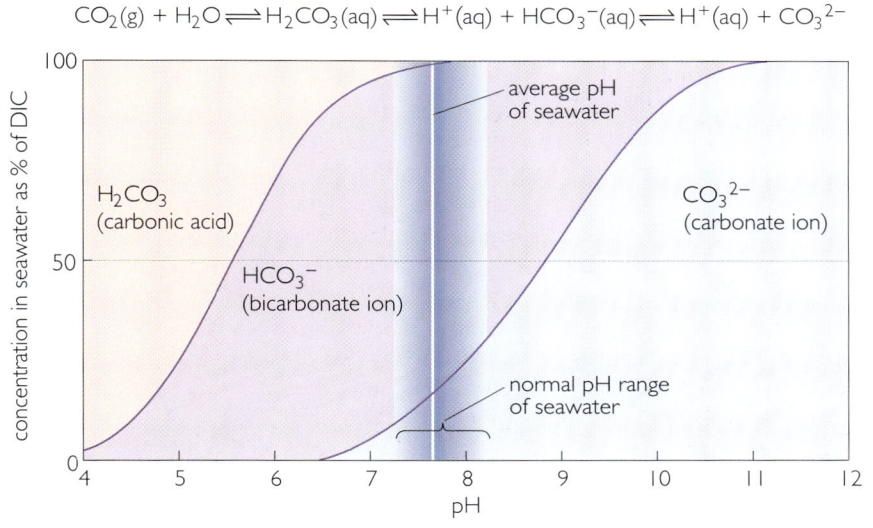

- According to Figure 3.9, what are the main contributions to dissolved inorganic carbon in seawater (pH around 7.7) and rainwater (pH 5.6 or less)?

- In seawater, most (~85%) of the DIC is as bicarbonate (about 10–15% is as carbonate ion, CO_3^{2-}, and less than 1% is as H_2CO_3), whilst in rainwater, about half the DIC is in the form of H_2CO_3, and about half as HCO_3^-. (Remember, however, that the positions of the curved lines can vary; see caption.)

Now let's briefly return to the land.

From rain to soil to river to sea

Imagine weakly acid rain falling onto soil where microbes are decomposing organic matter in respiration reactions that use oxygen and release CO_2 (Equation 3.1). Plants are also releasing CO_2 from respiration – most of it through their roots. As a result of both plant respiration and microbial activity pumping CO_2 into pore spaces in the soil, carbon dioxide concentrations there are typically 10 to 100 times higher than in the atmosphere. Because of this, concentrations of CO_2 in soil water are also raised, with the reactions in Equations 3.2–3.5 being moved to the right, so producing additional hydrogen ions. This, plus the addition of various organic acids (e.g. humic acid and fulvic acid) from the decomposition of plant remains, means that soil water is often significantly more acid than rainwater.

Acidic soil water is the main agent of chemical weathering of minerals in soil and rock. In general, carbonic acid plays a larger role in tropical forests where lower concentrations of organic acids remain after surface organic litter has decomposed, while organic acids dominate the weathering processes in cool temperate forests where weathering processes are slow and incomplete. Thus, especially in cool conditions, weathering rates are increased by the activities of living organisms, even those apparently as insignificant as small flowering plants (Figure 3.10), fungi and lichens.

Cations such as Na^+ and Ca^{2+}, dissolved from rocks, are essential for plant growth, but humic material which collects in a developing soil also plays an important part because essential elements – nitrogen (as nitrate), phosphorus (as phosphate) and other nutrients – are retained in the soil, and made available to plants largely through its organic content. As every good gardener knows, a high content of humus also helps retain water in the soil; in addition, the relatively dark colour of soil helps it absorb heat, and a warm soil generally increases plant growth rates.

As discussed in Section 2.3.2, compared with rainwater, river water contains high concentrations of HCO_3^-, Ca^{2+} and SiO_2 which have been released from rocks by chemical weathering. Most of the rocks making up the Earth's crust are made up of silicate or carbonate minerals, and so we can use just two equations to represent the two basic chemical weathering reactions. Equation 3.6a represents the reaction of rainwater with calcite, a form of calcium carbonate commonly found in sedimentary rocks (especially limestone and chalk), to release calcium and bicarbonate ions in solution. In Equation 3.6b, rainwater reacts with albite, a common mineral in igneous and metamorphic rocks, to produce a clay mineral, along with sodium and bicarbonate ions in solution, plus silica which is only partly in solution. (Albite is a sodium-rich silicate mineral, but similar equations could be written for other silicates, such as calcium-rich anorthite, $CaAl_2Si_2O_8$, potassium-rich orthoclase, $KAlSi_3O_8$, or magnesium–iron olivine, $(Mg,Fe)_2SiO_4$.)

First, for the weathering of carbonates (note that we have picked out the carbon atoms in bold):

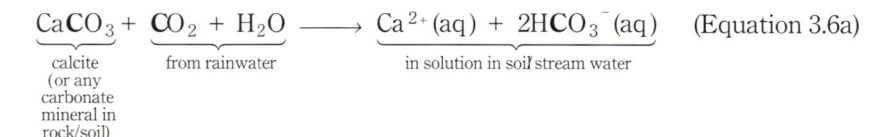

Figure 3.10
Soil forming in a limestone crevice around a newly established thrift (*Armeria maritima*) plant.

$$\underbrace{Ca\mathbf{C}O_3}_{\substack{\text{calcite} \\ \text{(or any} \\ \text{carbonate} \\ \text{mineral in} \\ \text{rock/soil)}}} + \underbrace{\mathbf{C}O_2 + H_2O}_{\text{from rainwater}} \longrightarrow \underbrace{Ca^{2+}(aq) + 2H\mathbf{C}O_3^-(aq)}_{\text{in solution in soil/stream water}} \quad \text{(Equation 3.6a)}$$

For the weathering of silicates:

$$\underbrace{2NaAlSi_3O_8}_{\substack{\text{albite} \\ \text{(or any silicate mineral} \\ \text{in rock/soil)}}} + \underbrace{2\mathbf{C}O_2 + 3H_2O}_{\text{from rainwater}} \longrightarrow \underbrace{Al_2Si_2O_5(OH)_4}_{\substack{\text{kaolinite} \\ \text{(a clay mineral)}}} + \underbrace{2Na^+(aq) + 2H\mathbf{C}O_3^-(aq)}_{\text{in solution in soil/stream water}} + \underbrace{4SiO_2(aq)}_{\substack{\text{silica,} \\ \text{partly} \\ \text{in solution}}}$$

(Equation 3.6b)

Each of these two weathering reactions result in two bicarbonate ions ($2HCO_3^-$) on the right-hand side.

- How do the two reactions differ in terms of where the two carbon atoms in the $2HCO_3^-$ come from?

- In the weathering of *carbonates* (Equation 3.6a), one carbon atom comes from atmospheric CO_2 and one comes from the carbonate mineral itself. In *silicate* weathering (Equation 3.6b), *both* carbon atoms come from the atmosphere.

You should try to ensure that you understand this distinction, because it will be crucial later.

The carbon cycle

Soil water carrying ions that have been released by weathering collects in streams, which run into rivers which eventually reach the sea.

- ■ Apart from being a much more concentrated solution, how, generally speaking, does seawater differ from river water?

- ■ The relative proportions of dissolved ions in seawater are different from those in river water (and are, in fact, closer to those in rainwater).

As discussed in Section 2.3.2, river water has relatively high concentrations of HCO_3^-, Ca^{2+} and SiO_2 (cf. Equation 3.6a and b), but the most abundant ions in seawater are Cl^-, Na^+ and SO_4^{2-}. As you will see shortly, the apparent shortfall of HCO_3^-, Ca^{2+} and SiO_2 in seawater is intimately related to the cycling of carbon in the ocean.

Carbon in the oceans

Figure 3.11 shows the marine carbon cycle in semi-pictorial form. Unlike the terrestrial carbon cycle, which is dominated by the formation, decomposition and recycling of *organic* (mainly plant) material (made of molecules built up of carbon, hydrogen and oxygen), the carbon cycle in the sea is dominated by the chemistry of *inorganic* carbon, that is, minerals (or dissolved ions) containing carbon. However, the ocean is a soup of living organisms – particularly in the sunlit surface layers – and the chemical transformations involved in the marine carbon cycle take place

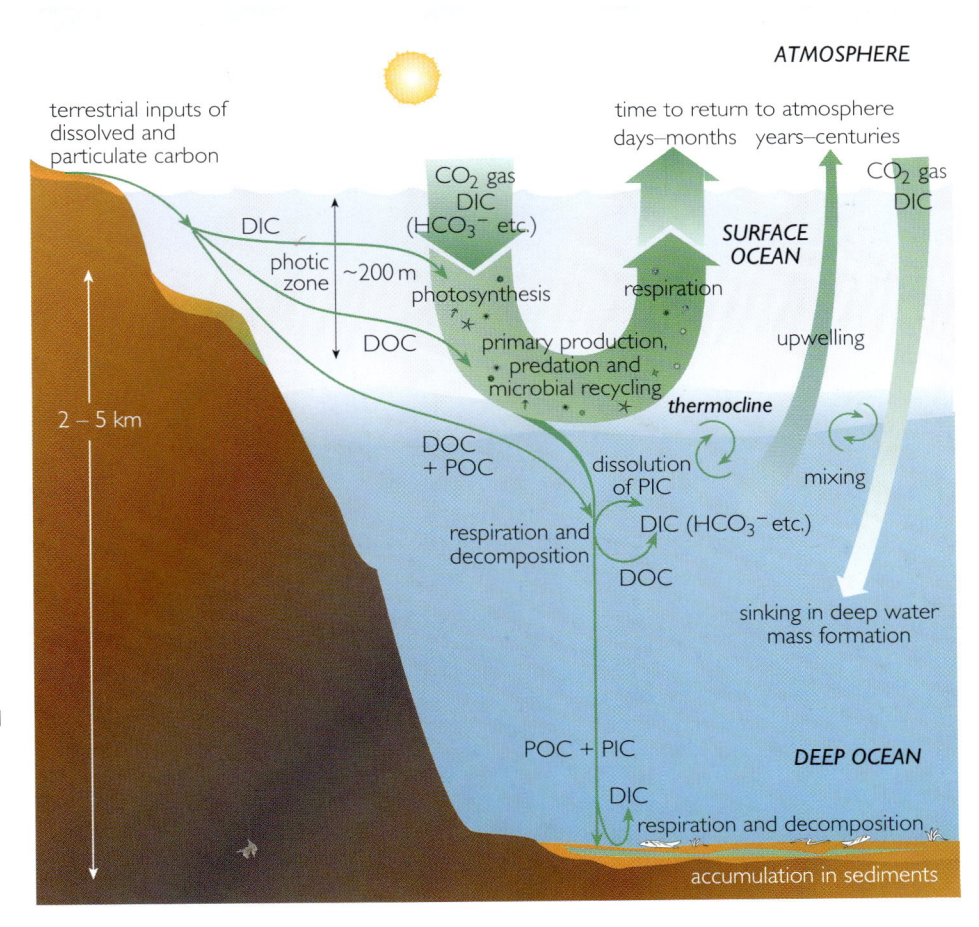

Figure 3.11
Pictorial representation of processes contributing to the marine carbon cycle (not to scale). Remember that while both physical and biological processes affect the fluxes of CO_2 across the air–sea interface, they occur through the action of *chemical equilibria* (cf. Figures 3.7 and 3.8). DIC and PIC = dissolved and particulate inorganic carbon; DOC and POC = dissolved and particulate organic carbon, respectively. Aspects of this diagram not discussed here will be addressed later.

largely through their agency. Nevertheless, the flux of carbon to and from the atmosphere – i.e. the direction of the net exchange of CO_2 across the air–sea interface (cf. Figures 3.7 and 3.8) – is not primarily regulated by photosynthesis (as it is on land) but by carbonate equilibrium reactions (Equations 3.3–3.5) which are affected by both biological and physical factors, both of which vary from place to place and season to season.

As discussed in Section 2.2 (Figure 2.29), the ocean can be divided into two main 'layers': the cold deep ocean, and above that, the warm, well-mixed waters. There is a continual exchange of CO_2 across the air–sea interface (cf. Figure 3.7) depending on its concentration in the atmosphere and its concentration in the mixed surface layer – the uppermost waters are effectively in contact with the atmosphere, particularly in windy conditions when waves churn air down below the surface. In Figure 3.11, you can think of the upper parts of the big arrow, which cross the sea-surface, as representing the exchanges of CO_2 across the air–sea interface, brought about by shifts in chemical equilibria.

Let's look at some experimental results which clearly show the effect of biological activity in surface waters. Figure 3.12 shows data collected in the North Atlantic in the spring of 1989, when light levels had risen sufficiently for phytoplankton to have begun to multiply (cf. Figure 2.40). The green plot shows how the concentration of chlorophyll pigment in surface water – an indication of phytoplankton biomass – varied along the ship's track; the highs and lows correspond to fronts, eddies and so on. The black plot shows the difference in the partial pressure of carbon dioxide between the atmosphere and the underlying ocean along the same track. Given the complications inherent in quantifying the amount of carbon dioxide gas in water, you do not need to worry about the way in which the pressure difference is expressed – all you need to know is that the negative values correspond to concentrations in the atmosphere being higher than those in the ocean.

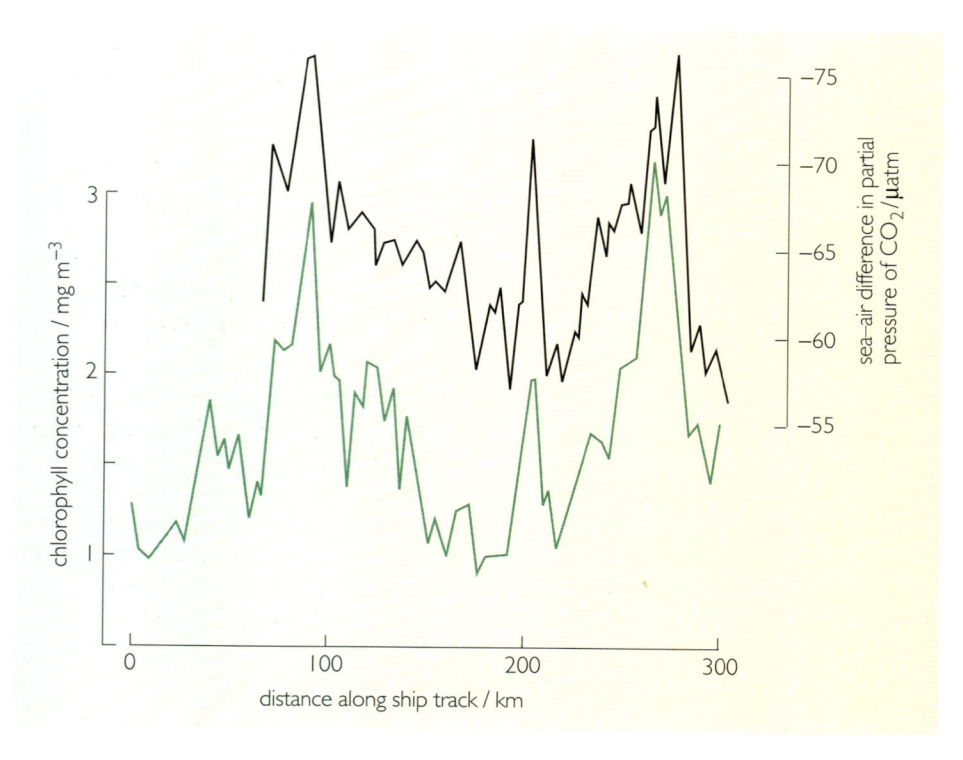

Figure 3.12 Variation in the concentration of chlorophyll in surface water along a ship's track in the North Atlantic (green plot). The difference between the partial pressure of carbon dioxide in surface waters and that in the overlying air (black plot), expressed in micro-atmospheres (μatm or 10^{-6} atm), along the same track. Negative values correspond to concentrations in the atmosphere being higher than those in the surface ocean. (The data were collected during the North Atlantic Bloom Experiment, which was part of BOFS, the Biogeochemical Ocean Flux Study.)

The carbon cycle

Question 3.4

(a) By reference to the right-hand axis of Figure 3.12, explain whether the CO_2 dissolved in the surface ocean was in equilibrium with the CO_2 in the overlying atmosphere at the time in question (cf. Figures 3.7 and 3.8). Was there a net flux of CO_2 across the air–sea interface, and if so was it into or out of the ocean?

(b) Why do you think the shapes of the two plots are similar?

So phytoplankton populations act as *sinks* for atmospheric carbon dioxide, as do microscopic benthic (i.e. bottom-living) algae and seaweeds (sometimes referred to as 'macro-algae'). Table 3.2 includes data on carbon fixation and productivity for both phytoplankton and benthic plants.

Table 3.2 Estimates of marine primary production (standing stock) and net primary productivity; cf. Table 3.1 for terrestrial biomass. (You are not expected to remember the details of this table.)

Region	Area/ 10^6 km^2	Mean plant biomass per unit area/ kgC m^{-2}	Total plant biomass/ 10^{12} kgC	Mean NPP per unit area/ kgC m^{-2} yr^{-1}	Total NPP/ 10^{12} kgC yr^{-1}
open ocean	332.0	0.0014	0.46	0.057	18.9
upwelling areas	0.4	0.01	0.004	0.225	0.1
continental shelf	26.6	0.005	0.13	0.162	4.3
algal beds; reefs	0.6	0.9	0.54	0.900	0.5
estuaries	1.4	0.45	0.63	0.810	1.1
total (or global mean)	361	(0.0049)	1.76	(0.069)	24.9

■ Which of the regions listed in Table 3.2 will *not* support any benthic plants?

■ Benthic plants can only grow where the bottom is within the photic zone, so the row for 'open ocean' (depth 2000–5000 m) cannot include data for benthic plants.

The photic zone may be as much as 200 m deep in the open ocean, but in turbid coastal waters it is a lot less. Algal beds and reefs only grow within the photic zone and support largely benthic production, while the data for 'estuaries', 'upwelling areas' and 'continental shelf' will include a variable amount of benthic plants, depending on the clarity of the water; if they are relatively clear, nutrient-rich coastal waters can support highly productive stands of vegetation (notably kelp forests; Figure 3.13).

It is difficult to define satisfactory categories for compilations such as Table 3.2 (e.g. many upwelling areas are over continental shelves) and, as for land vegetation, the estimates are subject to considerable error. Nevertheless, it is clear from Table 3.2 that the *net* amount of carbon converted each year into *new* plant (mainly phytoplankton) tissue, i.e. the net primary productivity (NPP), is less than that for terrestrial plants: only about 20×10^{12} kgC yr^{-1}. Furthermore, the actual amount of carbon in the marine plant reservoir at any one time is very small, only about ~ 1–2×10^{12} kgC (see fourth column), indicating a rapid turnover (i.e. a short residence time) of carbon in marine plants.

Figure 3.13
A kelp forest, characteristic of nutrient-rich near-shore waters under the influence of coastal upwelling.

Phytoplankton are consumed by zooplankton, and both phytoplankton and zooplankton are consumed by larger animals, and all of these organisms respire, releasing CO_2 back into the surrounding water, mainly as HCO_3^-. On death, soft tissues of these organisms are colonized by bacteria which decompose them, releasing organic molecules into solution. Bacterial respiration oxidizes both tissue and dissolved organic molecules, returning carbon dioxide into solution; in other words, decomposition returns both inorganic and organic carbon to solution (i.e. produces both DIC and DOC), to be available for re-use by organisms. In these various ways, most of the carbon in organisms of the upper ocean is recycled many times into dissolved inorganic and organic carbon and back again.

■ So how is it that the carbon fixed by photosynthesis does not all find its way back into the atmosphere in a very short time via respiration?

The reason is that the recycling system is not 100% efficient and some $4-5 \times 10^{12}$ kgC yr^{-1} escapes from surface waters, mostly as particles of organic debris and zooplankton faecal pellets (particulate organic carbon, or POC) and skeletal remains. Most of the detritus consists of very small particles – dead and dying algal cells and bacteria – and would take months to sink to the sea-bed were it not for the fact it forms clumps or aggregates of fluffy debris, often referred to as **marine snow** (Figure 3.14a,b). These aggregates sink to the sea-bed in a matter of days or weeks, and arrive more-or-less below where production occurred in surface waters, rather than being dispersed by currents. As it sinks, and after it arrives on the deep sea-bed, the particulate organic carbon is decomposed by bacteria, and both bacteria and organic remains provide food for animals in the water column (i.e. *pelagic* animals) and on the sea-bed (i.e. benthic, or bottom-dwelling, animals) (Figure 3.14c); all of these organisms respire, returning CO_2 to solution. Generally, soft tissue is almost completely decomposed before it reaches the sea-floor (Figure 3.15), but if productivity is very high, as during the spring bloom in the North Atlantic (Figure 2.40a), much phytoplankton debris aggregates into marine snow, sinks out of surface waters without being consumed, and provides a food bonanza for bottom-living animals (Figure 3.14d). This transfer of carbon from surface waters to the deep ocean (which means that more has to be supplied from the atmosphere to support plankton growth) is known as the **biological pump**.

Before moving on to look at the *physical* mechanisms that control the flux of carbon into and out of the ocean, we must consider another biological process that uses carbon in seawater. Many, though by no means all, planktonic organisms in the ocean have hard parts, in the form of protective shells or skeletons, as well as the soft tissues we considered above (as indeed do other larger marine organisms). Many of these hard parts are composed of calcium carbonate ($CaCO_3$) made from bicarbonate ions (HCO_3^-) and calcium ions (Ca^{2+}) abstracted from seawater. The chemical reaction for the precipitation of calcium carbonate shells is:

$$\underbrace{Ca^{2+}(aq) + 2HCO_3^-(aq)}_{\text{in solution in seawater}} \longrightarrow \underbrace{CaCO_3(s)}_{\substack{\text{precipitated} \\ \text{by organisms}}} + H_2O + CO_2 \quad \text{(Equation 3.7)}$$

Note that Equation 3.7 is Equation 3.6a in reverse. The reaction proceeds exothermically in only one direction; the other direction requires energy, i.e. it is endothermic.

Figure 3.14 (opposite)
(a) A particle of marine snow being consumed by copepods (small planktonic crustaceans). Aggregation of organic material into marine snow speeds up the descent to the sea-bed, so that there is much less time for it to be decomposed (or eaten) *en route*. The field of view is ~1–2 cm across.
(b) An individual 'snowflake', an isolated aggregate of marine snow ~1 cm across.
(c) The ultimate fate of most marine snow is to be consumed by benthic animals. Here you can see phytoplankton remains actually within a benthic foraminiferan (which is about 200 µm across).
(d) Marine snow carpeting the sea-bed and partially burying a mound made by an animal. Some benthic animals time their reproductive cycle so that their young can take advantage of organic debris from the spring bloom in surface waters. The field of view is about half a metre across.

The carbon cycle

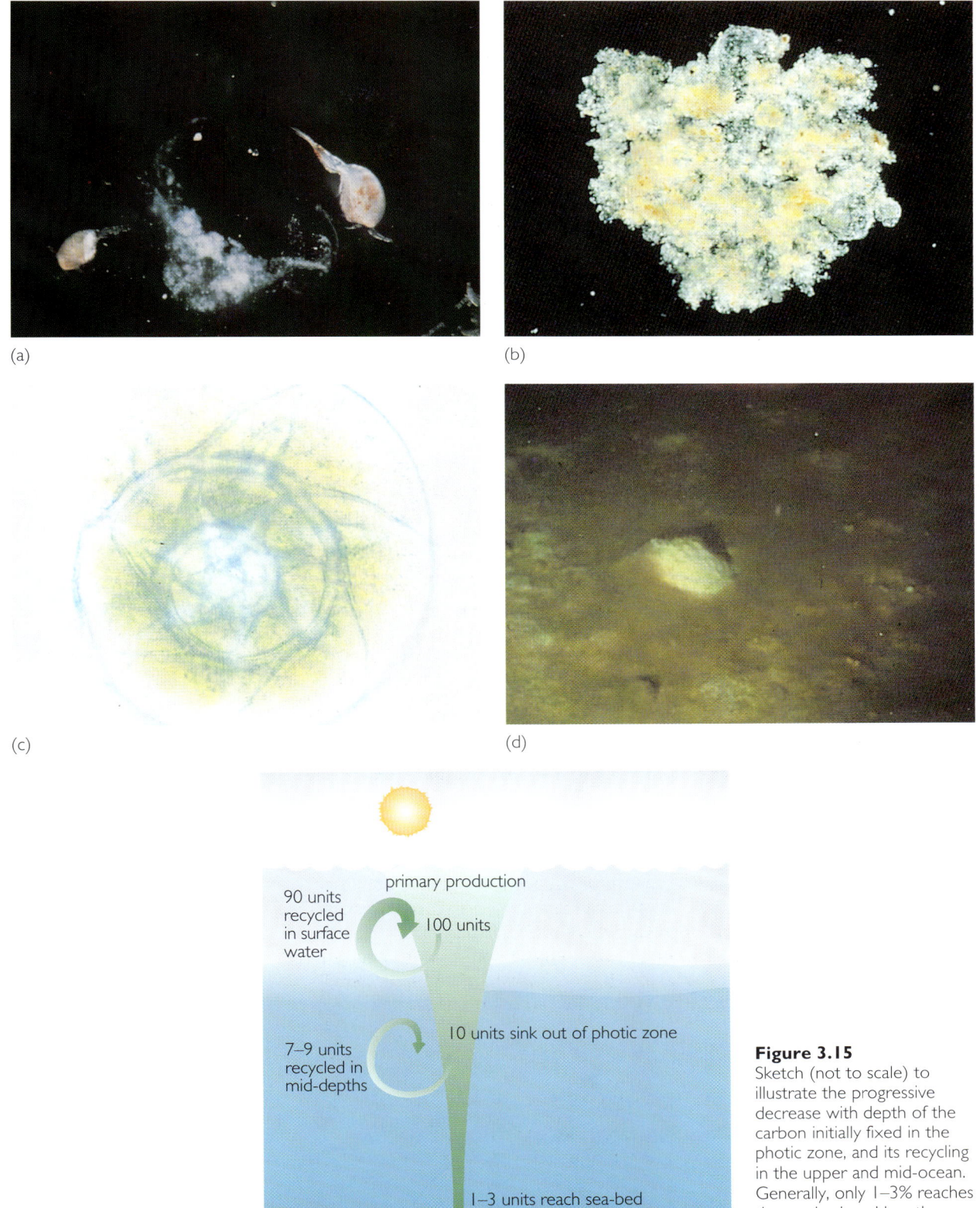

93

(a)

(b)

(c)

(d)

Figure 3.15
Sketch (not to scale) to illustrate the progressive decrease with depth of the carbon initially fixed in the photic zone, and its recycling in the upper and mid-ocean. Generally, only 1–3% reaches the sea-bed, and less than a third of that might be preserved in sediments.

primary production
100 units
90 units recycled in surface water
10 units sink out of photic zone
7–9 units recycled in mid-depths
1–3 units reach sea-bed
< 1 unit preserved in sediments

- Which direction proceeds exothermically and which endothermically?

- The weathering reaction (Equation 3.6a) proceeds spontaneously (think of crumbling limestone walls) and so must be exothermic; the biological reaction (Equation 3.7) requires energy – the organism is investing energy for its defence – and so is endothermic.

Two important examples of organisms that secrete $CaCO_3$ are the planktonic algae known as coccolithophores, which secrete plates ('coccoliths') of calcium carbonate (Figure 3.16a,b), and the zooplanktonic foraminiferans (Figure 3.16c). Other planktonic organisms precipitate *silica* from solution to form their shells and skeletons. These include *diatoms*, which are phytoplankton (see Figure 2.39a), and radiolarians, which are zooplankton. While (as you saw above) organisms that build hard parts of calcium carbonate use Ca^{2+} and HCO_3^-, these organisms remove SiO_2 from solution in seawater. Precipitation of shells and skeletons, along with precipitation of similar minerals in chemical reactions with the sediments and rocks at the sea-bed, is sometimes referred to as *reverse weathering*. (It is processes such as these that help maintain the difference in composition between seawater and river water.)

Despite their long journey to the deep sea-floor, the calcium carbonate remains of planktonic organisms are found in sediments over much of the sea-bed (Figure 3.16b). Whether calcium carbonate is eventually dissolved or incorporated into sea-floor sediments depends on the chemistry of the seawater (its acidity and the extent to which it is saturated with calcium carbonate), along with the speed of sinking of remains and the depth of the sea-floor. Calcium carbonate debris generally begins to dissolve some distance above the sea-bed, but by that stage of its descent it has already been incorporated into marine snow and is sinking fairly fast, so there is insufficient time for it to dissolve *en route*. Most dissolution therefore occurs at the sea-bed, and the extent to which remains are dissolved before being protected by further layers of sediment (including more calcareous debris) increases with increasing depth. The depth at which the proportion of calcium carbonate remains falls to less than 20% of the total sediment (biogenic sediments plus terrigenous clays) is known as the **carbonate compensation depth**, or **CCD**. Under areas of high surface productivity, the 'rain' of calcareous debris is such that carbonate in sea-bed sediments is preserved at greater depths than it would otherwise be, and so the CCD is depressed.

Of course, planktonic organisms are not the only ones that abstract constituents from seawater to make hard parts, and this process is not confined to the deep ocean. In shallow nearshore waters, bivalves such as mussels and oysters often grow together, forming large accumulations of shells of calcium carbonate; and in clear tropical waters, certain algae form carbonate-rich accumulations, while coral skeletons build up to form substantial reefs.

Now let's consider the physical factors that control fluxes of carbon dioxide across the air–sea interface, and its solubility within the ocean.

- From your reading of Section 2.3.2, does the solubility of CO_2 increase or decrease with (1) increasing temperature, (2) increasing pressure?

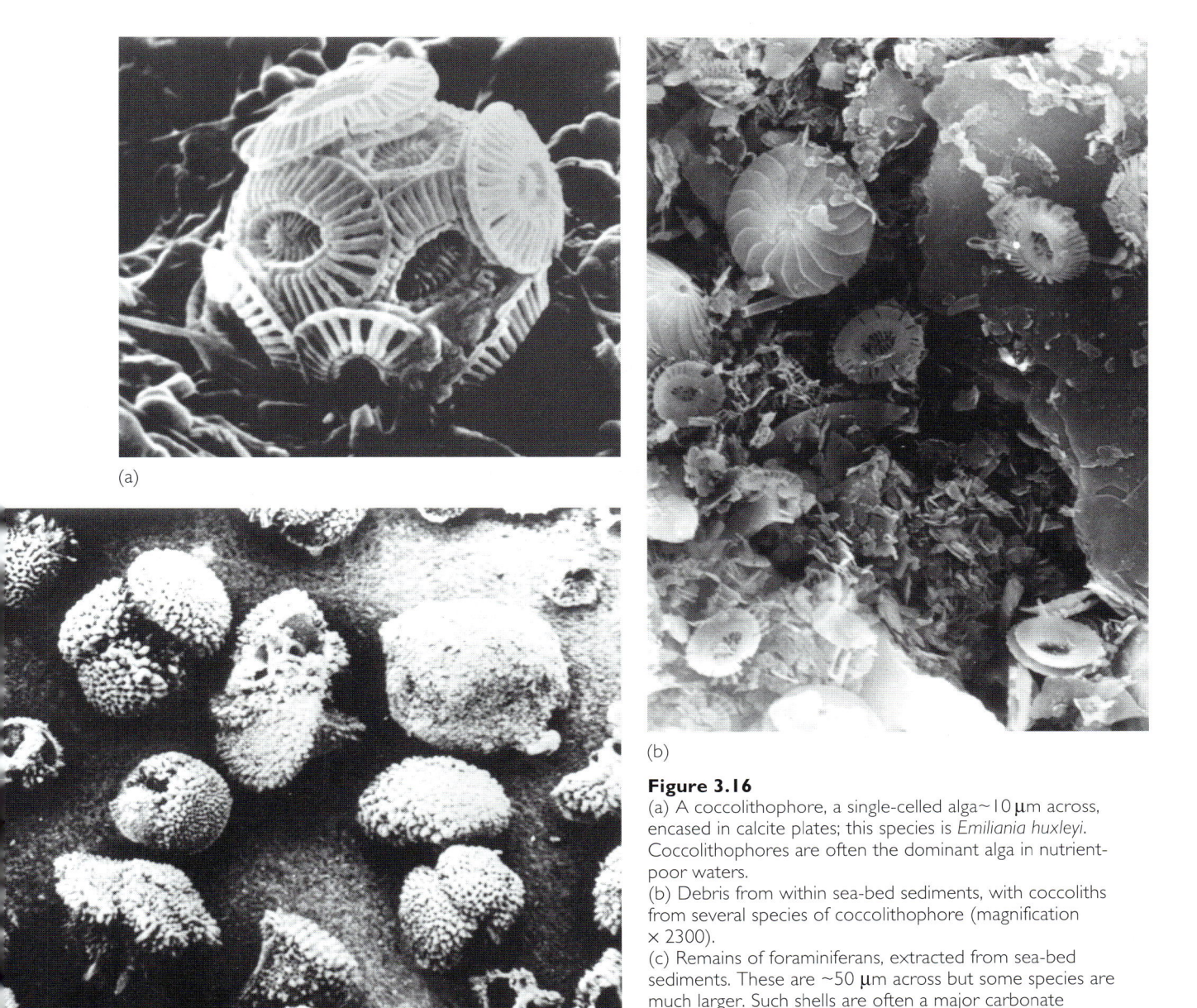

The carbon cycle

Figure 3.16
(a) A coccolithophore, a single-celled alga ~10 μm across, encased in calcite plates; this species is *Emiliania huxleyi*. Coccolithophores are often the dominant alga in nutrient-poor waters.
(b) Debris from within sea-bed sediments, with coccoliths from several species of coccolithophore (magnification × 2300).
(c) Remains of foraminiferans, extracted from sea-bed sediments. These are ~50 μm across but some species are much larger. Such shells are often a major carbonate component of sediments below areas of high productivity.

■ As discussed in connection with Figure 2.38, the solubility of gases decreases with increasing temperature and increases with increasing pressure.

If you have trouble remembering which way these controls on solubility act, think of the following everyday examples. (1) If you want to check that an electric kettle is working, you can look at the submerged element: if bubbles of gas are collecting there, air is being driven out of solution as the water warms, and the kettle is working. (2) When you open a can or bottle of gasified drink, it fizzes as gas comes out of solution: the solubility is reduced as pressure is reduced.

Now try Activity 3.1, which brings together much of what you have been reading about the roles of both biological and physical factors in determining fluxes of carbon dioxide into and out of the ocean.

Activity 3.1

Figure 3.17 shows fluxes of CO_2 across the air–sea interface in January–March and April–June, in the Northern Hemisphere. Areas acting as a net sink for atmospheric CO_2 (net flux of CO_2 from atmosphere to ocean) are shown in green and blue, and areas acting as a source (net flux of CO_2 from ocean to atmosphere) are shown in brown, red and yellow. Study the map, concentrating particularly on the Atlantic.

(a) By reference to Figure 2.29 if necessary, explain the direction of the CO_2 flux in the north-eastern North Atlantic during January–March.

(b) Bearing in mind when the data were collected, discuss possible reasons for changes in the CO_2 flux in the North Atlantic between January–March and April–June. (*Hint*: Look at Figure 2.28 and Figures 2.40 and 3.11.)

(c) Can you explain the direction of the flux in tropical and equatorial areas?

Figure 3.17
Fluxes of CO_2 across the air–sea interface at two seasons of the year: (a) averaged over the months January–March; (b) averaged over the months April–June.
Fluxes are given in moles of CO_2 m^{-2} yr^{-1}. Positive values (brown, red and yellow) correspond to a net flux of CO_2 out of the ocean. Negative values (greens and blues) correspond to a net flux into the ocean.

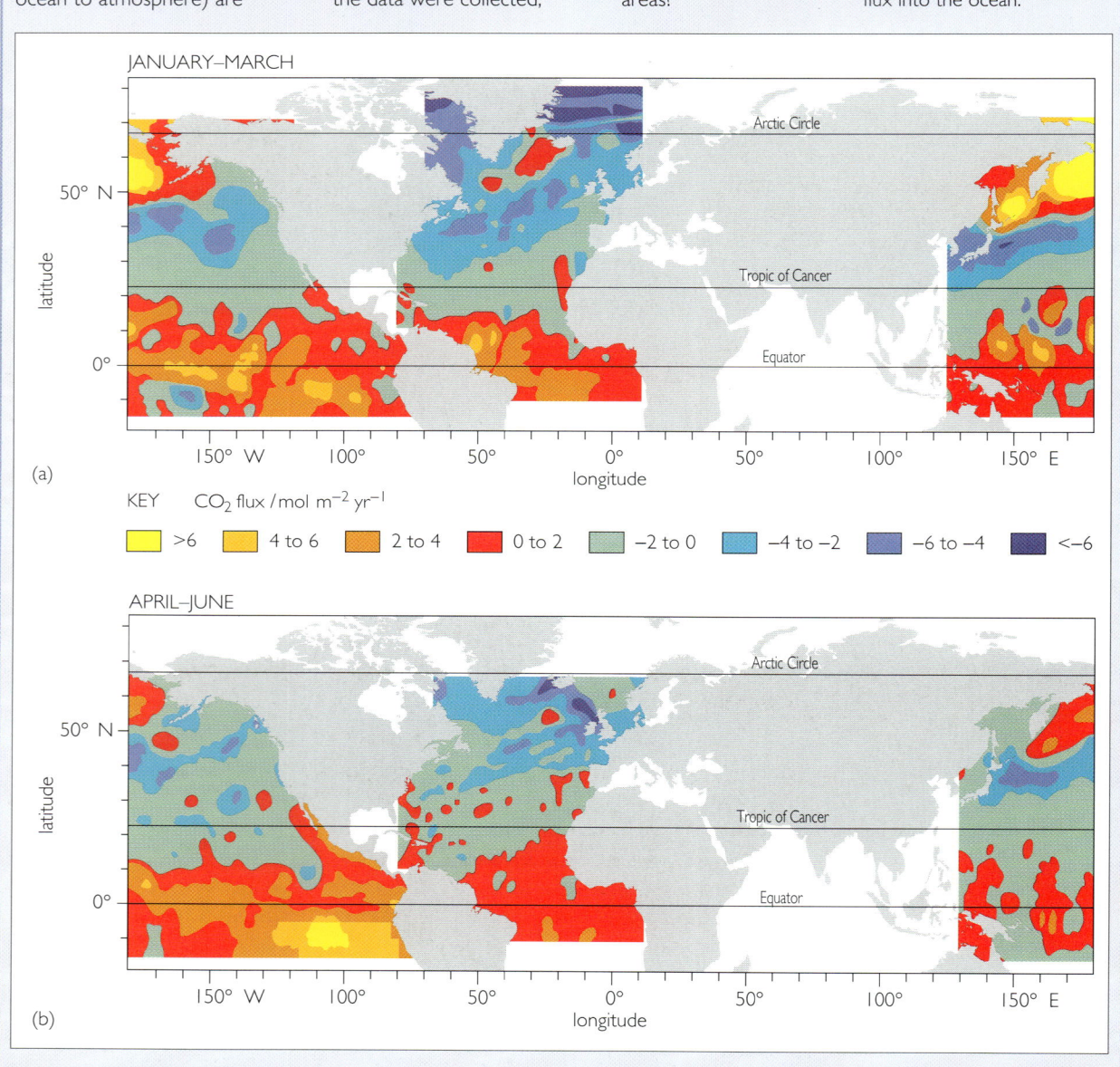

The carbon cycle

The answer to Activity 3.1 is illustrated in Figure 3.18 which extends the area of consideration to include the Southern Hemisphere. Do not worry if you did not manage to answer this all correctly: the main point to appreciate is that net fluxes of carbon dioxide are both into and out of the ocean, and that while regions of deep water mass formation generally act as sinks, other regions may be sinks at some times and sources at others, depending on the time of year, and the physical and biological processes occurring in surface waters.

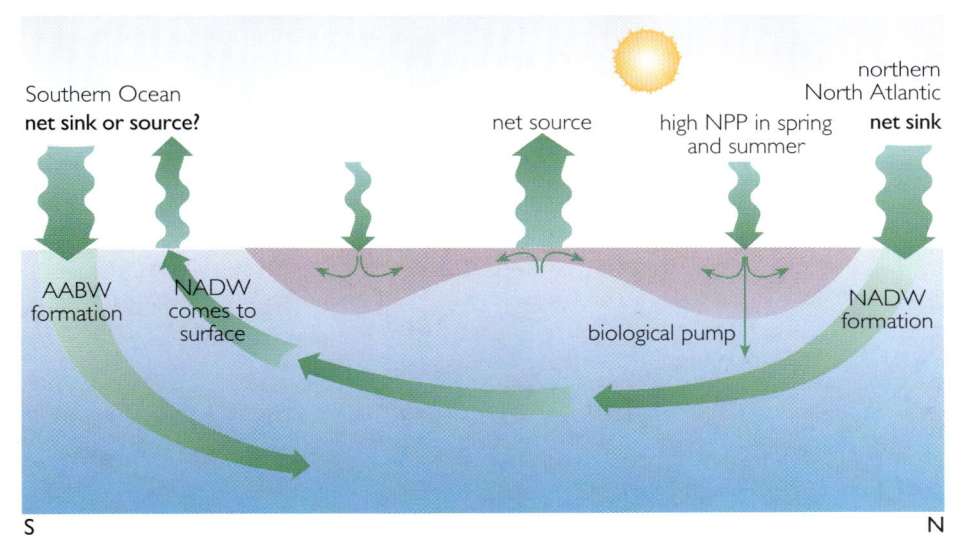

Figure 3.18
Sketch cross-section of the Atlantic Ocean to illustrate the carbon dioxide fluxes across the air–sea interface, including those discussed in Activity 3.1. Northern and southern high latitudes act as sinks for atmospheric CO_2 because they are regions of deep and bottom water mass formation (for North Atlantic Deep Water, NADW, and Antarctic Bottom Water, AABW, respectively). However, North Atlantic Deep Water eventually comes to the surface again at the Antarctic Divergence (cf. Figure 2.29), and it is not clear whether the Southern Ocean is a net source or a net sink of atmospheric CO_2. (The diagram is drawn for the northern summer and so high primary productivity, contributing to drawdown of CO_2, is shown for the North Atlantic.)

While surface waters may be in equilibrium with the overlying atmosphere as far as its gas content is concerned (cf. Figure 3.7), the cold waters filling the deep ocean were last in contact with the atmosphere when they were at the surface hundreds of years earlier.

- In general, therefore, would you expect deep waters to contain higher or lower concentrations of CO_2 (as dissolved inorganic carbon) than surface waters?

- Deep water generally contains higher concentrations of dissolved inorganic carbon. When deep water masses were last at the surface, they were at high latitudes where, because of the low temperatures, large amounts of CO_2 could dissolve from the atmosphere.

This is not the whole story, however. The decomposition of organic remains as they sink into the deep ocean, and the dissolution of skeletons and shells, also supply dissolved inorganic carbon to deep ocean waters, making them slightly more acid than surface waters. This increased acidity, along with increased pressures, partly explains why calcium carbonate debris dissolves more readily at depth.

Of course, deep ocean waters are not an endless sink for carbon dioxide, because there is exchange of water between the surface and deep ocean. As well as localized regions of downwelling and upwelling (mostly from relatively shallow depths, but also from greater depths at the Antarctic Divergence), there is mixing between the water masses flowing at various depths in the ocean. To compensate for the large volumes of water sinking at high latitudes, the net direction of mixing in the oceans as a whole is upwards. Because of the general movement of deep water towards the northern Pacific and Indian Oceans in the 'global thermohaline conveyor' (Figure 2.31), a relatively large proportion of carbon-rich water eventually comes to the surface in these regions (as can be seen by the yellow and red tone there in Figure 3.17). As a result of all the various physical, biological and chemical processes acting on it, on average a carbon atom will reside in the oceans for about 10^3 years, and most of this time will be spent in the deep ocean below the thermocline.

Figure 3.19 shows how the marine carbon cycle links with the terrestrial carbon cycle (cf. Figures 3.2 and 3.3). As you can see, the deep ocean is by far the largest reservoir of carbon, storing some $38\,000 \times 10^{12}$ kgC at any one time. The other carbon reservoirs in the ocean are the surface waters, which at any one time contain approximately 1000×10^{12} kgC (including 2×10^{12} kgC in phytoplankton) and the uppermost sea-bed sediments, which store some 3000×10^{12} kgC. These sediments are sometimes called 'reactive sediments' because carbon in them may undergo physical, biological or chemical reactions, so releasing it into the overlying water. Also shown in Figure 3.19 are estimates of annual fluxes between the various reservoirs.

If we wanted to calculate the residence time of carbon in ocean waters (reservoir size ~ $38\,000$–$39\,000 \times 10^{12}$ kgC, see above), it might seem that the best course of action would be to use the input from rivers as our flux. After all, although small – only ~ 0.4×10^{12} kgC – the river flux of carbon is one-way, while fluxes across the air–sea interface are in both directions.

■ What answer would we get for the residence time, using the input from rivers as our flux?

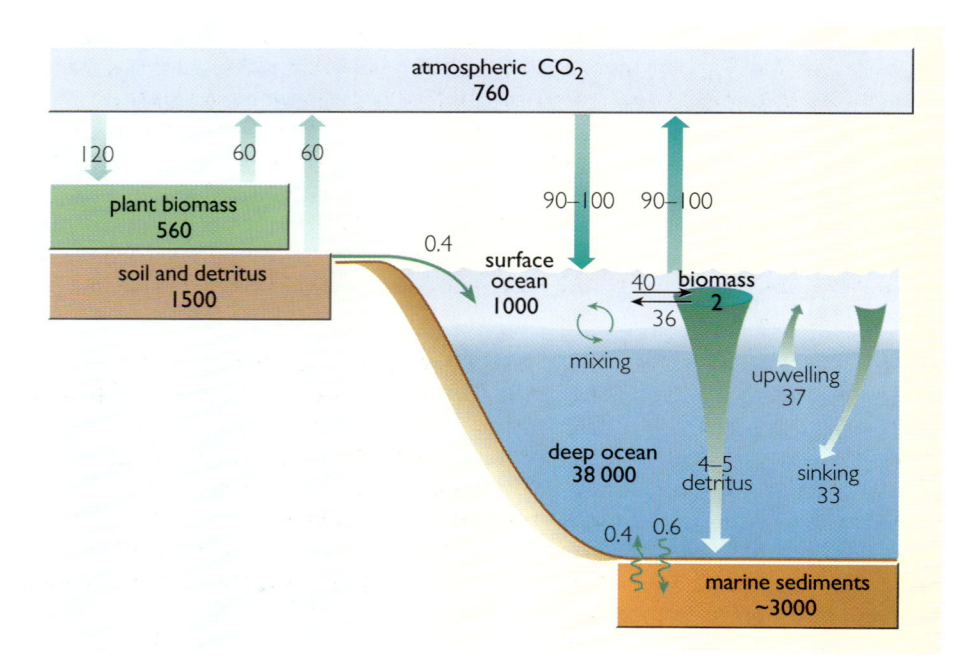

Figure 3.19
The carbon cycle (terrestrial plus marine), including those processes acting over time-scales of up to hundreds of thousands of years. Carbon reservoirs are given in 10^{12} kgC and fluxes in 10^{12} kgC yr^{-1}.

■ We would get $\dfrac{\sim 39\,000 \times 10^{12}\,\text{kgC}}{0.4 \times 10^{12}\,\text{kgC}\,\text{yr}^{-1}} \approx 10^5$ years.

But, as mentioned above, a carbon atom actually resides in ocean waters for about 10^3 years – a value obtained by observation and measurement, not residence-time calculation. We will come back to this conundrum shortly, but first let's look more closely at the fate of carbon in the ocean (shown schematically in Figure 3.11).

About half the carbon entering the ocean in run-off originates as organic carbon (DOC and POC) and half is inorganic carbon released in the weathering of carbonate and silicate minerals. Once in the ocean, both dissolved organic and inorganic carbon enters into biological cycling, and eventually becomes part of the bicarbonate–carbonate system in equilibrium with atmospheric CO_2, or is precipitated to form shells, skeletons, etc. Riverborne particulate material (remains of leaves, dead freshwater plankton, sewage) accumulates in estuaries or nearshore sediments whence the carbon may be released to the overlying water through biological oxidation. Some of the carbon accumulating in these sediments may become deeply buried and eventually isolated from the biospheric carbon cycle. These deposits, together with organic remains buried in deep-sea sediments and buried organic deposits on land (in swamps, etc.), account for the removal of $\sim 0.05 \times 10^{12}\,\text{kgC}\,\text{yr}^{-1}$ (not shown on Figure 3.19).

We can also put some approximate values on the fluxes of carbon to the deep ocean, discussed a little earlier in connection with Figures 3.14 and 3.15. The biological pump transfers $4–5 \times 10^{12}\,\text{kgC}$ from surface to deep water every year, and downward mixing and sinking adds another $33 \times 10^{12}\,\text{kgC}$; every year, however, upward mixing and upwelling bring approximately $37 \times 10^{12}\,\text{kgC}$ in solution up to the surface ocean. In other words, there is a net transfer of about $4 \times 10^{12}\,\text{kgC}\,\text{yr}^{-1}$ from the deep to surface ocean, balancing the carbon transported to the deep ocean via the biological pump. Over intermediate time-scales (decades to centuries), therefore, the cycle is roughly in balance.

By now, it should be clear that ocean waters cannot really be simplified into one reservoir as far as the marine carbon cycle is concerned. It is a number of interlinked reservoirs with fluxes between them. Most important of all, the surface ocean not only has fluxes in and out across the air–sea interface but also has fluxes across its lower boundary, the thermocline; and within the surface ocean is the reservoir of living biomass where carbon resides for a matter of days, on average. Moreover, although fluxes across the sea-surface are $90–100 \times 10^{12}\,\text{kgC}\,\text{yr}^{-1}$ in both directions, at any one location at any one time fluxes into and out of the surface ocean are by no means equal. What is more, carbon entering the ocean as CO_2 dissolving in cold surface waters at high latitudes, is carried straight down into the deep ocean – in fact, at high latitudes the surface and deep oceans are effectively joined (Figure 2.29).

Because of marine-snow formation, the time between the fixation of carbon in phytoplankton and the arrival of organic remains at the sea-bed can be as little as a few weeks, so this aspect of the marine carbon cycle (i.e. the biological pump) can reflect what is happening in surface waters on very short time-scales. Of the $4–5 \times 10^{12}\,\text{kgC}\,\text{yr}^{-1}$ transported to the deep ocean in biological debris, about $4 \times 10^{12}\,\text{kgC}\,\text{yr}^{-1}$ is redissolved in deep ocean water, so that – as shown in Figures 3.11 and 3.15 – only a small percentage of the carbon originally fixed in surface waters is preserved in sediments. Of the $0.6 \times 10^{12}\,\text{kgC}\,\text{yr}^{-1}$ deposited or precipitated from deep ocean water within marine sediments, about $0.4\, 10^{12}\,\text{kgC}\,\text{yr}^{-1}$ eventually finds its way back into deep ocean water. This means that about $0.2 \times 10^{12}\,\text{kgC}\,\text{yr}^{-1}$

becomes decoupled from the deep ocean water, and as in the case of carbon buried in sediments on land or in shallow water, this small amount of carbon 'leaks' from the cycle. This $0.2 \times 10^{12}\,\text{kgC yr}^{-1}$ is the amount preserved at the present time, but because of the short time-scales involved, short-term changes in biological productivity at the surface are likely to be reflected in the amount of carbon preserved in sediments (though the chemistry of deep ocean waters would also be affected, and this would in turn affect the preservation of material). However, it can be argued that over the long term the rate of accumulation of carbon in sediments will reflect its rate of supply in rivers. For this reason, although it is small, the river flux of carbon may be a far from trivial player in the global carbon cycle. We shall now consider the fate of the carbon that is preserved.

3.3.3 Long time-scales: the geological carbon cycle

The carbon cycle as we have described it so far involves cycling over time-scales of years or decades (the terrestrial carbon cycle) to up to hundreds of thousands of years (the marine carbon cycle). As described above, both the terrestrial and marine carbon cycle 'leak', so that over long time periods carbon accumulates deep within sediments and is removed from the cycles.

As discussed in the previous section, there are two different types of carbon-rich accumulations.

- What are they?
- They are sediments containing carbonate (carbon in inorganic form), and those containing soft tissue (organic) remains.

The first of these types of sediment is often described as *calcareous*. Such sediments are generally marine deposits containing accumulations of calcium carbonate shells and skeletons, which have been made by organisms using dissolved inorganic carbon, mainly HCO_3^-. They usually also contain variable amounts of land-derived sediments (mainly clay) and, particularly below upwelling regions, the remains of shells of silica; if there is a significant siliceous component, they are referred to as *calcareous–siliceous* sediments. (There are also sediments that are predominantly clay or predominantly siliceous, and these contain negligible amounts of carbonate.)

Over millions of years, because of high pressures due to increasing thicknesses of overlying sediment, and high temperatures due to heat loss from within the Earth, chemical and structural changes occur in the sedimentary accumulations and they become lithified, i.e. converted into rock. The kind of rock that is formed depends on the initial composition of the sediments: carbonate-rich sediments become rocks such as chalk and limestone.

The second kind of carbon-rich accumulation, containing organic material (i.e. that made up of molecules of carbon, hydrogen and oxygen), is often described as *carbonaceous*. As organic-rich material is covered with more organic matter and sediment, the weight of the overlying deposits causes compaction, squeezing out water and residual air from the pore spaces. In the resulting oxygen-poor environment, a dense residue enriched in carbon, known as **kerogen**, is formed. Under continued deposition of organic matter and sediments (often related to the subsidence of continental crust, as below the North Sea), the original material may be buried to depths of several kilometres. As in the case of calcareous remains, high temperatures and pressures eventually cause the sediments to become lithified. Large accumulations of land plants (e.g. the remains of trees that have

The carbon cycle

Figure 3.20
Summary diagram of the carbon cycle (cf. Figure 3.19), now also showing burial and preservation of carbon in sediments (as both calcareous and carbonaceous remains), its eventual return to the atmosphere, and the different time-scales involved. Note that the 'carbonate rocks' include lithified remains of calcareous organisms that lived in shallow waters as well as those that lived in the oceans. The approximate mass of carbon in each reservoir is given in units of 10^{12} kgC, and the fluxes are given in units of 10^{12} kgC yr^{-1}. (Anthropogenic contributions and current *imbalances* in the overall system are not included.)

accumulated in anoxic swamps) may become coal; marine sediments containing very high concentrations of phytoplankton debris can produce petroleum. If the organic matter is dispersed through sediments, oil-shales may result.

As mentioned earlier, at the present time about 0.05×10^{12} kgC yr^{-1} accumulates in buried organic matter (on land and sea) and about 0.2×10^{12} kgC yr^{-1} accumulates in shallow waters and in the deep ocean as inorganic carbonates. Over the Earth as a whole, there is approximately $10\,000\,000 \times 10^{12}$ kgC of carbon in carbonaceous rock (including fossil fuels) and four times that in limestones, chalks, etc. (see Figure 3.20). Collectively, rocks are the largest reservoir of carbon on Earth. The residence time for carbon within this reservoir is of the order of 100–200 million years, but even carbon incorporated into rocks can eventually enter the atmosphere again, as geological processes, particularly mountain-building, bring deeply buried rocks to the surface. The cycle then begins again, as the rocks are weathered and eroded by wind, water or biological activity. As discussed in Section 3.3.2, weathering of rocks by rain and soil water produces dissolved inorganic carbon, particularly HCO_3^- (Equation 3.6). Carbonaceous sediments exposed to the air may oxidize directly to CO_2, may be oxidized by bacterial respiration, or may be chemically weathered by soil water and rainfall to release carbon (dissolved and particulate carbon, DOC and POC) into streams. Sometimes, the carbon in rocks finds its way back into the atmosphere in volcanic gases (of which more later). Either way, the carbon is eventually returned to the atmosphere and/or the ocean.

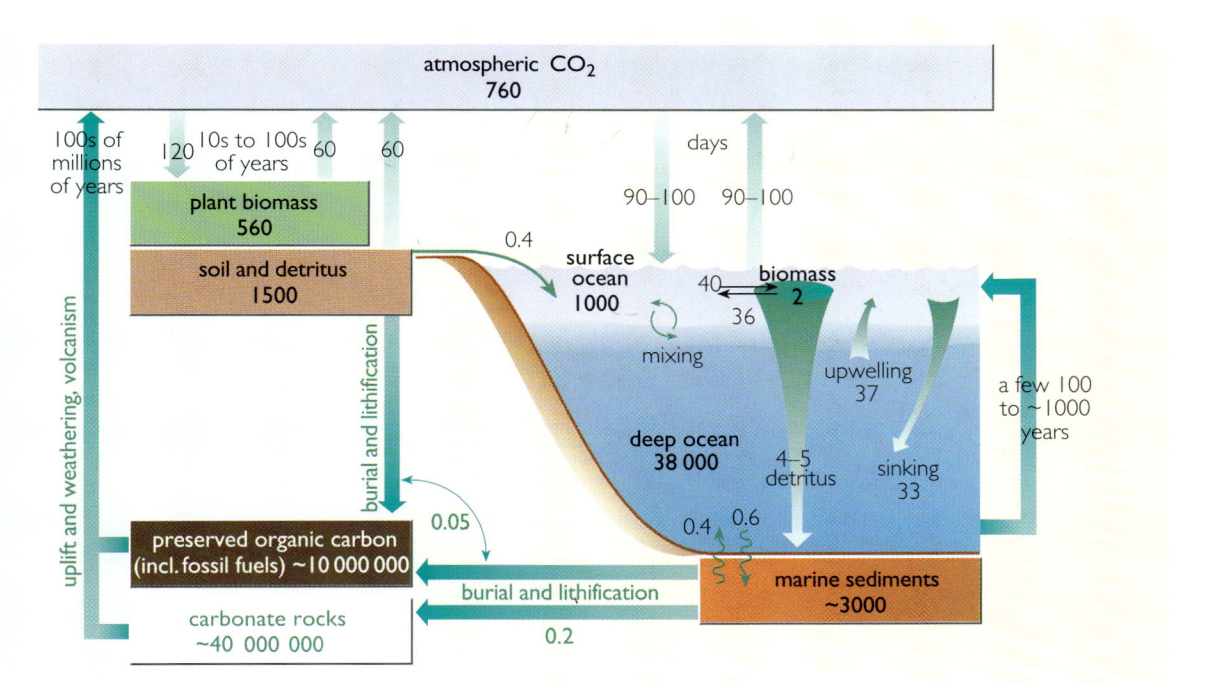

Long-term controls on atmospheric CO_2

Let us look at what affects the concentration of atmospheric CO_2 in more detail. As discussed in Section 3.3.1, the main flux of CO_2 into the atmosphere through the terrestrial carbon cycle is via respiration by plants and by decomposers, which is almost exactly balanced by that removed by photosynthesis (Figure 3.3), for which the chemical reaction is:

$$\underbrace{6CO_2}_{\text{from atmosphere}} + 6H_2O \xrightarrow{\text{energy}} \underbrace{C_6H_{12}O_6}_{\text{organic material}} + 6O_2 \qquad \text{(Eqn 1.2)}$$

If the current rate of burial of organic carbon into deep sediments roughly equals the rate of carbon release by oxidation and weathering of organic-rich sedimentary rocks (e.g. exposed coal seams), there will be no net gain or loss of atmospheric CO_2. We are fairly certain that the rate of burial does more or less match the rate of release by oxidation and weathering because a large imbalance would change the concentration of atmospheric oxygen, and such a change has not occurred in the recent geological past. The early Earth had very little, if any, oxygen in its atmosphere, and it was only the cumulative effect of burial of small amounts of organic carbon that allowed oxygen to build up in the atmosphere.

Whereas the flux of carbon into carbonaceous rocks can be balanced by the return flux from weathering, this is not the case for the flux of carbon into carbonate rocks. There is a discrepancy of about 0.03×10^{12} kgC yr^{-1} between the carbon that is stored in carbonate rocks (0.2×10^{12} kgC yr^{-1}, Figure 3.20) and that eventually returned to the atmosphere and ocean by weathering of carbonate rocks on land (currently estimated at 0.17×10^{12} kgC yr^{-1}).

- Remembering that *silicate* rocks, as well as carbonate rocks, are weathered (Equation 3.6a and 3.6b, and the associated text), can you suggest what the underlying reason for this discrepancy might be?

- The answer lies in the fact that in the weathering of carbonate minerals, only *one* of the carbon atoms going to form the bicarbonate $2HCO_3^-$ comes from the atmosphere, the other coming from carbonate mineral itself; in the weathering of silicate minerals, *both* carbon atoms come from the atmosphere.

So, weathering a carbonate mineral on land removes one atom of carbon from the atmosphere (in a molecule of CO_2) (Equation 3.6a). Precipitating a carbonate mineral in the ocean (e.g. as part of a shell or skeleton; p.92) returns that molecule to the upper ocean, which is in equilibrium with the overlying atmosphere:

$$\underbrace{Ca^{2+}(aq) + 2HCO_3^-(aq)}_{\text{in solution in seawater}} \longrightarrow \underbrace{CaCO_3(s)}_{\substack{\text{precipitated}\\\text{by organisms}}} + H_2O + CO_2 \qquad \text{(Equation 3.7)}$$

Weathering of a carbonate mineral followed by precipitation of a carbonate mineral therefore results in *no* net gain or loss of carbon (in effect CO_2) to the atmosphere.

Now, consider the case of silicate weathering. The reaction for weathering of a silicate mineral (e.g. $NaAlSi_3O_8$, as in Equation 3.6b) requires *two* molecules of atmospheric CO_2. Precipitating calcium carbonate in the sea returns *only one* of those molecules to the upper ocean–atmosphere equilibrium system (Equation 3.7). Burying this $CaCO_3$ in the sediments, so removing it from the terrestrial and marine carbon cycles, thus represents a *net depletion* of atmospheric CO_2. These arguments are summarized in Figure 3.21.

Silicate weathering on land followed by carbonate precipitation and burial in the sea removes about 0.03×10^{12} kgC from the atmosphere each year.

- There is presently (1996) a total of 760×10^{12} kgC in the atmosphere. How long would it take for silicate weathering / carbonate precipitation and burial to completely deplete the atmosphere of CO_2? (For the sake of this exercise, ignore all the other fluxes in the carbon cycle.)

- 760×10^{12} kgC divided by 0.03×10^{12} kgC yr^{-1} ≈ 25 000 years, i.e. all the CO_2 would be removed from the atmosphere in a very short time, geologically speaking.

The carbon cycle

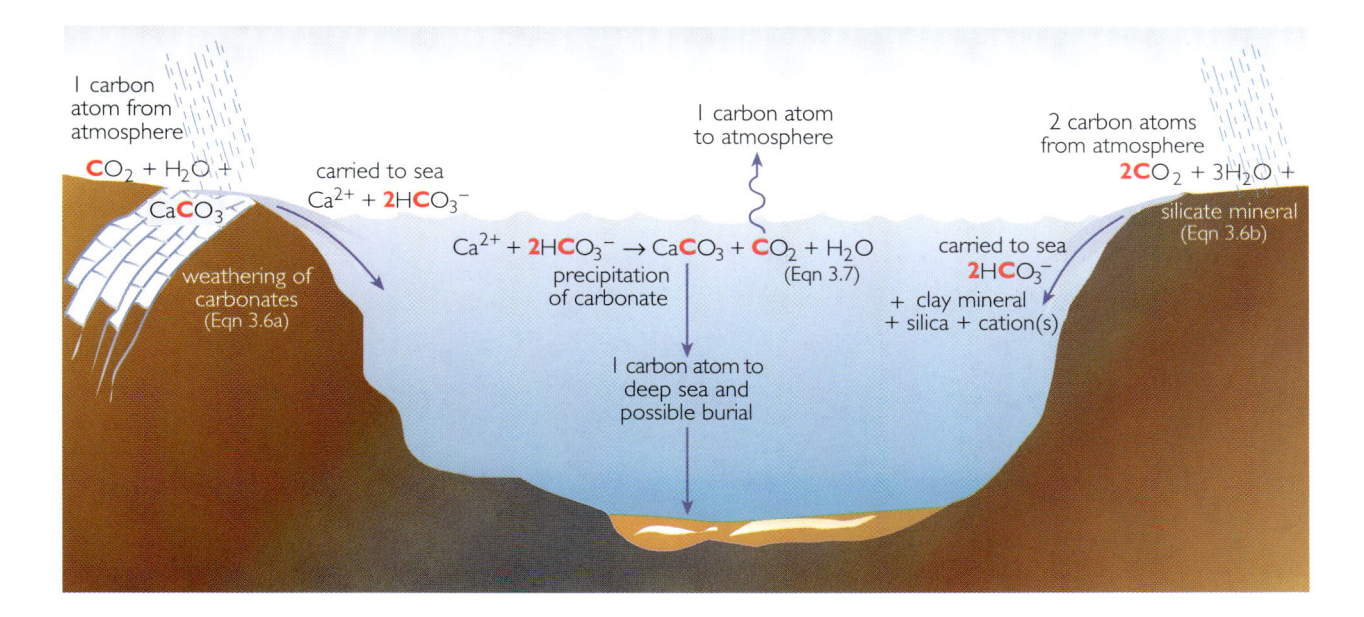

Figure 3.21 Why burial of sediments containing carbon from the weathering of silicates can result in a net removal of CO_2 from the atmosphere, but burial of carbonates does not. (Note that this diagram does not include processes involving *organic* carbon.)

The geological record shows that weathering of silicate minerals has occurred on Earth for at least the last 3.8 billion years and that the atmosphere has always contained some CO_2. There must be some return flux of atmospheric CO_2, and this is thought to be via volcanoes, components of the global 'conveyor belt', which recycles the entire crust over time-scales of hundreds of millions of years. Indeed, about 0.02×10^{12} kgC yr^{-1} is estimated to be released to the atmosphere by volcanoes at the present time. We will come back to this topic in later chapters.

3.4 A system in balance?

We have seen that the global carbon cycle is highly complex: carbon may be in gaseous, dissolved or solid form, and may reside in various reservoirs on time-scales ranging from decades (or less) to hundreds of million of years. With such complexity, is it possible to determine whether the cycle is in balance?

To explore this question, we need some kind of reference: a reservoir which we can use as an 'indicator' of any changes in the cycle.

- Would changes in the cycle show up first in a large reservoir or a small one? (Imagine a box containing a thousand marbles and a box containing ten marbles. If five marbles are added to or removed from each of these boxes, in which one would you notice it?)

- The addition or removal of five marbles would be much more noticeable in the box with only ten marbles initially. By extension, changes in the carbon cycle would show up more clearly in a small reservoir.

Atmospheric carbon dioxide is just such a small reservoir. Exchange reactions between the atmosphere and the other carbon reservoirs tend toward equilibrium, with the result that atmospheric CO_2 responds rapidly to changes in the larger reservoirs; in other words, changes in the larger reservoirs are detectable via the atmospheric CO_2 record long before they are noticeable in the sources and/or sinks themselves. Furthermore, the atmosphere is the only reservoir with direct links to the short, intermediate *and* long-term cycles of carbon (cf. Figures 3.19 and 3.20).

You calculated at the end of the previous section that the CO_2 in the atmosphere would in theory be depleted in 25 000 years if there was no return flux of CO_2 via volcanism. Over that time period, the amount of carbon in the carbonate rock reservoir would *increase* (in theory) by 760×10^{12} kgC, the mass of carbon originally in the atmosphere.

By this point you may be feeling a little suspicious.

- Can you suggest why the above calculations would not be accurate, even if volcanism *were* to stop entirely?

- The calculations effectively assume that all the other carbon equilibrium reactions would be unaffected, but this is unrealistic. For example, decreasing atmospheric CO_2 would cause the ocean carbonate equilibria to shift so that more oceanic CO_2 would be released to the atmosphere (cf. Figures 3.7 and 3.8, of which the latter illustrates the opposite situation), and this would slow the rate of depletion of atmospheric CO_2.

It seems that the concentration of CO_2 in the atmosphere has remained at a few hundreds to several thousands of p.p.m. for the past few hundred million years or so. This relative stability of atmospheric CO_2, and of the carbon cycle in general, is a result of the fact that changes in one reservoir cause repercussions on other reservoirs, and there are feedback loops between them. As mentioned in Section 2.2.3, in connection with physical interactions between atmosphere and ocean, positive feedback loops may be destabilizing but *negative* feedbacks are stabilizing.

A property of atmospheric CO_2 that could provide negative feedbacks within the carbon cycle is its role as a greenhouse gas.

Question 3.5
Imagine that the supply of CO_2 to the atmosphere was suddenly increased (by, say, enhanced volcanic activity). While CO_2 was being added to the atmosphere much faster than it was being removed, its concentration there would be increasing.

(a) From your reading of this chapter, and by reference to Figure 2.34, can you suggest at least two ways in which greenhouse warming resulting from such an increase could indirectly affect other processes forming part of the global carbon cycle?

(b) Could any of these processes result in extra carbon being preserved, and so removed from cycling for hundreds of millions of years?

Thus, if CO_2 levels on the Earth were, say, ten times higher than they are today we could say with confidence that the rate of CO_2 *removal* from the atmosphere would be increased (Equations 1.2 and 3.7). Precipitation and accumulation of carbonate resulting from silicate weathering, and preservation of organic material (mainly in the sea but also on land) both act as long-term sinks of carbon. Although we do not know how rapidly silicate weathering and global net primary productivity may respond to changes in atmospheric CO_2, we can be reasonably sure that a 'greenhouse' feedback loop, something like that investigated in Question 3.5, keeps fluctuations of atmospheric CO_2 within a relatively narrow range.

The carbon cycle

3.4.1 Short-circuiting the geological carbon cycle

If we change the scale of our observation from millions of years to decades, however, it is clear that the carbon cycle today is not in balance: there is more carbon entering the atmosphere than there is being removed from the atmosphere. This is one of the few certainties in the global carbon cycle! To see how we know this, read Box 3.3.

Box 3.3 The record on the mountain

In 1957, the scientist Charles David Keeling, then still a student, set up two stations for the continuous monitoring of CO_2 in the atmosphere: one in Hawaii on the volcanic peak known as Mauna Loa and the other at the South Pole. The results were dramatic. The CO_2 concentration of the atmosphere at Hawaii averaged around 317 p.p.m. but oscillated around that value over the year, increasing in the winter and decreasing in the summer. This provided the first clear evidence that life on Earth profoundly influences the atmosphere.

But even more was in store. Within about five years it became obvious that the annual average CO_2 concentration was steadily increasing (Figure 3.22). No natural phenomenon could be found to account for this spectacularly rapid (geologically speaking) rate of increase. The most likely source was the release of CO_2 from fossil-fuel combustion. Not only was the growth and death of plants recorded in the atmosphere, so too was the industrial activity of humans! The Mauna Loa and South Pole records are continuing and have, over the years, been augmented by a world-wide network of measurement stations. All the records of CO_2 concentration show the same rising trend. By 1996, the global average atmospheric CO_2 concentration stood at about 360 p.p.m. and it is still rising.

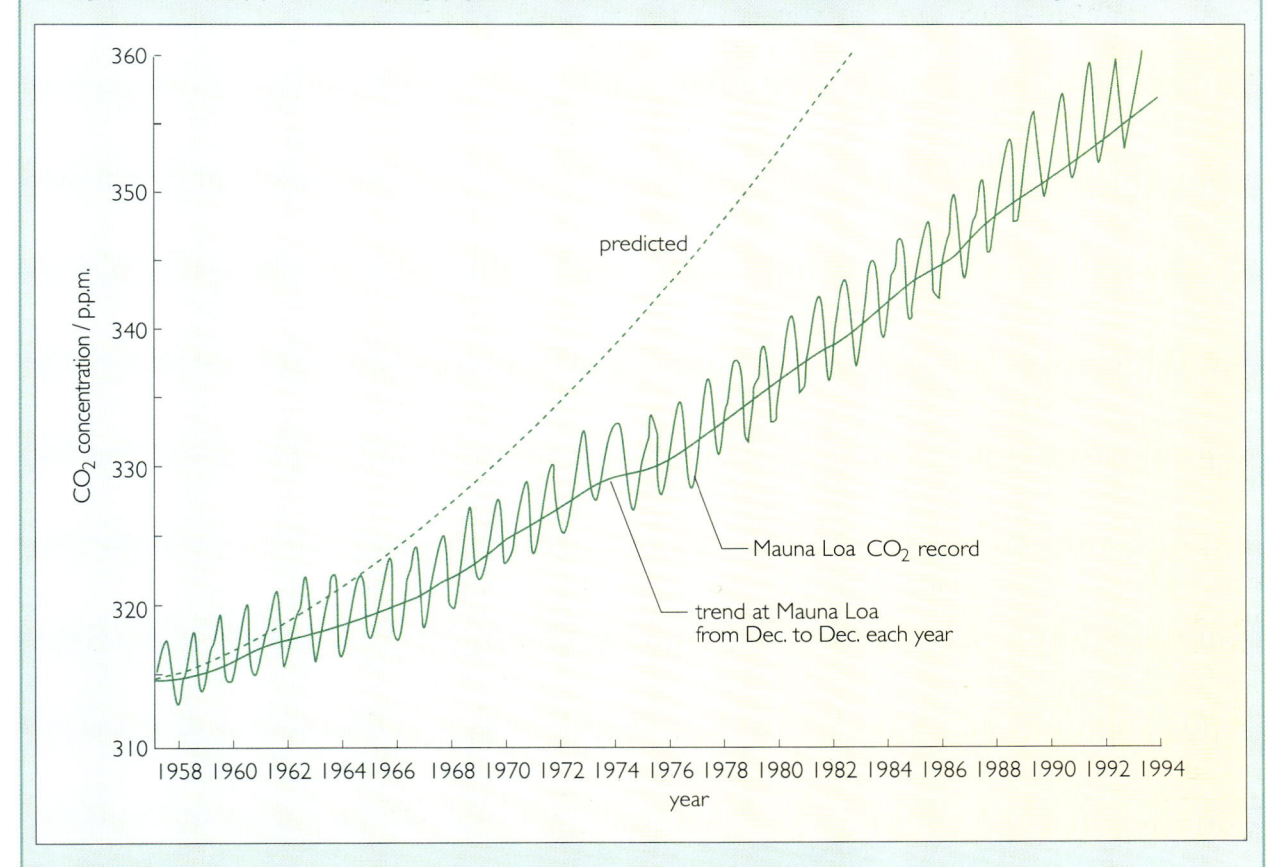

Figure 3.22
The increase in the concentration of CO_2 in the atmosphere, as measured at the Mauna Loa observatory in Hawaii. Also shown is the increase predicted on the basis of fossil-fuel combustion (discussed later).

Whether they are from Mauna Loa, the South Pole or the newer monitoring stations, all records of atmospheric CO_2 concentration show the same striking features (see Figure 3.22):

- A strong trend of increasing atmospheric levels of CO_2 over time, with the current rate of increase at about 1.5 p.p.m. per year or about 0.4% per year.
- Seasonal oscillations, with a CO_2 peak in the winter and a CO_2 trough in the summer.

However, all monitoring stations do not produce exactly the same results. Figure 3.23 is a composite diagram showing the variation of atmospheric CO_2 concentrations over the course of four years (1981–1984) over all latitudes.

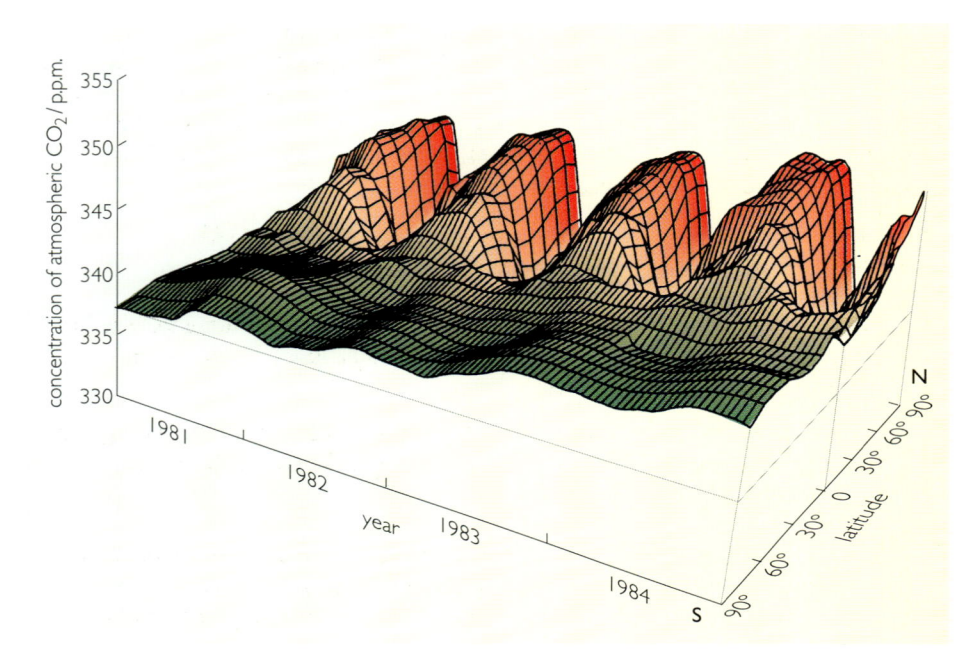

Figure 3.23
Seasonal fluctuations in atmospheric concentrations of CO_2 from 1981 to 1984, as a function of 10° latitude bands. Note that lows in the Northern Hemisphere correspond to (small) highs in the Southern Hemisphere.

■ Apart from the peaks and troughs, what is the most striking aspect of this diagram?

■ The seasonal oscillations are marked in the Northern Hemisphere, but extremely damped in the Southern Hemisphere with only slight peaks occurring during minima in the Northern Hemisphere.

The peaks and troughs and their global variations are both considered in Question 3.6.

Question 3.6
(a) What is the reason for the seasonal fluctuations in atmospheric CO_2 concentration?

(b) The patterns shown in Figures 3.22 and 3.23 have been dubbed 'the Earth breathing'. Do you think this is an appropriate description?

(c) By reference to any global map (e.g. Figure 3.4), can you suggest why this pattern is so damped in the Southern Hemisphere?

The carbon cycle

This annual cycle of CO_2 uptake and release is a striking example of how living organisms can influence geochemical cycles on Earth. The seasonal variations in CO_2 fluxes into and out of the atmosphere can be quite significant: at Hawaii, for example (latitude 19° N), they are sufficient to cause the atmospheric concentration of CO_2 to fall by about 7 p.p.m. over the course of spring and summer, whilst in Barrow, Alaska (latitude 71° N), the winter to summer decrease is double that. If seasonal decreases and subsequent increases remained equal in magnitude from year to year, the atmospheric reservoir would remain constant.

However, the rising trend apparent in Figures 3.22 and 3.23 indicates that there is a persistent disequilibrium in fluxes into/out of the atmospheric CO_2 reservoir. We know from analyses of air trapped in ice cores that this rapid increase in atmospheric CO_2 is a modern phenomenon; it began approximately 100 years ago, and has greatly accelerated in recent decades. The current rate of increase is about 1.5 p.p.m. per year: i.e. 3×10^{12} kgC are added to the atmosphere each year. This is rapid enough to make textbook values for atmospheric CO_2 obsolete: in 1958, atmospheric CO_2 stood at 315 p.p.m., by 1994 it was about 357 p.p.m., and over the lifetime of this Course it may exceed 370 p.p.m.

The overwhelmingly likely cause for the increase in atmospheric CO_2 concentrations is the extraction and burning of fossil fuels.

- By reference to Figure 3.19 (and to the title of this section!), explain briefly *why* extraction and burning of fossil fuels would cause an increase in atmospheric CO_2 concentrations.

- By extracting fossil fuels (organic carbon) and burning them, so adding CO_2 to the atmosphere, we are 'short-circuiting' the geological carbon cycle (see Figure 3.24 overleaf).

The important point here is not that carbon is being returned to the atmosphere (this would eventually happen anyway), but the *rate* at which it is occurring. Concentrations in the atmosphere of CO_2 (and other greenhouse gases – notably methane, which also contains carbon) are increasing faster than fluxes into the other reservoirs (the major short-term buffering processes) can accommodate them. Simply speaking, humans have caused an increase in the rate of return to the atmosphere of carbon previously stored in the Earth's crust. As we have been directly responsible for this flux through our own activities, we know its size to a high degree of accuracy: according to global production figures for coal and oil, it is $\sim 5 \times 10^{12}$ kgC yr^{-1}.

- How does the current yearly rate of increase of CO_2 in the atmosphere (see above) compare with the flux from fossil fuels?

- The current annual increase is only about 3×10^{12} kgC – i.e. between a half and two-thirds of that being added from fossil fuels.

So there is actually more carbon being *released* into the atmosphere by human activities than is *accumulating* there – hence the difference between the observed and predicted increases shown in Figure 3.22. In fact, when we add the amount of carbon added to the atmosphere each year by logging and burning of forest vegetation (poorly known, but currently estimated at $\sim 1.8 \times 10^{12}$ kgC yr^{-1}) the discrepancy is even greater.

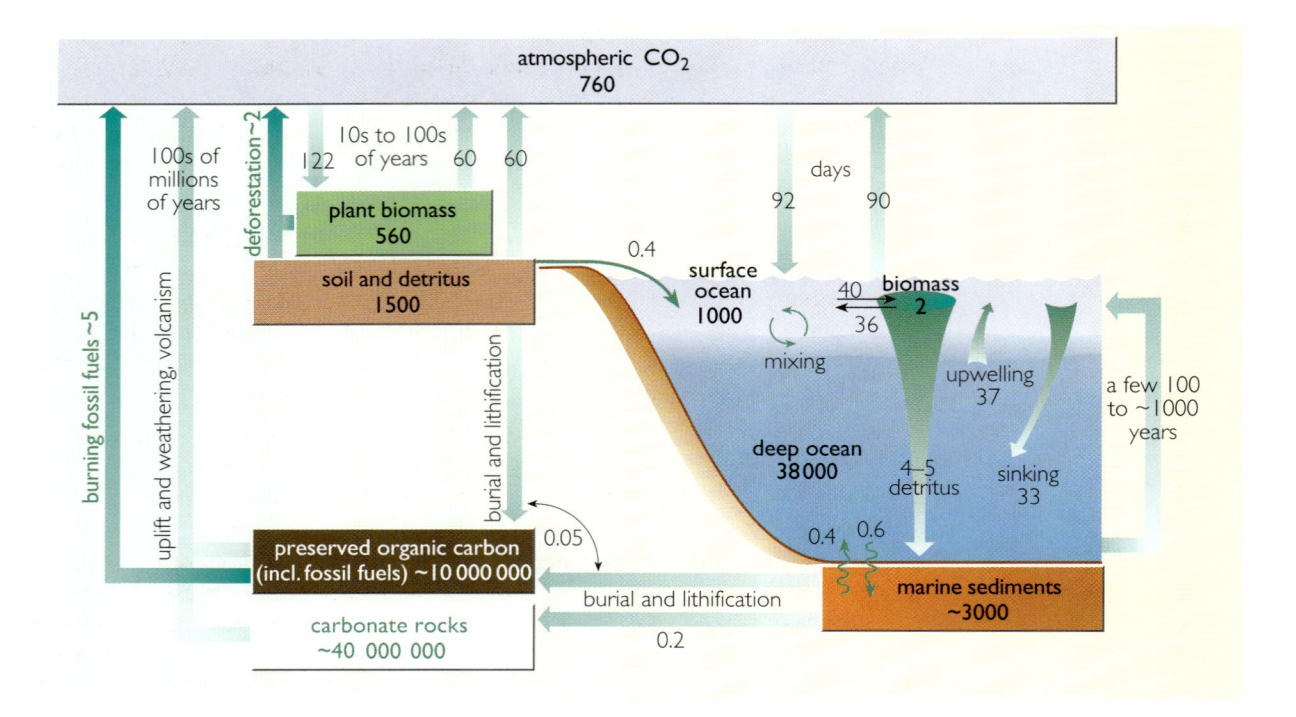

Figure 3.24 shows the global carbon cycle as perturbed by human activity: it is similar to Figure 3.20 but now includes not only estimates of the fluxes directly resulting from human activities (fossil-fuel burning and deforestation), but also the changes that might feed through into other fluxes.

■ By reference to Figure 3.24, can you suggest where the 'missing' carbon might have gone?

■ The only possibilities seem to be into the ocean or into plant biomass (terrestrial or marine, or both). (Figure 3.24 shows possible increased fluxes into these reservoirs from the atmosphere.)

Figure 3.24
Summary diagram of the carbon cycle (cf. Figure 3.20), now also showing the fluxes back to the atmosphere caused by human activities. Carbon reservoirs are given in 10^{12} kgC and fluxes in 10^{12} kgC yr^{-1}.

As illustrated in Figure 3.8 (and mentioned above), an increase in the concentration of atmospheric CO_2 will push the equilibria shown in Figure 3.7 to the right and cause an increase of the flux of CO_2 into the ocean (though it would take a long time for the oceans to equilibrate with the atmosphere). Computer models of ocean circulation and calculations of carbonate equilibria suggest that an extra 1.6×10^{12} kgC yr^{-1} could eventually be removed from the atmosphere to the oceans in this way. That still leaves us with $6.8 - (3.0 + 1.6) = 2.2 \times 10^{12}$ kgC yr^{-1} unaccounted for.

Conclusive evidence as to the identity of this 'missing sink' has so far eluded scientists, partly because, with the exception of the atmosphere, none of the reservoirs in the carbon cycle is small enough to detect a change of the order of 10^{12} kgC. However, as mentioned above, increased rates of fixation of carbon in plant material is a possibility. As stated in the answer to Question 3.5a, small-scale experiments have shown that an increase in the concentration of atmospheric carbon dioxide can (at least under experimental condition) increase rates of carbon fixation (i.e. net primary productivity), thus storing some of the missing carbon in vegetation or soil organic matter. Until recently, the favourite site for such increased fixation has been the temperate forests of the Northern Hemisphere. It now seems, however, that – somewhat surprisingly – the forests of Amazonia may be responsible for taking up a large proportion of the 'extra' carbon, both into continuing growth of existing trees and regrowth in cleared areas. Of course, the missing sink may turn out to be a combination of sites. Wherever it is, we should be grateful for it – human societies seem to have great difficulty in adapting to changing environmental conditions.

The rapid increase in atmospheric CO_2 we are observing today simply reflects the fact that the carbon cycle is currently not in balance. Eventually, however, coal and oil deposits will be exhausted and the rock-to-atmosphere CO_2 'shunt' will be closed. Some time after this, the various reservoirs of the carbon cycle will re-establish equilibrium with one another and atmospheric CO_2 will stop increasing, re-establishing itself at a higher concentration than previously.

We should not leave the carbon cycle without mentioning another component, the gas methane, CH_4, whose atmospheric concentration is some two orders of magnitude less than that of CO_2 (Table 2.1). Methane is in fact a more powerful greenhouse gas than CO_2, but its very low concentration means that it has a negligible effect. However, methane is emitted from swamps, wetlands, rice paddies and cattle. It is also the principal constituent of the natural gas many of us use for domestic heating, and substantial amounts are lost during extraction of fossil fuels. Since methane is relatively quickly oxidized to CO_2 in the atmosphere, its residence time there is only about a decade, and at present it is not considered a significant agent of global warming.

Atmospheric methane concentrations may of course have been greater in the geological past, and could be so in the future. Large quantities are presently locked up as so-called gas hydrates in frozen wetlands (tundra) and in some deep-sea sediments. It is possible that global warming could release this frozen methane into the atmosphere at rates sufficient to cause a positive feedback effect, which could exacerbate the warming because of the greenhouse gas characteristics of methane.

You may also have heard about the group of artificial carbon compounds (used as spray propellants and refrigerants) – the chlorofluorocarbons (CFCs). These are also greenhouse gases, but have a much more powerful effect in the stratosphere, where they are the principal cause of the ozone layer depletion that has received so much publicity in recent years. So far as the global carbon cycle is concerned, however, they can be neglected, not only because they are manufactured, but also because their atmospheric concentrations are even less than that of methane.

3.5 Summary of Chapter 3

1. The chemistry of the carbon atom means that it has a unique role in the living world. Its ability to share electrons with other carbon atoms allows the construction of large complex molecules of carbon, hydrogen and oxygen, of which organic material is built.

2. Carbon fulfils two essential roles in the biosphere: it is the primary component of living tissue (see 1 above) and, in its gaseous forms (CO_2 and CH_4), it allows the surface of the Earth to be warm enough to support life.

3. The main biogeochemical connection between Earth and life is the global carbon cycle: the movement of carbon through the atmosphere, biosphere, lithosphere and ocean. The global carbon cycle involves interlinking cycles over three major time-scales: (i) the terrestrial carbon cycle, driven by biological processes and acting over time-scales of months/years to decades;

(ii) the marine carbon cycle involving chemical, biological and physical components, and acting over an intermediate time-scale of up to hundreds of thousands of years; (iii) the geological carbon cycle, involving rocks and sediments, and acting on time-scales of up to hundreds of millions of years.

4   The terrestrial carbon cycle is driven by the fixation of atmospheric carbon into organic matter via photosynthesis, and involves large fluxes of carbon between the atmosphere and fairly small organic reservoirs (living organic matter and organic debris). The average residence time of carbon in plant biomass is about nine years. Estimates of annual net primary production (flux of carbon into the terrestrial biomass reservoir) and of standing stock are problematic, and involve determining the areas occupied by different biomes.

5   The marine carbon cycle is linked to the terrestrial carbon cycle through the carbon carried to the sea in rivers in organic and inorganic form (both dissolved and particulate). The surface ocean is, on average, in equilibrium with the overlying atmosphere and so the fluxes between the two more or less balance. Fluxes of carbon (CO_2) across the air–sea interface may be driven by physical processes (particularly sinking of cold water masses in high latitudes, and upwelling), and by biological activity (i.e. algal photosynthesis which, when combined with rapid sinking of organic matter into the deep sea, constitutes the *biological pump*). In both cases, however, the mechanism whereby CO_2 is 'pushed out' or 'drawn into' the ocean is chemical. The linked equilibria determine the relative proportions of CO_2 gas and the various forms of *dissolved inorganic carbon*: H_2CO_3 (carbonic acid), HCO_3^- (bicarbonate ion, the main constituent in seawater) and CO_3^{2-} (carbonate ion); these equilibria are together known as the carbonate system. Biological activity produces inorganic particulate carbon in the form of shells and skeletons which eventually sink to the sea-bed; whether they are dissolved or accumulate depends on the chemistry of the deep ocean water and the rate of supply of remains. The depth at which the proportion of calcium carbonate remains falls to less than 20% of the total sediment is known as the *carbonate compensation depth*.

6   In the geological carbon cycle, carbon in the organic and inorganic products of weathering on land is carried to the sea in rivers, takes part in the marine carbon cycle, and – in the case of a very small proportion is preserved and buried, eventually to be returned to the atmosphere or ocean via volcanism or via weathering and/or oxidation. The geological carbon cycle acts on a time-scale of up to hundreds of millions of years, and involves small fluxes between large reservoirs; carbonate and organic sedimentary rocks together store more than 99.9% of the carbon on the Earth.

7   Extracting and burning fossil fuels shortcuts the geological-scale return flux of carbon and has currently brought the global carbon cycle into disequilibrium. As a result, the concentration of CO_2 in the atmosphere is rising (as are those of other greenhouse gases), but it is not rising as much as expected. Some of the 'extra' CO_2 has almost certainly been taken up by the ocean; the rest (or some of it) is being taken up by increased net primary productivity of land plants, notably in forests.

8   Long-term stability of the carbon cycle is controlled by negative feedbacks acting on the atmospheric carbon reservoir via rates of silicate weathering on land followed by deposition and preservation of carbonates in marine sediments, and (possibly) the deposition and preservation of organic carbon, mainly in marine sediments.

Now try the following questions to consolidate your understanding of this chapter.

Question 3.7
The important chemical processes in the global carbon cycle are in approximate balance. What are the equivalent 'reverse reactions' for (a) photosynthesis, (b) weathering of limestones on land?

Question 3.8
Is the following statement true? 'Limestone deposits are examples of inorganic carbon, but skeletal remains and shells are examples of organic carbon.'

Question 3.9
(a) According to Table 3.2, what approximately is the total mass of carbon in marine plant material (i.e. the standing stock), and how does it compare with the average standing stock of plant material on land?

(b) According to Figure 3.19, what, on average, is the residence time of carbon in living phytoplankton? Does carbon cycle through the marine biomass reservoir faster or slower than through the terrestrial biomass reservoir?

Chapter 4
Volcanism and the Earth system

Like the workings of the carbon cycle, most of the processes occurring on and in the Earth are quiet and unspectacular. Volcanism can be very different: Earth's history has been frequently punctuated by massive explosive eruptions whose effects are discernible world-wide. Predictably, these occasionally catastrophic episodes have been linked with everything from the death of the dinosaurs to the triggering of Ice Ages. While the more extravagant speculations are hard to sustain, volcanism has played, and continues to play, a crucial role in the evolution of the Earth system. In this chapter, we will attempt to quantify some of the effects that volcanic emissions can have on climate, on both short and long time-scales. However, we should begin by establishing why volcanoes occur where they do.

4.1 Introduction

During volcanic activity, heat and material from the Earth's interior come to the surface, and it is probably not surprising that volcanoes owe their existence to the same internal processes that drive the motions of the Earth's lithospheric plates. Figure 4.1 shows the locations of current or recent volcanic activity in relation to plate boundaries.

- Volcanic eruptions related to plate boundaries can be divided into two main groups. What are they?

- The two groups are:
  1. Volcanic eruptions at mid-ocean spreading ridges.
  2. Volcanic eruptions in the vicinity of subduction zones.

These eruptions are an integral part of **plate tectonics**, the continual formation and destruction of the Earth's outer layer, and the resulting motion of the lithospheric plates over the surface of the Earth. The relationship between volcanoes and plate movements is outlined in Box 4.1, which you should read if you need to be reminded about the mechanisms of plate tectonics. Subduction zone eruptions often take place above sea-level, and so they attract more attention than submarine eruptions, particularly as they are intrinsically more explosive (Figure 4.2). However, both types of eruption are important in the Earth system.

Also shown on Figure 4.1 are the positions of 'hot spots', places where volcanism is triggered by **mantle plumes**, upcurrents of magma rising from the boundary between the mantle and the core, deep within the Earth. This source remains fixed as the lithosphere moves over it, so that over time a chain of volcanoes is generated, rather in the way that a linear scorch mark appears as a sheet of paper is moved over a candle flame (see Figure 4.5). Such hot-spot volcanism is responsible for linear island chains such as those in the northern Pacific, of which the best known is the Hawaiian group (visible in Figure 4.1) with the island of Hawaii itself at the 'young' end of the chain.

Figure 4.1 (opposite) The distribution of active (or recently active) volcanic activity in relation to the margins of the lithospheric plates, of which there are three types: constructive margins (mid-ocean spreading ridges), destructive margins (subduction zones), and conservative margins (transform faults). Note that at mid-ocean ridges, volcanism is sporadic on short time-scales (decades), but over time-scales of hundreds of years ridges are continually active. A third type of volcanic activity, unrelated to plate margins, occurs at 'hot spots' resulting from mantle plumes. The arrows indicate the relative motions of the plates, assuming the African Plate to be stationary; the arrow shown in the key corresponds to a relative velocity of 5 cm yr^{-1}.

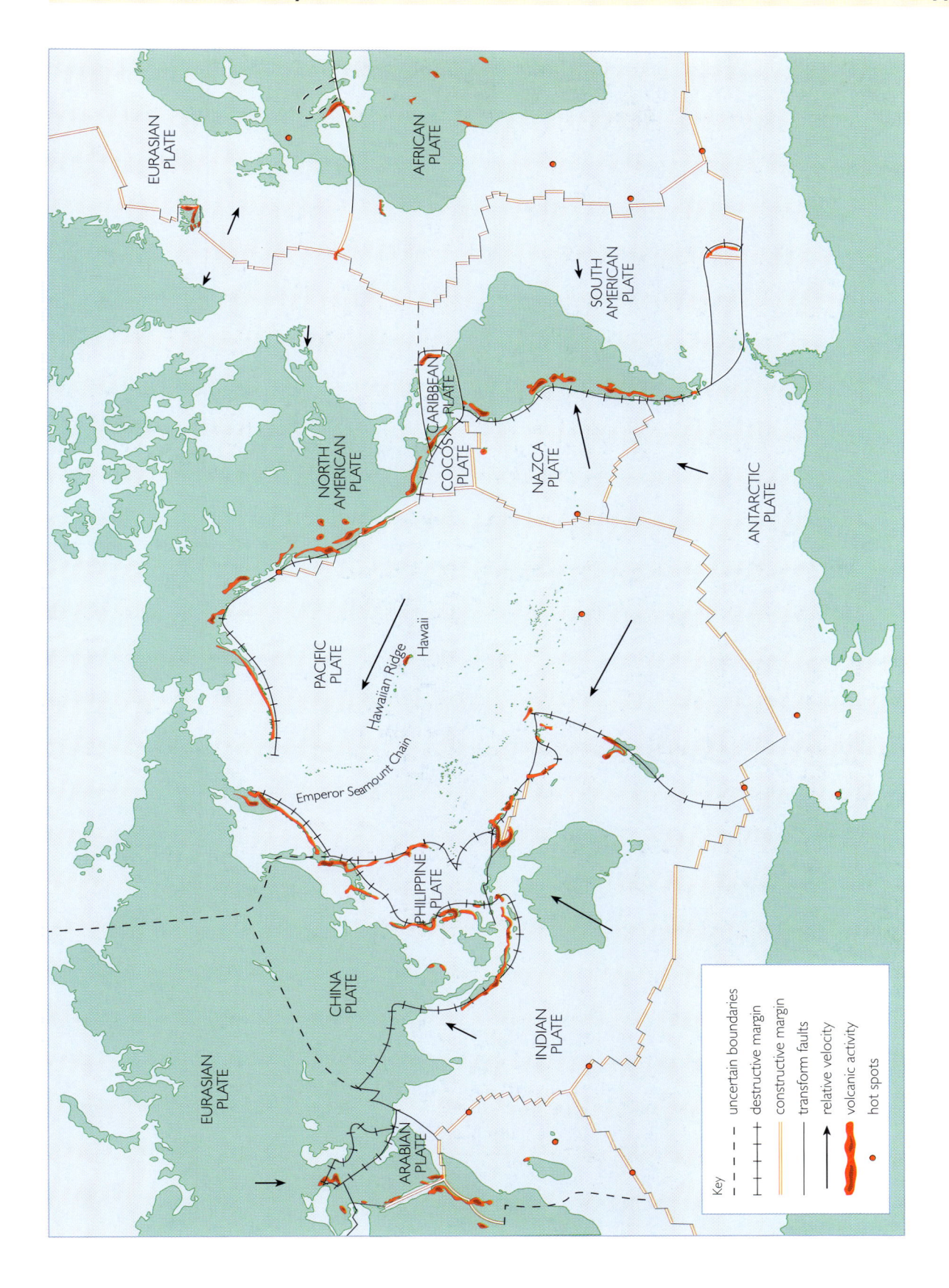

Volcanism and the Earth system

Figure 4.2
(a) A drawing of an early phase of the great Krakatau eruption, based on a photograph taken by the last party of people known to visit the island, on 27 May 1883. Stumps of trees, their foliage stripped by falling ash, are scattered over the flanks of the island. No photographs exist of the catastrophic phase of the eruption.
(b) A view from the *Space Shuttle* of an explosive eruption in progress. A plume of ash, gas and aerosols forms a banner hundreds of kilometres long downwind (eastwards) from the Klyuchevskaya volcano on the Kamchatka Peninsula, Russia, which erupted in 1994. The area covered by the photograph is about 250 km across.

Box 4.1 Plate tectonics

For over a century, the Earth has been considered as consisting of crust on the outside, mantle beneath, and the core at the centre. But the Earth can equally well be subdivided into layers on the basis of how rocks behave under stress — their rheology. The crust and a good portion of the upper mantle beneath move around the surface of the Earth as rigid plates. This rigid outermost layer — the **lithosphere** — is allowed to move by a layer that behaves in a ductile fashion; this underlying layer is known as the *asthenosphere*. Only the upper part of the lithosphere is made up of crustal rocks; plate tectonics therefore describes the movement of lithosphere — not just the crust — over the asthenosphere.

As shown in Figure 4.1, today the Earth's lithosphere is made up of six or seven large plates and a multitude of smaller ones.

■ To what extent do plates consist entirely of continental crust, or entirely of oceanic crust, or of both types of crust?

■ The Pacific Plate and some small plates are entirely oceanic, but most plates have both continental and oceanic components.

Throughout most of the Earth's history, the movement of lithospheric plates has been responsible for many of its important features, including the distribution of its earthquake and volcanic zones, the disposition of ocean basins and continents, mountain ranges, and the topography of the sea-floor.

The submarine volcanic ridges snaking through the oceans mark the sites where lithosphere is created, and so are referred to as *constructive margins* (Figure 4.3). These also provide the setting for hydrothermal fluids to be expelled onto the sea-floor as mentioned in Chapter 2 (cf. Figure 2.37). Because new lithosphere moves outwards in both directions from the ocean ridges (Figure 4.3), the process of plate creation leads to **sea-floor spreading**. New ocean floor is formed at ridges, and older lithosphere moves sideways

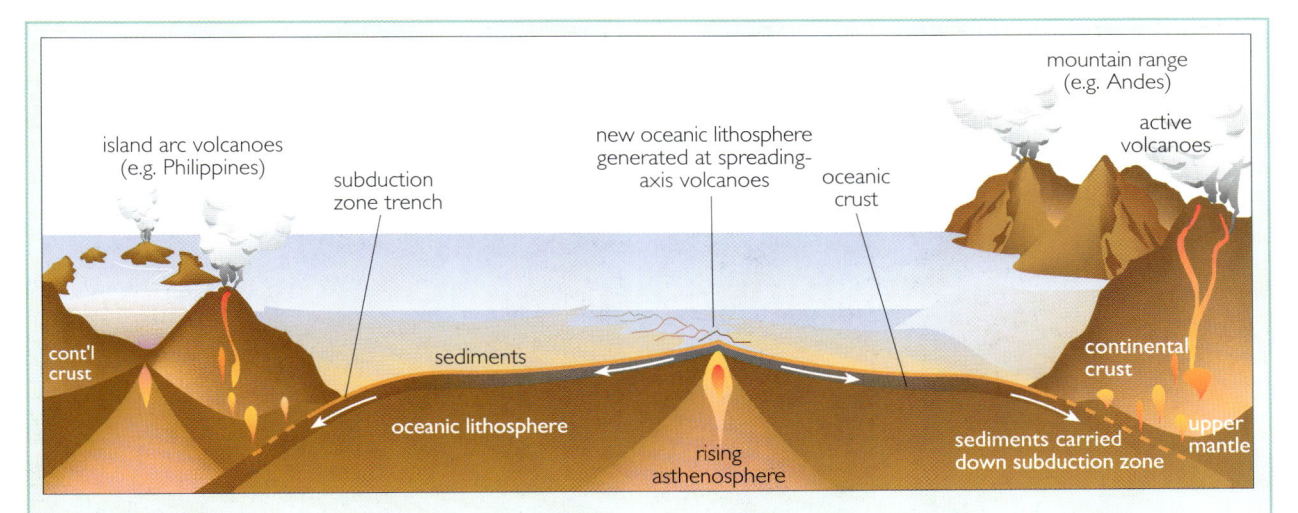

Volcanism and the Earth system

Figure 4.3
The plate-tectonic setting of mid-ocean ridge volcanism and subduction zone volcanism. New oceanic lithosphere is created by magma rising from the asthenosphere beneath the ocean ridge. As the lithosphere moves away from the ridge it cools, contracts and sinks, eventually becoming covered by a few kilometres of sediment (cf. Figure 2.37); the arrows show direction of sea-floor spreading. The right-hand side shows oceanic crust being subducted below a continent. Most of the sediment is scraped off and accreted onto the continental margin, but a small proportion may be subducted. The subducted oceanic crust melts, and there is also partial melting of the mantle above the subduction zone. The resulting magma is both erupted as lavas at volcanoes and intruded into the continental crust, where it solidifies, thus increasing its thickness.
The left-hand side shows the result of oceanic crust being subducted beneath crust which is also oceanic: eruption of the magmas formed by melting along the subduction zone results in the formation of a volcanic island arc, the crust of which is also thickened by intrusion and solidification of magma within it. Such island arcs are common in the western Pacific; indeed, you can imagine this hypothetical cross-section as being a simplified slantways section across the Pacific, from the Philippines to the coast of South America. Note that the vertical scale is greatly exaggerated, though not as much as in Figure 2.37, with which this diagram should be compared.

to accommodate it. So the further it is from the ridge, the older the ocean floor.

Plate creation is balanced by plate destruction. Where plates converge, oceanic lithosphere is drawn down beneath the margin of the other plate, back into the asthenosphere where it is resorbed. This process is called **subduction**. Where oceanic crust is subducted below a continent, a volcanic mountain range forms along the edge of the continent; an example of this is the Andean coast of South America (see right-hand side of Figure 4.3). Where oceanic crust is subducted below crust that is also oceanic, an arc of volcanic islands is formed (e.g. the Philippines; see the left-hand side of Figure 4.3). Offshore from the volcanic arc lies a deep ocean trench which marks the zone where the oceanic lithosphere is being subducted.

The energy required for driving the lithospheric plates comes from the heat energy being released by the Earth. The Earth behaves as a heat engine, losing most of its heat at ocean ridges through the formation of new lithosphere.

- Figure 4.4 shows how ductile flow may occur in the mantle. What type of flow patterns are these?

- The flow patterns are convection cells, analogous to those occurring in the atmosphere and ocean (cf. Figure 2.2).

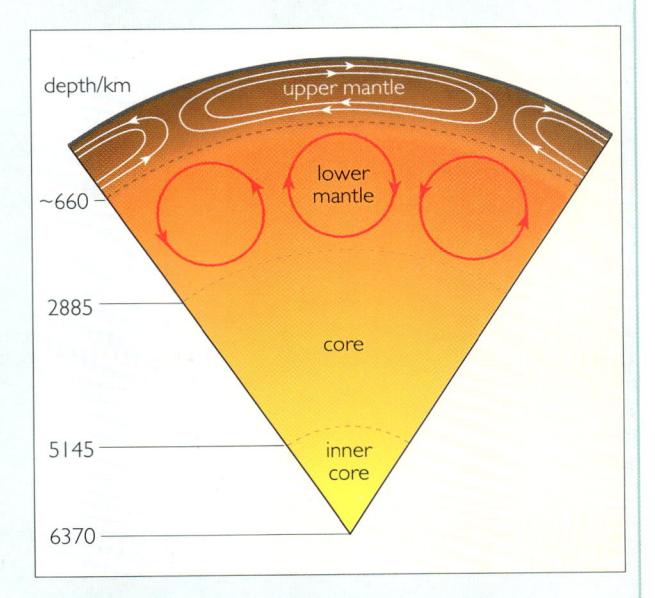

Figure 4.4
Sketch cross-section (not to scale) showing how ductile flow may occur in the mantle. The size and location of convection cells remain a matter for debate, and you may see other values quoted for the boundary between the upper and lower mantle.

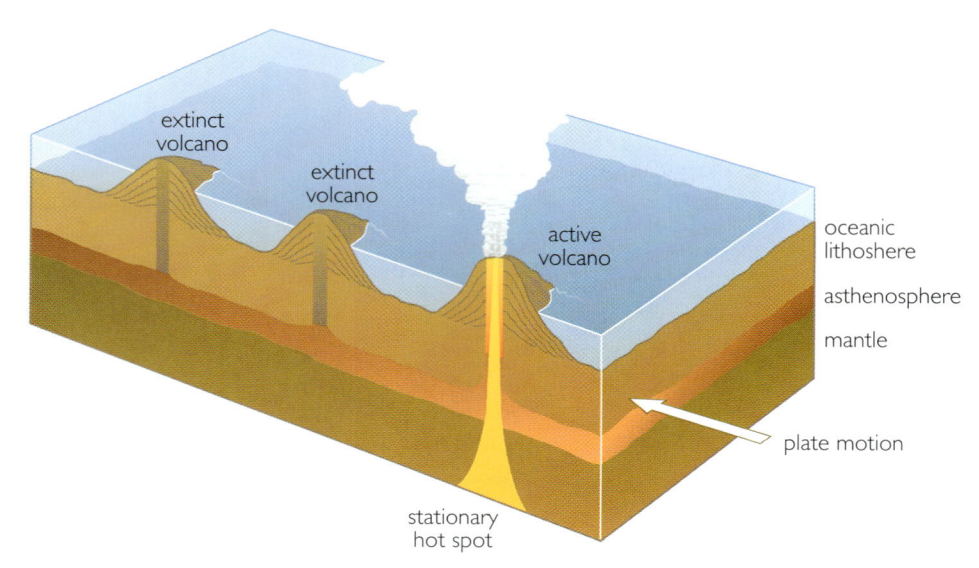

Figure 4.5
Schematic diagram (not to scale) illustrating how a volcanic island chain such as the Hawaiian Chain forms over a stationary hot spot. The volcano on the right (the equivalent of Hawaii) is still active, and the age of the volcanoes increases towards the left.

Hawaiian eruptions have been well documented, so they offer a useful starting point for our studies. Hawaii contains two major active volcanoes, Mauna Loa and Kilauea; a third volcano, Mauna Kea, is inactive, but is well known for its array of massive astronomical observatories. The peak of Mauna Kea is 4206 m above sea-level and, rising almost 9 km from the Pacific Ocean floor, it is the world's tallest mountain. Kilauea is the world's most persistently active volcano, steadily spewing molten lava into the ocean at about five cubic metres per second. Such a production rate makes it is easy to appreciate the role that volcanism plays in simply forming new land – the island of Hawaii has been built up from the sea-floor over the past 1–2 million years (Ma) (Figure 4.6). But rocks are only part of the story. Table 4.1 lists the composition of a typical Hawaiian lava, just on the point of eruption.

(a)   (b)

Figure 4.6
(a) Mauna Kea volcano seen from the flanks of Mauna Loa. The buildings in the middle distance are part of a weather station on Mauna Loa; as discussed in Box 3.3 and related text, the rising concentrations of atmospheric CO_2 recorded at this weather station have played a fundamental role in current debates on the contribution of human activities to global climate change.
(b) A plume of steam rises to mark the spot where lava is flowing into the ocean on the south-east coast of Hawaii. When this photograph was taken in 1994, the flow of lava had been virtually continuous for more than ten years.

Volcanism and the Earth system

Table 4.1 Chemical composition of the solid and gaseous components of a typical Hawaiian basalt lava, erupted from Mauna Loa in 1859.

Non-volatile material (elements present as silicate or oxide minerals in magma)	
Element	Weight %
silicon, Si	24.15
aluminium, Al	7.31
iron, Fe	8.23
magnesium, Mg	4.35
calcium, Ca	7.55
sodium, Na	1.71
potassium, K	0.35
manganese, Mn	0.13
titanium, Ti	1.24
phosphorus, P	0.10
oxygen, O	44.26
Dissolved gases	
water vapour, H_2O	0.27
carbon dioxide, CO_2	0.32
sulfur dioxide, SO_2	0.18

- Look at the analysis of the magma in Table 4.1. Which of the ingredients are most likely to affect the environment some distance from the volcano?

- Water, carbon dioxide and sulfur dioxide, because they can escape as volatiles (gases) and mingle freely with the atmosphere. On the other hand, the first 11 components (Si to P), combined with oxygen in silicates and oxides, all contribute to the rocky lava.

Some of the water may be 'secondary', that is, recycled groundwater that has been heated up by the volcanic rocks, rather than being derived from the Earth's interior. At destructive plate margins, some of the carbon preserved in sea-floor sediments and subsequently lithified (discussed in Section 3.3) may be returned to the atmosphere as carbon dioxide, which escapes from solution in the magma as it is erupted at volcanoes above the subduction zone (Figure 4.3).

At Hawaii, however, the magma originates deep within the mantle so that the carbon dioxide exhaled represents a net addition to the atmosphere. Early in the history of the Earth, almost all of the first stable atmosphere formed by volcanic outgassing in this manner. Sulfur dioxide (a poisonous gas used in the past in warfare), has a range of effects on the atmosphere, which we will review shortly. First, though, let us estimate the volcanic input of carbon dioxide (which, like water vapour, is a greenhouse gas) to the atmosphere.

Question 4.1
At the present time, Mauna Loa has a volume of $2.3 \times 10^{14}\,m^3$, and a mass of about $6.6 \times 10^{17}\,kg$. Every kilogram of lava that has been erupted has liberated 0.0032 kg of carbon dioxide. How much carbon dioxide has been so far been released into the atmosphere during the construction of Mauna Loa?

This impressive number is more meaningful if we consider the *rate* of emission. If Mauna Loa has been building up over 1.5 million years, the annual flux into the atmosphere would have been 1.4×10^9 kg (about the same as a 200 MW coal-fired power station).

■ What is the total annual human contribution to atmospheric CO_2 (expressed as mass of carbon) at the present time, according to Figure 3.24?

■ The total annual human contribution of carbon to CO_2 in the atmosphere is about 6.8×10^{12} kg – about 5×10^{12} kg from burning fossil fuels and about 1.8×10^{12} kg (rounded up to $\sim 2 \times 10^{12}$ kg on Fig. 3.24) from deforestation.

To obtain the corresponding mass of CO_2 we need to multiply this value by the relative molecular mass of CO_2 divided by the relative atomic mass of carbon, i.e. by $(12 + 16 + 16)/12 = 44/12$. This gives us 6.8×10^{12} kg $\times 44/12 = 24.9 \times 10^{12}$ kg $\approx 2.5 \times 10^{13}$ kg.

■ How would the annual flux of CO_2 from Mauna Loa compare with this?

■ If Mauna Loa's annual contribution is only about 1.4×10^9 kg, it is

$$\frac{1.4 \times 10^9 \text{ kg} \times 100}{2.5 \times 10^{13} \text{ kg}} \approx 0.006\%$$

of the annual human contribution.

Table 4.2 shows the global annual production of both CO_2 and sulfur dioxide, SO_2, compared with the human-generated flux (you can now fill in that for CO_2), as well as those for Kilauea (which is currently producing 1.32×10^9 kg of CO_2 a year – slightly less than the value we assumed above for Mauna Loa).

Table 4.2 Annual production of SO_2 and CO_2 from Kilauea volcano and the total global volcanic production, compared with the global anthropogenic (human-generated) fluxes (that for CO_2 has been left for you to complete, see above).

	Kilauea outgassing rate / kg yr^{-1}	Global volcanic outgassing rate / kg yr^{-1}	Global anthropogenic flux / kg yr^{-1}
CO_2	1.32×10^9	6.5×10^{10}	
SO_2	4.3×10^8	1.8×10^{10}	1.9×10^{11}

Some volcanoes – Mount Etna, for example – exhale more CO_2 than Kilauea, but the total global volcanic contribution to the atmosphere is only thought to be about 6.5×10^{10} kg yr^{-1} – less than 0.3% of the total anthropogenic (human-generated) flux (Table 4.2). This figure illustrates strikingly how human activity is perturbing the natural environment. It is important to note, however, that although volcanoes presently account for the addition to the atmosphere of a relatively small amount of CO_2, at certain times in the past the impact of volcanic CO_2 was much greater (of which more later in this chapter).

■ Of the three types of volcanism shown on Figure 4.1 – mid-ocean ridge, subduction zone or mantle plume – which do you think will generally be responsible for the release of most CO_2 into the atmosphere?

■ It is likely to be the volcanism associated with subduction zones, which is much more widespread, and overall produces much more lava, than the volcanism associated with hot spots.

You may well have thought that the volcanism associated with mid-ocean spreading ridges would have contributed a significant amount of CO_2 to the atmosphere. At first sight this seems a reasonable supposition, given the enormous lengths of spreading ridges, but in fact very little of the gas released by mid-ocean ridge volcanism enters the atmosphere *directly*. Mid-ocean ridges are at about 2.5 km depth, and therefore under pressures of several atmospheres, enough to prevent gases escaping from the lavas. Nevertheless, CO_2 is an important component of the gases dissolved in the fluids expelled at hydrothermal vents, which are found wherever there is submarine volcanism. It will eventually get into the atmosphere after circulating in the ocean in solution for anything up to 1000 years.

So far, we have tended to concentrate on volcanic emissions of CO_2, but *on short time-scales* SO_2 has far more dramatic effects, and it is to these that we now turn.

4.2 Volcanic aerosols and climatic change

> *The ashes now began to fall upon us, though in no great quantity ... darkness overcame us, not like that of a cloudy night, or when there is no moon, but of a room when it is all shut up and all the lights extinguished.*
> *Pliny the Younger*

This brief extract from Pliny the Younger's classic account of the eruption of Vesuvius in AD 79 reminds us that volcanic ash clouds can be highly effective at blocking out sunshine. Major explosive eruptions (still called plinian eruptions to commemorate Pliny's observations) blast huge amounts of volcanic ash high into the atmosphere, often reaching the stratosphere. Awe-inspiring though these eruption columns are, it is not the ash particles that cause the most significant climatic effects. Ash fragments are relatively large (ranging from shards 1–2 μm across to fist-sized lumps of pumice) and they fall back to the ground in hours or days. It is volcanic gases that have the longest-lasting environmental effects.

As mentioned above (Tables 4.1 and 4.2), volcanoes produce large quantities of sulfur dioxide as well as carbon dioxide. After eruption, sulfur dioxide reacts with water vapour in the atmosphere to form aerosols (i.e. tiny airborne droplets) of sulfuric acid, H_2SO_4. Like those produced from the oxidation of DMS (Section 2.3.2), these acid aerosols form by a complex series of photochemical reactions which may continue for months, replenishing the aerosol cloud so that new particles form while older, larger ones settle out of the atmosphere. Aerosol particles are very tiny – initially only 0.1–1 μm in diameter – so any reaching the stratosphere may remain suspended there for months or years, far longer than solid ash fragments.

Volcanic aerosols often produce gloriously colourful sunsets (and sunrises) which can be enjoyed around the world (Figure 4.7). These were so spectacular after the 1883 eruption of Krakatau that they were the inspiration for the imagery in a poem by Alfred Lord Tennyson:

> *Had the fierce ashes of some fiery peak*
> *Been hurl'd so high they ranged about the globe?*
> *For day by day, thro' many a blood-red eve,*
> *In that four-hundredth summer after Christ*
> *The wrathful sunset glared against a cross ...*
> *Saint Telemachus, by Alfred Lord Tennyson*

Tennyson's 'ashes' were, in fact, aerosols as these were the cause of the 'blood-red eves', through their tendency to scatter sunlight. During the day, the sky looks blue because gas molecules scatter sunlight in all directions. These molecules are smaller (~ 0.03 μm) than the wavelength of light and they scatter blue wavelengths (~ 0.4 μm) more effectively than red (~ 0.7 μm). Thus, blue light from the Sun actually reaches your eye even when you are not looking directly at it, whereas the other wavelengths don't. When the size of particles is about the same as the wavelength of light, as in aerosols and smoke, the scattering process is more complex but the dominant effect shifts to longer visible wavelengths, around orange to red. It is less intense than the blue-sky effect, and is most noticeable at sunset and dawn when the Sun is low in the sky and its rays have to traverse a much longer atmospheric path (cf. Figure 1.6). In fogs and mists, the water droplets are much larger (> 1 μm) than the wavelength of light, and so visible wavelengths are not differentially scattered. Thus, the Sun seen through fog looks white.

In detail, the physics of light-scattering by aerosols is complex, but the overall effect is that when a layer of volcanic aerosols gets between the Earth and the Sun, a fraction of the Sun's radiation is scattered back to space. In other words, the presence of the aerosols effectively increases the Earth's albedo (cf. Figure 2.33). However, to deduce the climatic consequences, we have to consider the balance between incoming solar radiation, and outgoing long-wave (thermal) radiation from the Earth and atmosphere. As Figure 2.33 showed, a fraction of the incoming radiation is reflected back to space; of the remainder, some passes directly through

Figure 4.7
During the autumn of 1883, after the eruption of Krakatau, spectacular sunsets around the world attracted the attention of artists. This is one of a series of six paintings by William Anscom of the Sun setting on a November evening seen from Chelsea, London. It was published in the contemporary Royal Society report on the great eruption.

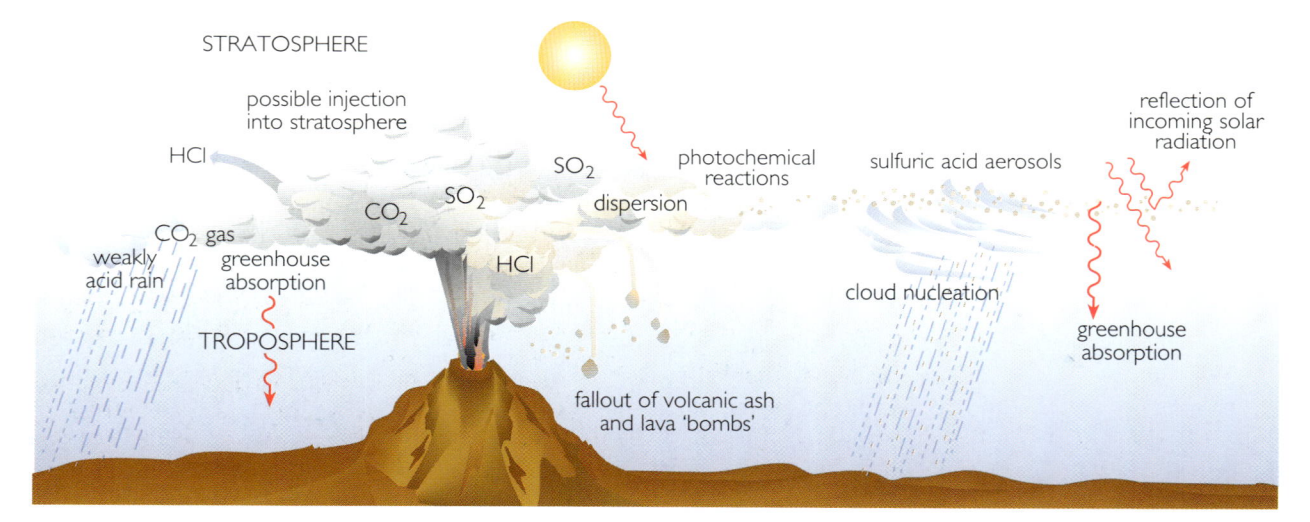

Figure 4.8
Physical and chemical interactions of a volcanic eruption with the atmosphere. The eruption releases sulfur dioxide (SO$_2$), which is eventually converted to sulfuric acid (through reaction with atmospheric oxygen and water), producing a mist of droplets, or aerosols. These reflect a fraction of the incoming solar radiation, so cooling the troposphere; they also absorb some solar radiation, warming the stratosphere. If the aerosols are sufficiently large, the aerosol layer may absorb and re-radiate thermal energy, warming the lower atmosphere (see Figure 4.9); note that they are sometimes referred to as *sulfate* aerosols, rather than sulfuric acid aerosols. The eruption also produces large amounts of CO$_2$, which mixes in the atmosphere, forms a much weaker acid with atmospheric water, and does not form aerosols; smaller amounts of hydrogen chloride (HCl) are also released into the atmosphere, forming hydrochloric acid.

Volcanism and the Earth system

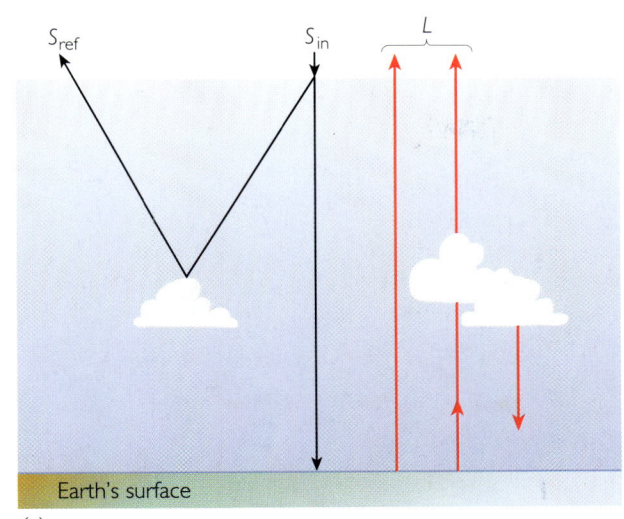

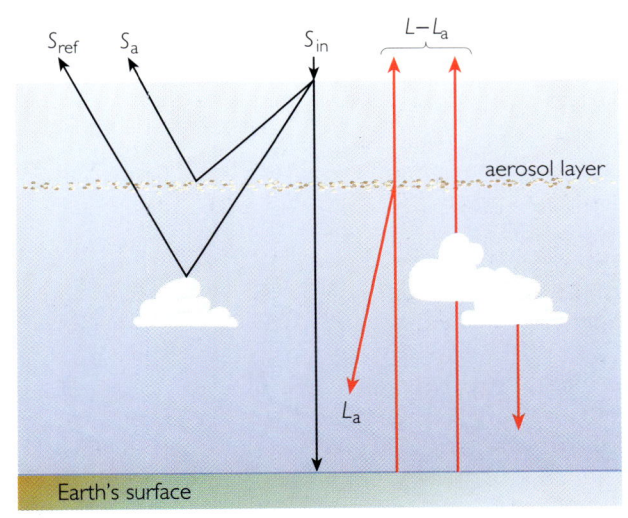

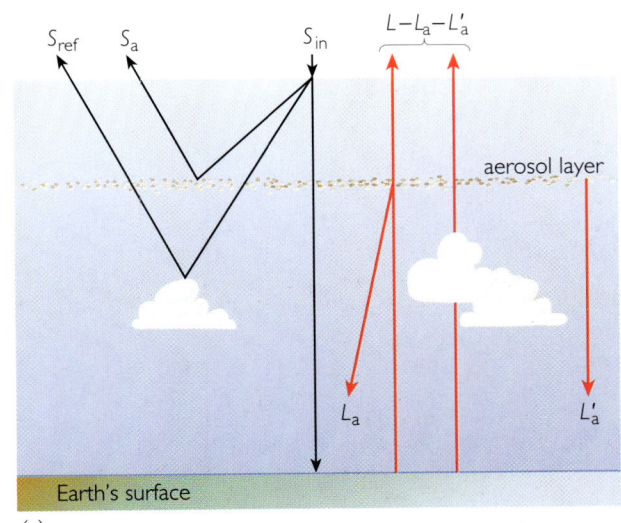

to the surface of the Earth, and some is forward-scattered, also reaching the ground. The ground heated by the sunshine gets warm, and re-radiates long-wave thermal energy, some of which is re-radiated back to Earth by the atmosphere and clouds, while the remainder escapes to space. As incoming radiation (S_{in}) and outgoing (reflected short-wave, S_{ref}, plus thermal long-wave, L) radiation must balance, we can write:

$$S_{in} = S_{ref} + L \qquad \text{(Equation 4.1a)}$$

Figure 4.9a applies these symbols to a simplified version of Figure 2.33: part (b) of Figure 4.9 is the same as part (a), but also includes the effect of aerosols, which not only scatter incoming solar radiation back to space (as mentioned above) but also scatter long-wave thermal radiation back towards the Earth's surface. If we now include in our equation S_a for solar radiation scattered back into space by the aerosol layer and L_a for long-wave thermal radiation back-scattered to Earth by the aerosol layer, we can now modify this equation to read:

$$S_{in} = S_{ref} + S_a + (L - L_a) \qquad \text{(Equation 4.1b)}$$

S_a will add an additional component to the sunlight already normally reflected by clouds, etc., while L_a will reduce the amount of long-wave thermal radiation escaping to space (Figure 4.9b).

- What would be the overall effect of increasing the term $S_{ref} + S_a$, bearing in mind that the radiation budget would eventually reach a state of balance?

- In order to maintain balance, the term $L - L_a$ would have to decrease.

Figure 4.9
(a) Simplified version of Figure 2.33 where S_{in} represents incoming solar radiation, S_{ref} represents the solar radiation reflected by the Earth (i.e. by clouds, the Earth's surface and the atmosphere), and L represents the long-wave (i.e. thermal) radiation that escapes to space.
(b) If a layer of aerosols is added to the situation shown in (a), the amount of solar radiation reflected back to space becomes $S_{ref} + S_a$, and the amount of long-wave radiation that escapes to space becomes $L - L_a$.
(c) If the aerosol layer also *absorbs* outgoing thermal radiation and *re-radiates* an amount L'_a towards the Earth, the amount of long-wave radiation escaping to space becomes $L - L_a - L'_a$. Whether the net result is global cooling or global warming depends on the relative sizes of L_a and L'_a; in general, though, $L - L_a$ is greater than L'_a and the net result is global cooling.
Note Although the diagram suggests that solar radiation is all short-wave, it does include some infrared (albeit at relatively short wavelengths, Figure 2.32a).

Now, $L - L_a$ is the thermal radiation escaping from the Earth; if this term is decreasing, the Earth's global temperature must be decreasing. It follows that increasing the term S_a (and hence $S_{ref} + S_a$) will lead to global cooling. In other words, *prima facie*, volcanic aerosols will cause global cooling.

Things are more complicated in detail, and for a given aerosol layer, the fraction of solar radiation that is actually scattered back into space depends on both the amount of aerosol present and the size of the droplets. An increased albedo caused by aerosol-scattering certainly cuts down the sunlight reaching the surface of the Earth, causing cooling. However, while droplets of all sizes may scatter radiation, they can also absorb it if their diameters are of the same order as the wavelength of the radiation. As re-radiated thermal radiation is of relatively long wavelength, it can be absorbed by large aerosols, which will emit long-wave radiation, some of it back towards the Earth (Figure 4.9c). An aerosol layer can therefore act as a 'greenhouse', preventing some thermal radiation escaping, and causing surface warming.

It is difficult to make precise predictions about the effects of an aerosol layer on the Earth's surface temperature. Generally, only if the radius of the aerosols exceeds 2 μm does the global average greenhouse warming effect exceed the cooling effect. Most long-lived volcanic aerosols are so small that only net cooling results.

4.2.1 Climatic effects of the eruption of Mount Pinatubo

The eruption of Mount Pinatubo on 15 June 1991 (Figure 4.10) was the second largest this century. Fortunately, the climax of the eruption was correctly predicted by volcanologists, and a great catastrophe for the people of the Philippines was averted. For climatologists, the eruption provided an ideal opportunity to study the effects of a major eruption using the whole panoply of modern space-borne remote-sensing instruments. Specifically, they were able to measure the total mass of aerosol injected into the atmosphere and the fraction of incident solar radiation that was scattered back into space by the aerosol layer (cf. Figure 4.9). This was an unexpected bonus for global climatologists working on an existing project known as the Earth Radiation Budget Experiment (see Figures 4.11 and 4.12). They were also able to estimate the global average temperature decrease – perhaps the most difficult parameter of all to extract, given all the 'noise' of normal climatic variation.

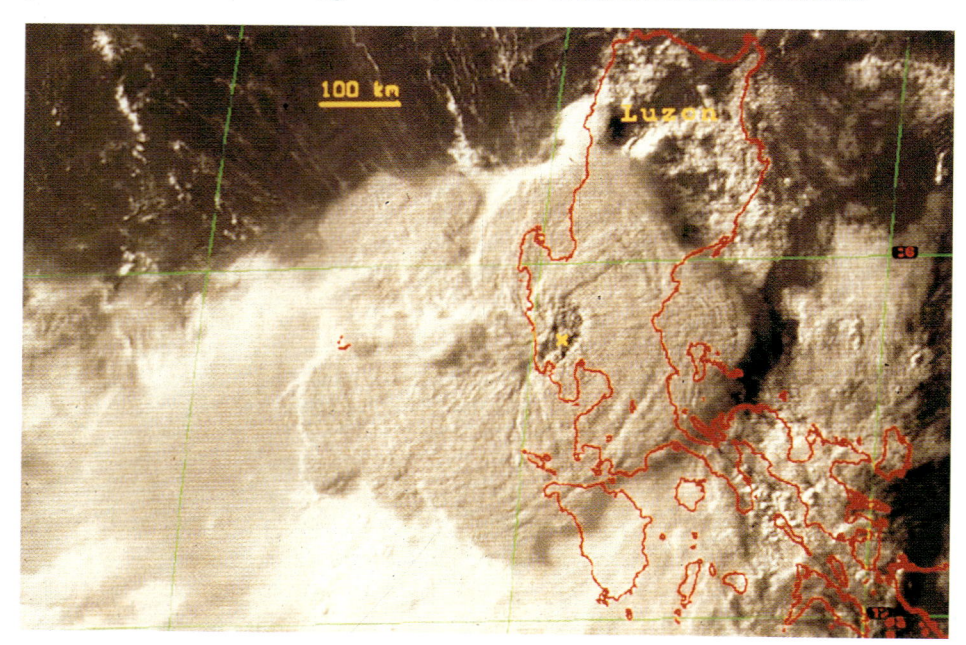

Figure 4.10
A weather satellite image showing the great cloud produced by the eruption of Mount Pinatubo, at 8.30 a.m. on 15 June 1991. The red outline shows the island of Luzon, where the explosion occurred, and other smaller islands at the northern end of the Philippine arc.

Volcanism and the Earth system

Figure 4.11
The cloud of sulfur dioxide gas in the stratosphere shortly after the Mount Pinatubo eruption, as recorded by an instrument on board a weather satellite on 16 June 1991. At this stage, the gas had drifted only a short distance from the Philippines. The diagram is contoured in units of sulfur dioxide concentration (white highest, brown lowest). The cloud was estimated to contain about 2×10^{10} kg of sulfur dioxide.

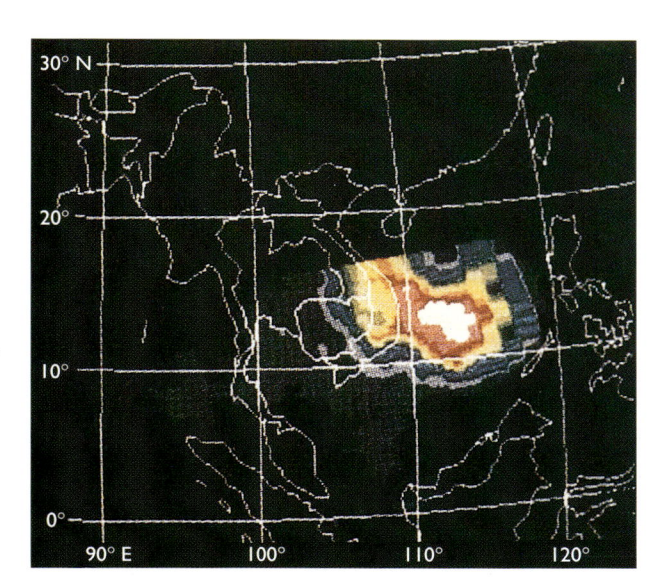

Figure 4.12
This image shows that by July, stratospheric aerosols from the Pinatubo eruption had spread right around the globe, reaching high latitudes in both the Northern and the Southern Hemisphere. Note that while Figure 4.11 shows sulfur dioxide *gas* concentrations, this figure shows the distribution of aerosols, and is contoured in terms of atmospheric opacity (white being highest, red intermediate and purple/black the lowest), effectively the extent to which the aerosols reduce incoming solar radiation.

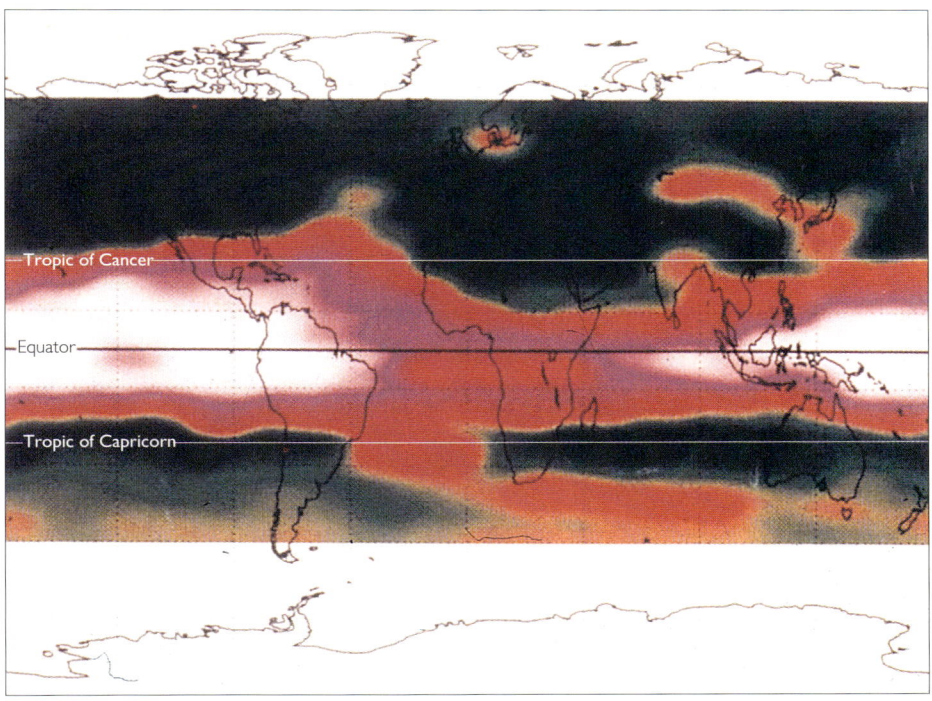

Most important for our discussion here, the Earth Radiation Budget Satellite measurements showed that as a result of the eruption of Mount Pinatubo the global albedo increased from 23.6% to 25.0%. This may not sound very much, but over normally cloud-free regions of the Earth the albedo increased by as much as 20%. Mount Pinatubo is at 15° N, and the satellite measurements showed that in tropical areas, where the aerosols were concentrated (Figure 4.12), the increase in solar radiation reflected back into space because of aerosols (S_a in Figure 4.9) was significant: between 5° S and 5° N it was equivalent to a reduction in the radiation reaching the Earth's surface of about 8 W m^{-2}, i.e. a radiative forcing of -8 W m^{-2}. At latitudes between 5° and 40°, the average forcing was about -4.3 W m^{-2}. In Chapter 1, we noted that the Earth's average climate sensitivity to radiative forcing is about 0.75 °C per W m^{-2} of forcing.

- On this basis, how much cooling – which we all lived through – might we expect to have followed the Pinatubo eruption?

- The climate forcing observed to result from Pinatubo was about −4.3 W m^{-2} (a minus is used because there was net *decrease* in radiation energy warming the Earth). We should therefore expect there to have been a cooling of 4.3 × 0.75 °C, or about 3 °C.

In fact, meteorological records showed that in 1992 Pinatubo caused a decrease in globally averaged temperature of only about 0.5 °C. The major reason for this discrepancy was probably that the aerosols from Pinatubo dispersed in a few years, and the cooling due to the aerosol haze did not last nearly long enough for feedback effects to come into play and permit the Earth's climate system to respond fully.

Mount Pinatubo was small beer compared to many historic and prehistoric eruptions. However, the detailed observations made at the time provide a useful benchmark for estimating the effects of some of those earlier eruptions, for which actual records are sketchy or absent. According to the observations following the Mount Pinatubo eruption, the net radiative forcing (R_{net}) due to volcanic aerosols can be estimated using:

$$R_{net} = -(1.43 \times 10^{-10}) \times m \ \text{W m}^{-2} \quad \text{(Equation 4.2)}$$

where m is the mass of erupted sulfuric acid aerosols in kilograms. This empirically deduced 'rule of thumb' enables us to investigate the atmospheric impact of some large historic eruptions, and the cooling they may have caused.

4.2.2 Historic eruptions

Table 4.3 summarizes data for some major eruptions within the past 200 years. The second and third columns list the masses of magma erupted and the masses of aerosols injected into the stratosphere. Both of these datasets are subject to considerable errors: it is difficult to estimate the total mass of material erupted, especially as much of it ended up in the ocean (and two of these eruptions took place more than 100 years ago); and the mass of aerosol erupted can only be inferred indirectly, for example from records of ice acidity in drill cores in the polar ice-caps.

Table 4.3 Data on the amount of magma erupted, aerosol loading and observed cooling for some selected recent eruptions (to be completed later).

Location of eruption	Mass of magma erupted / 10^{11} kg	Mass (m) of aerosols in stratosphere / 10^9 kg	Climatic forcing / Wm^{-2}	Observed Northern Hemisphere temperature change / °C
Tambora, Indonesia, 1815 (8° S)	2400	150	−21.4	−1.0
Krakatau, Indonesia, 1883 (6° S)	247	55	−7.8	−0.5
Agung, Indonesia, 1963 (8° S)	24	20		−0.3
Katmai, Alaska, 1912 (58° N)	240	25		−0.2
El Chichón, Mexico, 1982 (17° N)	7.5	12		−0.2
Pinatubo, Philippines, 1991 (15° N)	120	30	−4.3	−0.5

Volcanism and the Earth system

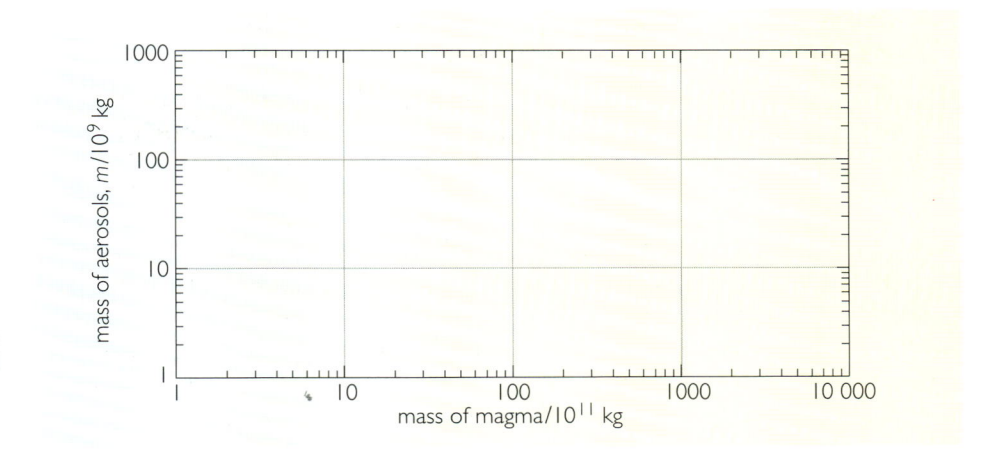

Figure 4.13
Graph for plotting the mass of aerosols injected into the stratosphere against the mass of magma erupted for several major historic eruptions (given in Table 4.3). Note that the axes are logarithmic.

Question 4.2
Using the axes in Figure 4.13 and the data in the second and third columns of Table 4.3, make a quick plot of the mass of aerosols produced against the mass of magma erupted. How would you describe the correlation between the mass of erupted magma and mass of aerosols?

The famous eruption of Mount St Helens in May 1980 produced about 6.5×10^{11} kg of magma, and 0.3×10^9 kg of aerosols. Plot this on Figure 4.13.

■ How do the data for El Chichón and Mount St Helens compare?

■ Despite erupting about the same amount of magma as Mount St Helens, El Chichón produced a far greater mass of aerosols – about 40 times more.

We can draw an important conclusion from this simple observation: eruptions vary greatly in their potential for climatic impact. Some volcanoes erupt far more sulfur dioxide per unit mass of magma than others, and hence have far greater effects on the stratosphere.

There are also some less obvious reasons why apparently similar eruptions have different effects. If the character of an eruption is such that the erupted gases are not injected into the stratosphere, then there is little chance of the resulting aerosols having any global effect, as they will be rapidly rained out of the troposphere.

Question 4.3
By reference to Figure 2.7, explain:
(a) Why the effect of aerosols injected into the troposphere from a volcano in (for the sake of argument) the Southern Hemisphere will be largely confined to that hemisphere.
(b) Why an eruption from a volcano at high latitudes is more likely to eject gases (and hence aerosols) into the stratosphere than a volcano at low latitudes.

Figure 4.12 illustrates that the situation is different if the volcanic gases get into the stratosphere. About 40 days after the eruption of Mount Pinatubo (at ~15° N), stratospheric aerosols had reached high latitudes in both hemispheres. However, an eruption sending ejecta into the stratosphere at *high* latitudes will have less effect in the other hemisphere than one taking place in the tropics, because stratospheric circulation patterns are mainly zonal (east to west, or west to east) and so inhibit the spread of aerosols from one hemisphere into the other.

Now return to Table 4.3 and, filling in the blank spaces, use Equation 4.2 to compute the climate forcings produced by the eruptions listed (your completed version should resemble Table 4.4, overleaf). Next, on Figure 4.14a plot the forcings you have just calculated against the Northern Hemisphere cooling that resulted, as given in the last column of the table. (If in doubt, you can compare your completed Figure 4.14a with Figure 4.14b, overleaf.)

- ■ Does the graph show a simple relationship between the forcing that you calculated and the temperature decrease observed in the Northern Hemisphere?

- ■ More or less. The plot shows that there is a rough correlation between forcing and climatic impact.

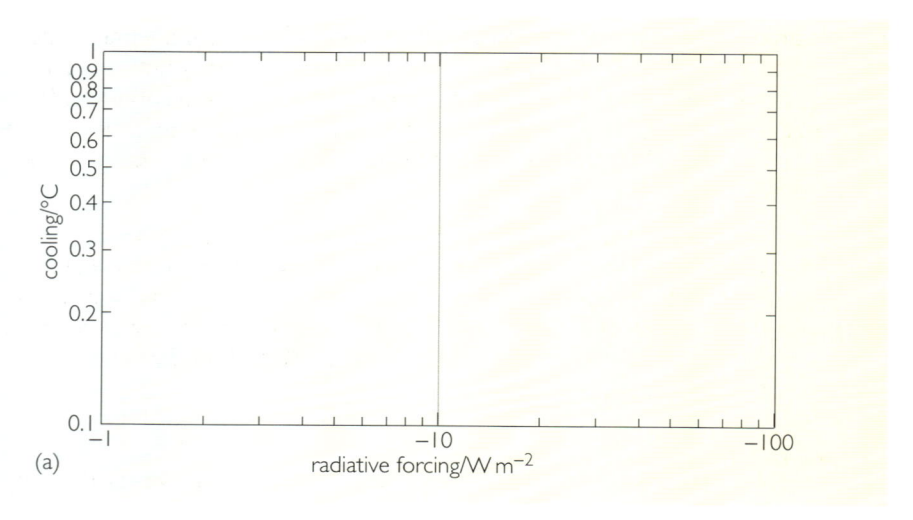

Figure 4.14(a)
Plot of observed cooling against radiative forcing for the eruptions listed in Tables 4.3/4.4 (to be completed).

You should bear in mind, however, that the plot is again logarithmic – the correlation would look less impressive if plotted on an ordinary graph. The shape of the plot is also heavily influenced by the data for Tambora, which was far bigger than any other recent eruption. Its climatic effects were so severe that the following year, 1816, went down in history as the '**Year without a Summer**'. Contemporary accounts of the weather that year paint a dismal picture of cool, rainy weather across much of the Northern Hemisphere, leading to widespread crop failures and famine.

There are several reasons why one should not expect to find a strong correlation between calculated volcanic forcings and their observed climate impacts. One reason is simply the poor quality of the data for historic eruptions – it is difficult to extract a clear 'signal' of global eruption-induced climate change from contemporary meteorological data, which were collected in a rather haphazard way for much of the 19th century. Another is that most recent eruptions have been rather small, so they do not provide useful yardsticks – Mount St Helens was so small that its stratospheric aerosol loading was almost negligible.

4.2.3 The Toba eruption

Although the Tambora eruption of 1815 was exceptional in the historical record, and produced sufficiently bad weather to enter folk history, it was modest on the scale of some earlier eruptions. An eruption about 75 000 years ago which excavated the 100-km long Toba caldera in Indonesia was the largest eruption that we know of in the recent geological record. About 7.5×10^{15} kg of magma are believed to have been erupted, accompanied by an estimated 3.3×10^{12} kg of stratospheric aerosols.

Question 4.4

(a) Taking these estimates together with the empirical relationship summarized in Equation 4.2, what would have been the climatic forcing resulting from the Toba eruption?

(b) Why is this answer clearly nonsense? (Refer to Figure 1.5, if necessary.)

The Toba data illustrate the problems of trying to scale up from small, well-observed phenomena to much larger ones: it shows that the relationship between the mass of aerosols produced and radiative forcing cannot always be the simple linear one expressed in Equation 4.2. This holds up reasonably well for eruptions of modest magnitude, but breaks down for much larger ones. In other words, provided that the magnitude of the eruption is within the range of that covered by historic data, the equation gives a useful guide to the resulting climate forcing; extrapolations beyond this are unreliable.

Most scientists believe that there is a limit to the mass of aerosols that the atmosphere can sustain. Pumping sulfur dioxide into the atmosphere will not only 'consume' water in the stratosphere through making sulfuric acid, but the aerosol particles may also coalesce to form larger droplets and fall out as acid rain (Figure 4.8). Thus, the relationship between eruption magnitude and the formation of aerosols may be self-limiting.

Question 4.5

This is an opportunity for you to work out for yourself the effects of a large volcanic eruption on stratospheric water. The mass of water in the stratosphere (above ~ 20 km) is about 8×10^{11} kg. The Toba eruption injected about 2.1×10^{12} kg of SO_2 into the stratosphere. If three molecules of water are consumed for each molecule of sulfur dioxide converted to sulfuric acid, in theory how much stratospheric water could the Toba acid gases consume? (You will need to use the following relative atomic masses: H = 1, O = 16 and S = 32.)

Of course, in real life things are much more complex than this simple calculation suggests. In particular, a large eruption actually injects a lot of water into the stratosphere, as well as SO_2. However, your calculation emphasizes how dry the stratosphere is, and how easy it is to overload it.

No-one knows what the climatic effects of the largest volcanic eruptions might be, but they are certainly more serious than those caused by Tambora in 1815. Contemporary accounts of an eruption in AD 536 state that 'the Sun gave forth its light without brightness, like the Moon, during this whole year …'. Some scientists believe that, at worst, a Toba-sized eruption would cause such extensive dimming of the Sun that daylight would resemble a brightly moonlit night. They speak of a global catastrophe they call 'volcanic winter'; this may overstate the case, but it seems likely that there would be global cooling by several degrees for at least a year or two.

Table 4.4 Completed Table 4.3, showing computed climatic forcings and the average cooling that was observed to result.

Location of eruption	Mass of magma erupted / 10^{11} kg	Mass (m) of aerosols in stratosphere / 10^9 kg	Climatic forcing / W m^{-2}	Observed Northern Hemisphere temperature change / °C
Tambora, Indonesia, 1815 (8° S)	2400	150	−21.4	−1.0
Krakatau, Indonesia, 1883 (6° S)	247	55	−7.8	−0.5
Agung, Indonesia, 1963 (8° S)	24	20	−2.86	−0.3
Katmai, Alaska, 1912 (58° N)	240	25	−3.5	−0.2
El Chichón, Mexico, 1982 (17° N)	7.5	12	−1.7	−0.2
Pinatubo, Philippines, 1991 (15° N)	120	30	−4.3	−0.5

This raises an interesting issue. During the peak of the last glacial period, global temperatures were about 5 °C lower than at present – could the onset of glaciation have been triggered by a massive eruption like Toba?

■ What is the strongest argument *against* the notion that a glacial period could be triggered by a large explosive eruption?

■ It is the issue of response time. Although a single large volcanic eruption can provide a significant climate forcing, its effects are not sustained sufficiently long for feedback processes to come into play.

In fact, Antarctic ice-cores show that the Toba eruption *did* coincide with a period of global cooling, but that this cooling had already been initiated prior to the eruption. However, it is possible that the Toba aerosols amplified a cooling trend that was already underway; as you may have gathered from Chapter 1, warming and cooling of the Earth has occurred cyclically in response to changes in its orbit and tilt (i.e. the Milankovich cycles) for at least the past 800 000 years. We will come back to this topic in the next chapter.

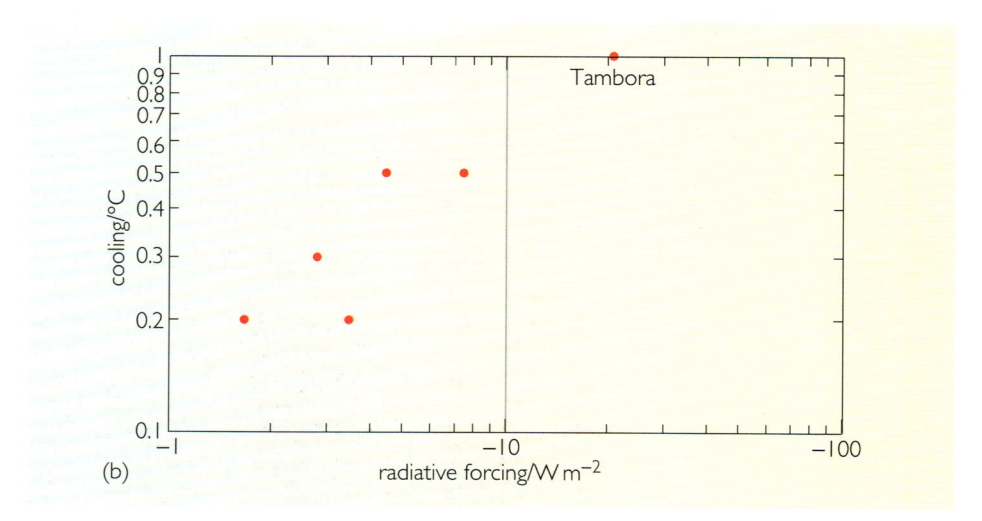

Figure 4.14(b) Completed Figure 4.14a.

Volcanism and the Earth system

4.3 Flood basalts and their climatic effects

So far, we have been discussing short-lived, violently explosive eruptions such as that of Mount Pinatubo. Although its effects were perceptible for years, the eruption itself lasted only a matter of hours. A different kind of eruption may have more serious environmental effects than the explosive kind: the effusion of prodigious volumes of lava. These are termed **flood basalts**.

Accumulations of flood basalts cover large parts of the Earth. Some of the statistics are impressive: India's 65-million-year-old Deccan Traps currently cover half a million square kilometres (Figure 4.15), and may have covered a million-and-a-half square kilometres when first erupted. They have an average thickness of at least one kilometre, and locally two. Most of their huge volume may have been erupted in less than half a million years (although this is controversial) and they may be made up of many hundreds of individual lava flows. On erosion, they may be seen as a distinctive topography resembling flights of steps: the term 'traps' is an old Swedish word for stairs. In the once-volcanic region of the north-western USA known as the Columbia River Province, 170 000 km^3 of basalt flooded out between 17 and 15 million years ago, in individual flows 20–30 m thick, covering an area of about 165 000 km^2 (Figure 4.16); 99% of the volume of the entire province may have been erupted during those 2 million years. Then 14 million years ago, another 1200 km^3 of basalt erupted to form the Roza flow, which travelled more than 300 km westwards from its source, along the Columbia River gorge. The slightly younger Pomona flow travelled even further, reaching over 550 km from its source (Figure 4.17).

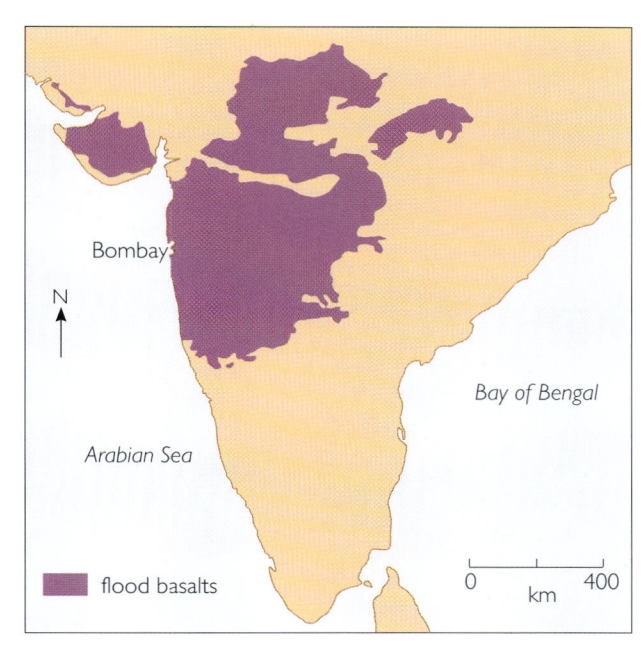

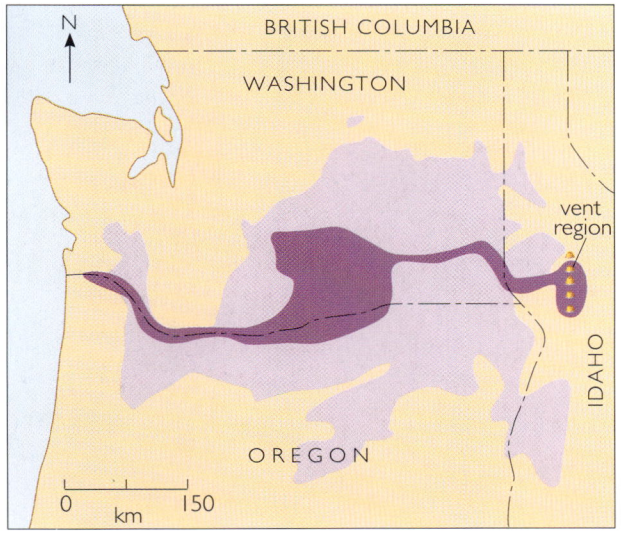

Figure 4.15
The areal extent of the Deccan Traps. A significant fraction of India's land surface is covered by the outcrop of these 65-million-year-old basalt lavas.

Figure 4.16
River valley and waterfall cutting through basalts of the Columbia River Province. Each of the horizontal layers represents a single basalt lava flow; the total thickness of lavas visible in the foreground is ~100 m. Cracks formed during cooling ('joints') give the flows a marked columnar appearance.

Figure 4.17
Columbia River Province covering parts of Washington, Oregon and Idaho. Some flows extended from their source region on the borders of Idaho and Washington as far as the Pacific coast. The darker area defines the extent of the 14-Ma-old Pomona lava flow in Washington and Oregon; also shown is the vent region for the Pomona flow. (Hooper, P.R. (1982) 'The Columbia River basalts', *Science*, **215**, pp.1463–68, copyright © 1982 American Association for the Advancement of Science.)

There may even be examples of submarine flood basalt provinces: recent research suggests that enigmatic **submarine plateaux** may be analogous to flood basalt provinces on the continents. Submarine plateaux are extensive elevated regions 2–3 km higher than the surrounding sea-floor, characterized by unusually thick oceanic crust (20–40 km) and containing huge volumes of lava which were erupted very rapidly, within 5–10 million years. In Alaska and British Columbia, sequences of flood basalt as much as 6 km thick may have formed about 240 million years ago as a submarine plateau, which about 100 million years ago accreted to the North American continent. We will come back to these large oceanic igneous provinces later; first, we should consider whether eruptions of the recent past can be used to deduce the effects of other, perhaps larger, eruptions.

4.3.1 Lessons from the Laki Fissure eruption

Perhaps fortunately, massive basalt lava eruptions are relatively rare. The most recent was the 1783 Laki Fissure eruption in Iceland, which caused one of the most serious environmental disasters in modern history. Lavas were erupted from a 27-km-long chain of more than 140 craters which cut across an older ridge called Laki; the lavas entered and drowned two river valleys (Figure 4.18). Apart from the 14 cubic kilometres of lava, large amounts of sulfur dioxide were pumped into the atmosphere over a period of months, forming a low-lying haze of acid aerosols. Some of the environmental and climatic consequences of the eruption are described in Box 4.2.

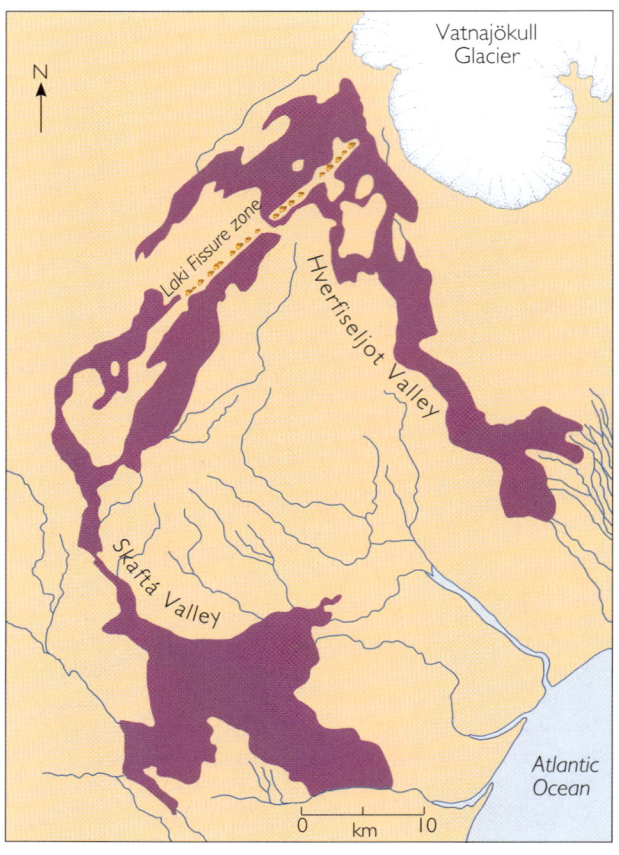

Figure 4.18
Areas of Iceland covered by lava from the 1783 Laki Fissure eruption. Local priests and farmers carefully documented the progress of the advancing flows. Some of the rivers shown now flow through or over the lava, while others flow around the edge of it.

Box 4.2 Environmental and climatic effects of the Laki Fissure eruption

The effects of the acid haze from Laki were widespread. Benjamin Franklin drew the connection between the 'dry fog' that crept southwards to blanket much of Europe, and the events unfolding in Iceland. In 1784, while serving in Paris as the first diplomatic representative of the newly formed United States of America, he wrote:

'During several of the summer months of the year 1783, when the effects of the Sun's rays to heat the Earth in these northern regions should have been the greatest, there existed a constant fog over all Europe, and great part of North America. This fog was of a permanent nature; it was dry, and the rays of the sun seemed to have little effect towards dissipating it, as they easily do a moist fog, arising from water. They were indeed rendered so faint in passing through it that when collected in the focus of a burning glass, they would scarcely kindle brown paper. Of course, their summer effect in heating the earth was exceedingly diminished.'

Some accounts suggest that the haze reached as far as Syria and western Siberia. In Europe, where its effects were most obvious, the haze was so dense that it dimmed the Sun. In Pavia, northern Italy, people could look at the Sun with the naked eye, while in southern France the setting Sun was sometimes invisible once it had sunk within 17° of the horizon. Acid rain fell in Norway and Scotland. In Iceland itself, the 'dry fog' was a dense bluish acid haze which destroyed most of the summer crops. Famine followed, leading to the deaths of 75% of Iceland's livestock, and 25% of the human population.

Apart from sulfur dioxide, many other gases, including hydrogen fluoride, were emitted. Fluorides caused the death of cattle that grazed on grass contaminated by the acid haze; subsequent experiments have shown that if the fluorine content of dry grass exceeds 250 parts per million, animals grazing on it die in two or three days. Fluorine is present in proportions of a only few hundred parts per

Volcanism and the Earth system

million in basaltic lavas, and its abundance in erupted volatiles as hydrogen fluoride is difficult to estimate. But the highly reactive gas binds firmly to the surface of erupted particles and aerosols, so when hungry cattle grazed on contaminated grass – as starvation compelled them – they ingested lethal doses of the element, even at considerable distances from the volcano.

While Franklin was enduring the bitter winter of 1783–84 in Paris, conditions were also bad on the other side of the Atlantic. Records show that an abnormally steep decline in temperature began in autumn 1783, with values hitting rock-bottom between December and February 1784, when the lowest-ever winter average temperature for this region was recorded: 4.8 °C below the 225-year average for minimum winter temperature (cf. Figure 4.19). Taking the Northern Hemisphere as a whole, a cooling of about 1 °C seems to have taken place.

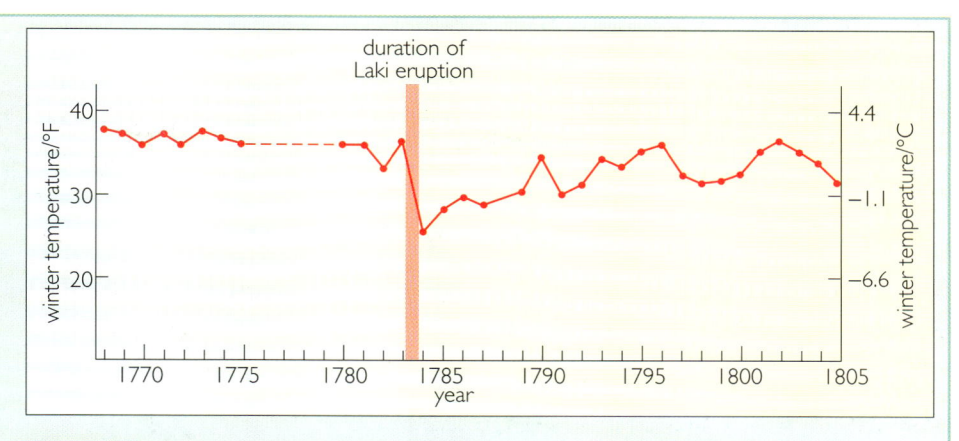

Figure 4.19
Winter (Dec.–Feb.) temperatures for the eastern United States from the late 1760s until 1805. Several distinguished scholars kept temperature records at the time, including Thomas Jefferson, third President of the USA, and Ezra Stiles, President of Yale College in Connecticut. (The observations were recorded in °F (left-hand axis), here also converted to °C (right-hand axis).)

In Britain, the renowned diarist and naturalist Gilbert White, author of *The Natural History of Selborne*, also noted that conditions during the summer of 1783 were unusual. He termed the summer 'amazing and portentous', but was unaware of the eruption in Iceland. White recorded the presence of a 'peculiar haze or smoky fog' from 23 June to 20 July, but also noted that the summer in England was exceptionally hot.

■ From what you have read earlier about the effects of aerosols, can you suggest one set of circumstances that might have accounted for the hot summer in England in 1783?

■ If the tropospheric aerosols that formed the 'dry fog' over Britain had been composed of large particles (1–2 μm), then they might have had a warming effect, by trapping long-wave infrared radiation (cf. Figures 4.8 and 4.9c).

Much of Scandinavia suffered more seriously than Britain, and crop harvests were ruined by the acid haze. For example, a letter of 4 September 1783 states that in the Eidsberg Parish of Sweden, in many fields it was not even possible to determine what crops had been planted there. The

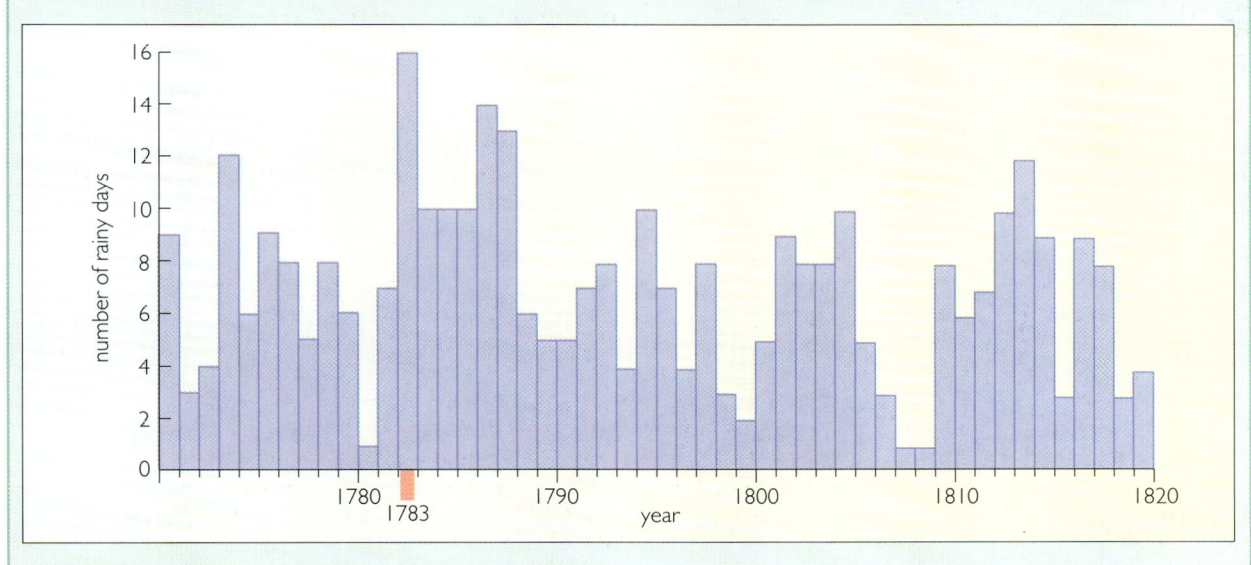

Figure 4.20
Number of August days with rain in Japan, in the period 1770–1820.

letter continues '... many farmers who normally harvest two barrels expect not even a single bushel this time, and the same is the case with the oats'. Effects were felt as far away as Japan, where the summer of 1783 was so unprecedently wet that there were widespread crop failures, leading to famine (Figure 4.20). (In 1816, by contrast, while Europe and North America were experiencing the 'Year without a Summer' after the Tambora eruption, the weather in Japan was unusually dry and sunny; see Table 4.5.)

Table 4.5 Exceptionally cold years and seasons in the period 1783–1817 in Brunswick, New Jersey. Laki erupted in 1783 and Tambora erupted in 1815.

Year	Year overall	Spring	Summer	Autumn	Winter
1783	average	warm	average/warm	average/cool	coldest by 1.6 °C
1784	5th coldest	coldish	average/warm	average/cool	2nd coldest
1785	9th coldest	3rd coldest	average/warm	average/cool	average
1816	average	normal	cool	average	average
1817	4th coldest	coldest	coldest by 1 °C	average	average

Because the Laki eruption has been closely studied, it can be used to infer the consequences of much larger, ancient eruptions. The Laki magma was rich in sulfur, containing initially about 1760 p.p.m. of the element by weight. 73% of the sulfur was released to the atmosphere during the eruption, while the remainder remained in the lavas. Thus, for every kilogram of rock erupted, 0.0013 kg of sulfur was produced. About 3.8×10^{13} kg of magma were erupted, leading to the production in the atmosphere of 1.5×10^{11} kg of sulfuric acid aerosols – a value that matches quite well with independent estimates derived from the mass of acid aerosols that fell on the Greenland ice-cap.

By working through Question 4.6 you can get some idea of the effects this great eruption had on climate.

Question 4.6

(a) Estimate the climate forcing due to the Laki eruption, using the relationship we introduced earlier: $R_{net} = -(1.43 \times 10^{-10}) \times m$ W m^{-2}.

(b) According to information given previously (in Table 4.4), with which recent eruption might the climatic forcing caused by Laki have been comparable?

So the cooling caused by the Laki eruption would have been comparable to that which gave rise to the 'Year without a Summer', in 1816. Laki, however, was different from Tambora in several important ways:

- it was an eruption of lava, not a major explosive event;
- the eruption continued for many months, rather than hours or days;
- it took place in Iceland, at a latitude of 64° N.

■ What is the significance of the latitude of the eruption, as far as climatic effects were concerned?

□ Because the eruption took place at a high northern latitude, global wind patterns would have dictated that its effects would be concentrated in the Northern Hemisphere. In the case of Tambora, located near the Equator, it was easier for aerosols to spread into both hemispheres, causing global effects.

Furthermore, because Laki was a lava eruption, not an explosive one like Tambora, it would have been more difficult for the eruption products to be injected into the stratosphere, even though the tropopause is lower at high latitudes than in the tropics. The vast majority of Laki's aerosols therefore remained confined to the troposphere in the Northern Hemisphere. Because the troposphere is where almost all 'weather' happens, the aerosols were easily flushed out of it as acid rain. However, this loss was offset by the long duration of the eruption, so that fresh supplies of aerosols were continually pumped into the atmospheric circulation, and were transported long distances by westerly winds.

Laki's long drawn-out eruption therefore provides us with three important lessons:

1. The insidious environmental effects of an eruption can be far more serious than the more immediately obvious manifestations.
2. An effusive eruption of modest dimensions can have climatic effects comparable with those of explosive eruptions of much larger volume.
3. The fact that the summer of 1783 was unusually hot in England, unusually wet in Japan, and average in the eastern USA, emphasizes that while the overall effect of volcanic aerosols is almost always to cause global cooling, their effect on regional weather is much more varied, and they may even cause warming locally.

Recent research suggests that because of the way weather systems are perturbed by *overall* global cooling, large explosive eruptions may actually lead to warmer winters in the Northern Hemisphere. This counter-intuitive finding demonstrates that one should never jump to conclusions about the way the Earth system works!

Scaling up the effects of the Laki Fissure eruption

Despite the problems inherent in extrapolating from one eruption to another, Laki provides some valuable clues to the effects of the far bigger eruptions which formed basaltic plateaux. In particular, it provides us with two values which we can use to make estimates about these eruptions:

- An average eruption rate over eight months of almost $700 \, m^3 \, s^{-1}$, or $1.8 \times 10^6 \, kg \, s^{-1}$.
- For every kilogram of rock erupted, 0.0013 kg of sulfur was sent into the atmosphere.

You can now examine the implications of these data for one of the huge basaltic flows in the Columbia River Province (cf. Figures 4.16 and 4.17), the Roza flow mentioned earlier.

Question 4.7

The Roza flow had a volume of *at least* 1200 km^3, probably much more.

(a) What would be an approximate lower limit for the mass of sulfur sent into the atmosphere? (Begin by calculating the minimum mass of rock produced by assuming that the density of basalt lavas is $2.7 \times 10^3 \, kg \, m^{-3}$.)

(b) Convert your answer to (a) into an estimate of the minimum mass of sulfuric acid aerosols produced. You will need to use the appropriate relative atomic masses (S = 32, O = 16, H = 1), and to begin by calculating the relative molecular mass of sulfuric acid, H_2SO_4.

Table 4.4 showed that the Tambora eruption of 1815 yielded only 1.5×10^{11} kg of sulfuric acid aerosols, roughly two orders of magnitude less than the Roza eruption. Thus, it seems that the Roza flow must inevitably have had a huge effect on the environment.

- The eruption of Mount Pinatubo lead to the production of about 30×10^9 kg of aerosols. Would we be justified in using the climate forcing data for Mount Pinatubo to scale up the effects of the Roza flow?

- Not really. Not only was the Roza flow an entirely different kind of eruption, but it was about 300 times bigger. As you saw in Question 4.4, the climatic effects of large eruptions do not in reality scale linearly with the mass of aerosols produced.

No-one knows for sure how long it took for a giant flow like Roza to be erupted, but recent research suggests that the process may have been more akin to a prolonged ooze than a giant, brief torrent.

- If the Roza flow was erupted at the same average rate as Laki, how long would the eruption have lasted?

- At a rate of $700 \text{ m}^3 \text{ s}^{-1}$, the eruption would have lasted about 50 years.

Thus, we can imagine large-scale environmental disruption resulting from the production of acid aerosols continuing over periods of many years, possibly decades. It is also unclear what proportion of the aerosols penetrated the stratosphere. Some volcanologists believe that powerful 'fire fountains' may have accompanied the eruption, injecting aerosols into the stratosphere, but this remains a matter for investigation. Whether or not the aerosols had long atmospheric residence times as a result of stratospheric injection, the duration of the eruption seems likely to have ensured long-lasting effects. As mentioned earlier, climate forcings sustained over long periods – tens or hundreds of years – provide the opportunity for feedback effects (such as increased cooling caused by increased snow and ice cover) to come into play.

We must also remember that the Roza flow is only a single component of the Columbia River Province. The plateau is up to 3 km thick, and is composed of several hundred major flows, one on top of another. There are clear time-breaks between some flows, but not between others. Thus, the possibility arises that huge amounts of basalt lavas were welling up to the surface of the Earth and producing aerosols continuously for periods of hundreds of years.

4.3.2 Are flood basalts implicated in mass extinctions?

At the end of the Cretaceous Period 65 million years ago, there was world-wide extinction of many animal species – a **mass extinction** – both in the sea and on land. The demise of the dinosaurs has captured the public imagination, but many less notorious species also became extinct, notably many species of planktonic foraminiferans. An increasing body of evidence indicates that at that time a major impact event covered the Earth in a film of iridium-rich ejecta, now seen in the sedimentary record at the Cretaceous–Tertiary boundary. Researchers have even zeroed-in on the site of the impact, at Chicxulub on the Yucatan peninsula of Mexico.

Volcanism and the Earth system

But at the time of the Chicxulub impact, at least a million cubic kilometres of basalts were being erupted to form the Deccan Traps of India (Figure 4.15). Were these two remarkable events connected? Did the impact extinguish the dinosaurs? Or was it the basaltic flood eruption – Laki on a global scale? Did the impact event perhaps trigger the basaltic flood? Or did the two extraordinary events reinforce each other in some lethal way to cause mass extinctions? These highly contentious issues have sparked vigorous debates.

As long ago as the early 1970s, some scientists argued that widespread basaltic volcanism might have been implicated in the extinction of the dinosaurs. A round-number estimate of the volume of the Deccan basalts is one million cubic kilometres. Simple linear scaling from Laki (always risky) suggests that a total of about 1×10^{16} kg of sulfuric acid was liberated. An alternative method based on observed eruption data suggests a figure of 1.7×10^{16} kg. It seems inevitable that the huge mass of acid aerosol produced by the Deccan eruptions had far-reaching consequences, and probably made more than one *Tyrannosaurus* retch.

An even greater mass extinction defines the Permian–Triassic boundary, the milepost at 248–245 million years ago, separating the Paleozoic ('ancient life') Era from the Mesozoic ('middle life') Era. Huge numbers of species, including trilobites (marine arthropods, probably related to the ancestors of the crustaceans), became extinct by the end of the Permian, while the Mesozoic saw the blossoming of the age of reptiles, including the dinosaurs. Is it purely coincidental that the Paleozoic–Mesozoic boundary coincides approximately with the vast outpouring of the Siberian flood basalts (shown on Figure 4.21)? More than 1.5 million cubic kilometres of basalt may have been erupted within a period of less than a million years, an average rate of more than 1.5 km^3 per year (about 50 m^3 s^{-1}).

- How many extinction events correspond with a flood basalt eruption, within the limits of error of the ages cited? Does this suggest that flood basalts are implicated in mass extinctions?

Table 4.6 Dates of continental flood basalt eruptions and mass extinctions. (You do not need to memorize the names of the various provinces or geological boundaries.)

Flood basalt province (most recent first)	Initiation of eruptions/ Ma ago	Stage in which mass extinction occurred (most recent first)	Age of upper boundary/Ma
Columbia River	17.0 ± 0.2	Pliocene	1.64 ± 0.03
Ethiopian	35 ± 2	Middle Miocene	10.5 ± 1
Brito-Arctic	61 ± 1	Upper Eocene	35 ± 1
Deccan Traps	66 ± 1	Maastrichtian	65 ± 1
Madagascar	90 ± 5	Cenomanian	91 ± 1
Rajmahal	116 ± 2	Aptian	110 ± 3
Ontong–Java	121 ± 3	Tithonian	141 ± 4
Serra Geral	133 ± 1	Bajocian	171 ± 4
Namibian	135 ± 5	Pliensbachian	190 ± 4
Antarctic (Kerguelen)	176 ± 2	Rhaetian	204 ± 4
South African	190 ± 5	Dzulfian	250 ± 3
West African	200 ± 3		
Eastern North America	201 ± 1		
Wrangellian	230 ± 5		
Siberian	247 ± 2		

- Eight of the eleven extinction events match well with dated basalt eruptions. Upper Eocene (corresponding to the Ethiopian eruption); Maastrichtian (Deccan Traps); Cenomanian (Madagascar); Tithonian (Namibian); Bajocian (Antarctic); Pliensbachian (South African); Rhaetian (West African and Eastern North American); and Dzulfian (Siberian).

  These correspondences indicate that it is at least plausible that the flood basalts may be implicated in the extinctions.

The issue is far from resolved, however – to assess the correlations adequately we require more data on individual eruptive and extinction events.

4.4 CO$_2$ from the Deccan Traps: a cause of greenhouse warming?

In considering the climatic effects of volcanism, we have so far focused mainly on the effects of sulfuric acid aerosols. However, during basaltic volcanism more carbon dioxide is produced than sulfur dioxide (Table 4.1) so it is natural to enquire whether the prodigious effusions that formed the Deccan Traps could also have contributed to global warming. As before, we will examine this possibility quantitatively.

Estimates of the volume of the Deccan Traps range from 1×10^6 to 2.6×10^6 km^3. Taking a mean value of 1.8×10^6 km^3 gives us a mass of 4.8×10^{18} kg. Studies of the basalts suggest that they contained about 0.2% by mass of carbon dioxide. Of this, about 60% was actually outgassed on eruption. The total mass of carbon dioxide outgassed was thus about $0.6 \times 0.002 \times 4.8 \times 10^{18}$ kg or about 5.8×10^{15} kg. Now the total mass of carbon dioxide in the atmosphere today is only about $\sim 2.8 \times 10^{15}$ kg, so at first sight it may seem that the eruption of the Deccan Traps should have caused a massive CO$_2$ greenhouse effect. But did it? To find out more about the effects of the eruption, do Activity 4.1.

Activity 4.1

The enormous pile of Deccan lavas was not erupted either instantaneously or continuously: geological studies show that there were some quite long breaks, long enough for soil horizons to develop.

(a) (i) Assuming for the sake of argument that the Deccan Traps were erupted spasmodically over a period of 500 000 years, how much CO$_2$ did the eruption release to the atmosphere annually, on average? What proportion would this have been of the total mass of CO$_2$ in the atmosphere, assuming for convenience that this was not very different from what it is today (see above)?

(ii) Explain whether you would expect the mass of CO$_2$ in the atmosphere to have increased annually by the amount you calculated in (i)?

(b) According to one estimate, the formation of the Deccan Traps could have led to an overall increase in atmospheric CO$_2$ of about 75 p.p.m. It has been calculated that a doubling of the concentration of atmospheric CO$_2$ from an initial level of 330 p.p.m. (the concentration in the mid-1970s) would result in a climatic forcing of about +4.4 W m^{-2}. Although it is a very crude method, we can use this information to estimate the climatic forcing resulting from CO$_2$ emission from the Deccan Traps.

(i) By means of a simple proportion calculation, estimate the climatic forcing resulting from an increase in atmospheric CO$_2$ of about 75 p.p.m., assuming for convenience that there were initially 330 p.p.m. of CO$_2$ in the atmosphere.

(ii) Assuming that the Earth's climate sensitivity is about 0.75 °C per W m^{-2} of forcing, would the Deccan volcanism have been likely to have caused significant warming? What other factors do we need to take into consideration?

(iii) Bearing in mind the residence time of carbon (mainly as CO$_2$) in the atmosphere, explain whether the climatic effects of the Deccan Traps would have fluctuated over the course of the thousands of years that they were being erupted.

Volcanism and the Earth system

So, the effect of the eruption of the Deccan Traps on climate was probably fairly small. But it seems highly improbable that these massive events, coinciding as they do with great mass extinctions, were environmentally neutral.

4.5 Taking a global view

We have been dealing so far with the effects of individual eruptions, or episodes of eruptions. There is another, simpler, way of looking at the issue. As you know from your reading of Chapter 3, at any given time in Earth history, the amount of carbon dioxide in the atmosphere is the result of the balance of fluxes between sources and sinks in the carbon cycle. If the average global flux of carbon dioxide to the atmosphere from volcanoes should increase for some reason, and if that increased flux is sustained, then the carbon cycle will settle into a new equilibrium, with important implications for climate (Section 3.4).

But is there any evidence that the flux of carbon dioxide into the atmosphere has varied over geological time? There are some pointers in this direction. Most attention has been focused on the Cretaceous Period (142–65 million years ago), which seems to have been an exceptionally productive period in terms of creation of ocean crust and eruption of massive basaltic plateaux (cf. p.130). So much new material was erupted in the Pacific basin during the Cretaceous that it is difficult to find any older oceanic crust or sediments there – they have been simply swamped by Cretaceous lavas. Two huge plateaux were formed at this time: the Ontong–Java Plateau about 122 million years ago and the Kerguelen Plateau about 112 million years ago (Figure 4.21). Both of these were probably above sea-level when they formed. The Ontong–Java Plateau is 25 times bigger than the Deccan Traps: so much new lava was involved in its formation that it elevated sea-level by about 10 m, displacing seawater like a body in a bath tub.

Estimates of the average global rate of ocean crust production over the past 150 million years range between 16 and 26 $km^3 \, yr^{-1}$. Estimates of the production rate of the Ontong–Java Plateau are around 12–15 $km^3 \, yr^{-1}$, the same order of magnitude.

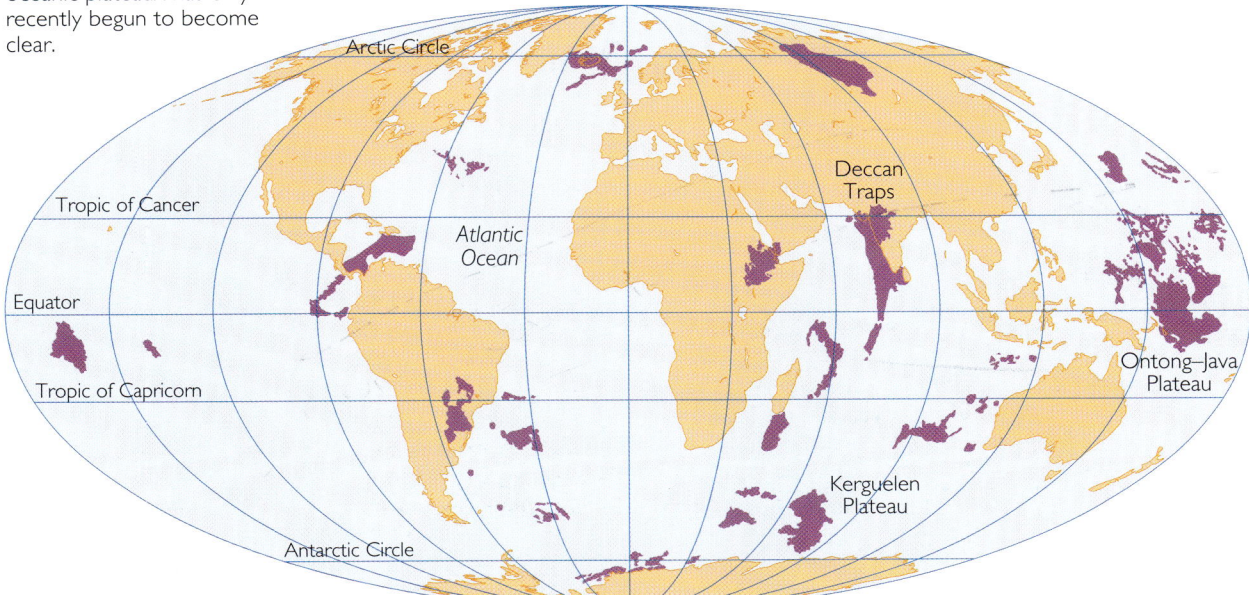

Figure 4.21
Global distribution of large igneous provinces. The significance of the large oceanic plateaux has only recently begun to become clear.

If we attempt to plot a graph of oceanic crust production rate through time (Figure 4.22), we can see that there was a distinct increase in rate about 120 million years ago, probably due to the combined effects of increased 'ordinary' oceanic crust formation, and the abrupt formation of the enormous Ontong–Java and Kerguelen Plateaux and some lesser ones in the south-west Pacific (plus a small contribution from hot-spot island chains).

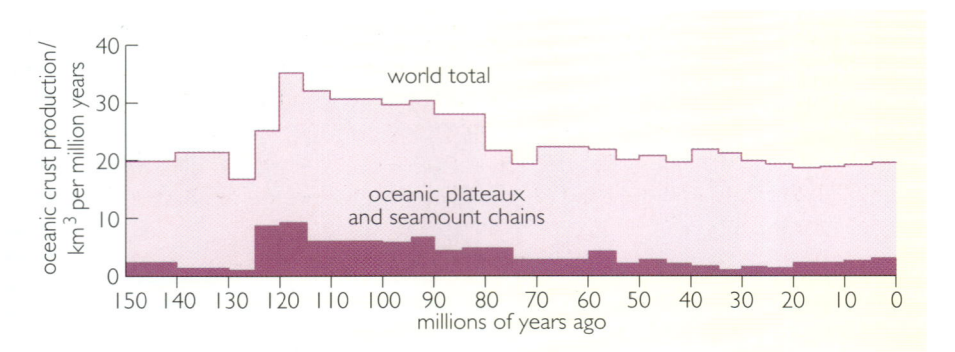

Figure 4.22
Variation with time of the estimated rate of production of oceanic crust. Note the large peak between about 120 and 80 million years ago.

What were the effects of all this volcanic activity on CO_2 in the atmosphere? There may have been both direct and indirect consequences. As discussed at the start of this chapter, as long as it remained below sea-level, the volcanism taking place at mid-ocean ridges, and leading to creation of new oceanic crust, would have had little immediate effect on the atmosphere. Increased sea-floor spreading, however, has to be balanced by increased subduction – if it were not, the total area of the Earth's lithosphere would increase, so that the Earth would have to be expanding. As discussed in Section 3.3.3, when oceanic crust is subducted, carbonate-rich sediments are also subducted, and their carbon may ultimately be exhaled at subduction zone volcanoes. More important, if 'hot spot' mantle plumes lead to lots of large volcanoes – like Hawaii – poking their noses above sea-level, then a *direct* flux of CO_2 from the mantle into the atmosphere could be sustained.

Estimates of the CO_2 production from the Cretaceous outburst of volcanic activity suggest that atmospheric CO_2 could have reached a level anywhere between four and 15 times the modern pre-industrial level of 285 p.p.m., leading to global greenhouse warming of between 3 and 8 °C. This would have been accompanied by other changes, including increases in sea-level and a rearrangement of the continents (both explored in Chapter 5) and climate modellers have suggested that the net effect of all the activity in the Cretaceous was a warming of between 8 and 13 °C. This is in good agreement with the 6–14 °C of warming inferred from fossil evidence.

The Cretaceous, then, provides evidence of climate change driven by internal processes. But what was different in the Cretaceous? What actually happened? To answer this, we have to look into the Earth's mantle. Geoscientists have for many years accepted that individual flood basalt provinces such as the Columbia River were the result of convective plumes arising from the core–mantle boundary: when the plume head reached the surface, rapid outpourings of basalt ensued. For the vast eruptions that formed the Kerguelen and Ontong–Java plateaux, and the generally enhanced level of Cretaceous oceanic crust production, the term **superplume** has been coined.

Is a 'superplume' just jargon for something which geoscientists don't really understand? Possibly. But underlying it all there may be a new vision of the way the world works, according to which the 'normal' plate tectonic cycle of ocean crust creation and subduction is occasionally interrupted by major mantle overturning events, originating at the core–mantle boundary (Figure 4.23). This hypothesis would account well for many puzzling aspects of the Earth's history – for example the coincidence between a decrease in the magnetic field reversal rate with the increase in oceanic crust production rate 120 million years ago (Figure 4.22). Thus, the 'superplume' concept offers a framework which links events in the Earth's core and mantle with those at the surface. Furthermore, it looks as though Venus may experience similar massive mantle overturning events: one of the major conclusions from the *Magellan* mission to the planet is that 300–500 million years ago virtually the whole surface of the planet was resurfaced through massive outpourings of basalt lava.

Figure 4.23
Contrasted modes of mantle convection. At left, normal plate tectonics operates, with opening and closing of oceans, and mantle convection characterized by isolated upper and lower mantle domains. At right, during 'superplume' events, accumulated cold material descends from the 660 km boundary into the lower mantle, and multiple major plumes rise from the core–mantle boundary to the surface, thus creating a major overturn.

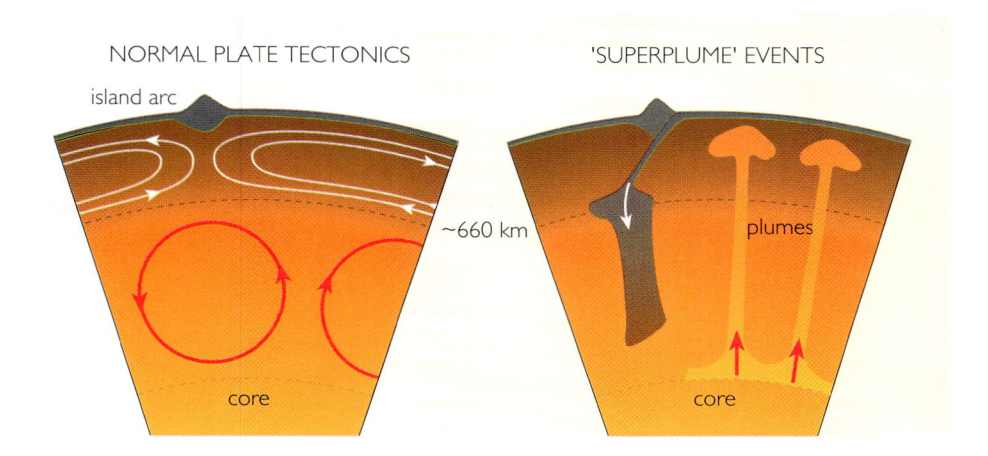

4.6 Summary of Chapter 4

1   Volcanoes are conduits linking deep mantle processes to atmospheric composition and hence climate; they exhale large amounts of water, CO_2 and SO_2. Over geological time-scales, volcanoes have an important role in the carbon cycle, but in the short term their contribution to atmospheric CO_2 has been overwhelmed by the anthropogenic flux. Major eruptions, like that of Mount Pinatubo in 1991, pump large amounts of SO_2 into the stratosphere. Over time, the SO_2 combines with water to form sulfuric acid aerosols (H_2SO_4). These change the Earth's radiative balance, reflecting a proportion of incoming solar radiation back into space.

2   The effects of an eruption on climate are dictated by:
    ◆   the magnitude of the eruption;
    ◆   the sulfur content of the magma;
    ◆   whether or not the volcanic gases get into the stratosphere;
    ◆   the location of the eruption site on the globe;
    ◆   the type of eruption.

    Eruptions taking place in the tropics are most likely to have global effects. Explosive eruptions are more likely than effusions of lavas to inject gases into the stratosphere.

3   The radiative effects of major explosive eruptions can be large, reducing the amount of solar radiation reaching the Earth's surface by more than 10%. The effects of most explosive eruptions, however, are short-lived (2–3 years), as aerosols fall out of the stratosphere. Because of the long response time of the Earth's climate system as a whole, brief volcanic events do not have a significant effect on climate. While the effects of some large historic eruptions such as that of Tambora in 1815 (followed by the 'Year without a Summer' in 1816) are well documented, it is not possible to scale directly up from these to deduce the effects of much larger prehistoric eruptions. For instance, there may be a limit to the aerosol loading the stratosphere can sustain.

4   Eruptions of flood basalts such as those of the Columbia River Province may involve effusion of more than 1000 km^3 of sulfur-rich basalt lava over periods of 10–100 years. Because these eruptions are sustained over longer periods than great explosive events, their radiative and environmental effects may be more profound. The 1783 Laki Fissure eruption in Iceland, which yielded 14 km^3 of lava, was followed by an exceptionally severe winter in North America and Europe, and widespread starvation of humans in Iceland and elsewhere. The close coincidence in timing between great episodes of flood basalts in Earth history with major mass extinctions – such as those at the Cretaceous–Tertiary and Paleozoic–Mesozoic boundaries – suggests that mass extinctions may be linked with the environmental and climatic effects of the flood basalts.

5   Major flood basalt provinces are themselves expressions of plumes rising through the Earth from the core–mantle boundary. Occasional massive convective overturning events in the mantle may lead to 'superplumes', expressed at the surface in the generation of large igneous provinces such as the Ontong–Java Plateau. The additional CO_2 released into the atmosphere by superplumes may be linked with the climatic warming which occurred during the Cretaceous Period.

Now try the following questions to consolidate your understanding of this chapter.

Question 4.8
Some volcanologists and climatologists argue that volcanic eruptions usually affect weather, not climate. What aspect of volcanic eruptions underlies this point of view? In what circumstances might it be appropriate to speak of the effects of volcanism on climate *per se*?

Question 4.9
In Section 4.4.1, we stated that the Laki Fissure eruption produced 1.5×10^{11} kg of sulfuric acid aerosols. Referring to Question 4.7 as necessary, show how this value was obtained, given that 3.8×10^{13} kg of magma were erupted, and for each kilogram of magma 0.0013 kg of sulfur was released.

Question 4.10
According to computer models of the climate, the amount of global warming that has occurred since the industrial revolution is *less* than ought to have occurred in response to the observed increase of greenhouse gases in the atmosphere. Bearing in mind that many industrial processes (including the burning of fossil fuels) produce sulfur dioxide, can you suggest a possible explanation for the lower-than-expected rise in global temperatures?

Chapter 5
Plate tectonics, carbon and climate

Up to now, we have been considering only the relatively recent geological past, so we may have given the impression of regarding the surface of the solid Earth as fairly static. In fact, on time-scales of millions of years, it is as dynamic as the oceans. In this chapter, we explore how plate-tectonic processes (summarized in Box 4.1) result in interactions between the outer shell of the solid Earth and the atmosphere, oceans and biosphere in ways that can cause long-term climatic change.

5.1 Continental drift and climate

Sea-floor spreading occurs at rates of a few centimetres a year, which means that major ocean basins open and close on time-scales of 100–200 million years. In other words, continents break up, are carried over the surface of the globe, and eventually are brought together again, at intervals of hundreds of millions of years. Not surprisingly, such movements cause the climate of individual continents to change through time. As an example, let us see how climatic conditions have changed for that part of the European landmass that is now southern Britain.

Figure 5.1 shows how the latitude of that region has changed over the past 450 million years or so. The latitudes shown on this plot were obtained by measuring the alignments of magnetic minerals in rocks of different ages, i.e. by using paleomagnetism. Many sedimentary and igneous rocks contain magnetic minerals that are, in effect, 'fossil compasses'. If a compass needle is free to move in three dimensions instead of just a horizontal plane, it will still align itself along the lines of force of the Earth's magnetic field and point towards the North Magnetic Pole.

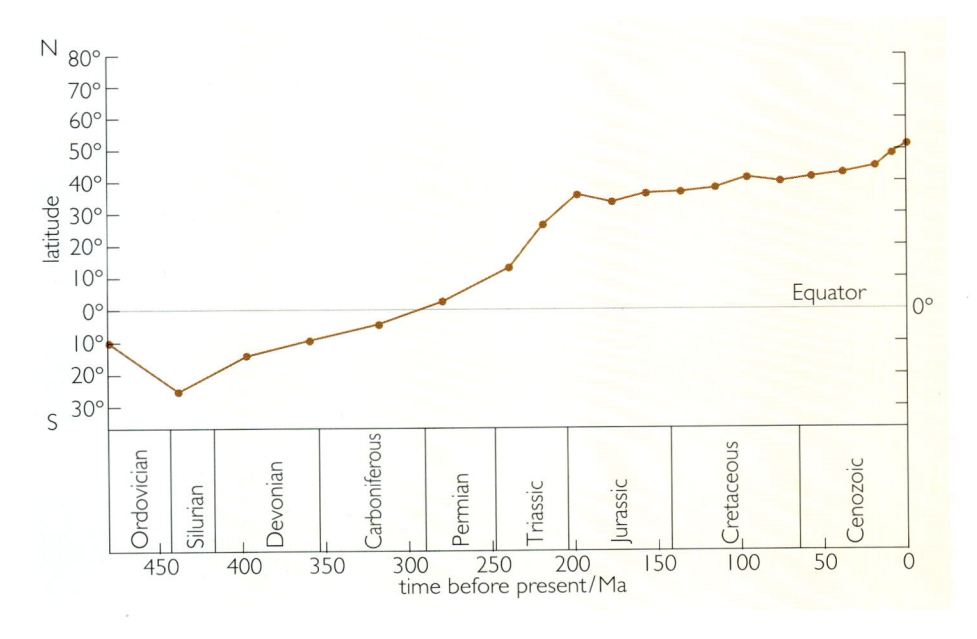

Figure 5.1
A plot of the generally northward drift of 'southern Britain' over approximately the last 450 million years or so, based on paleomagnetic measurements. (Note that paleomagnetism cannot tell us anything about eastward or westward movements.)

As a result, it will be near-vertical at the poles (90° latitude), and near-horizontal at the Equator (0° latitude). Its orientation therefore approximates to its latitude. Using this relationship, and assuming that the positions of the magnetic poles have remained approximately the same throughout much of geological time, the alignment of magnetic minerals within certain sedimentary and igneous rocks can be used to determine the approximate latitude at which they were deposited (in the case of sedimentary rocks), or cooled and solidified (in the case of igneous rocks); ancient latitudes determined in this way are often referred to as *paleolatitudes*.

We can use Figure 5.1 to make some general deductions about how the climate (and hence vegetation) of the area that is now southern Britain has changed over time. At the beginning of the Silurian, ~ 440 million years ago, the landmass concerned was at about 25–30° S.

- Bearing in mind what you read about the Earth's climatic zones in Chapter 2, how would you expect the climate of 'southern Britain' to have changed since the beginning of the Silurian and ~ 290 million years ago, the end of the Carboniferous Period?

- Around 440 million years ago, the region would have been in the dry subtropics, but by the end of the Devonian it would have been moving out of the subtropics into a more tropical climate. By 290 million years ago, it would have been in the vicinity of the Equator: at these lower latitudes, the climate would have been hot, and because of the influence of the Intertropical Convergence Zone, it would also have been wet.

440 million years ago plants had scarcely begun to colonize the land, so the vegetation cover of 'southern Britain' would have been sparse, particularly in the dry climate. By Devonian times, land plants had probably become established, so vegetation could have been the equivalent of savanna. By 290 million years ago, the vegetation would have been lush, like that in present-day tropical rainforests.

The rock record in Britain contains evidence consistent with these inferences. During the Devonian (i.e. before the Carboniferous), red sediments characteristic of hot deserts accumulated. At the beginning of the Carboniferous, when Britain lay just south of the Equator, life teemed in shallow tropical seas, where reef corals and large tropical shellfish contributed to the formation of the thick limestone deposits that today give rise to the characteristic scenery of the Mendips, as well as the Derbyshire Peak District and the Yorkshire Pennines. Towards the end of the Carboniferous, the shallow seas were replaced by broad coastal plains crossed by rivers flowing down from newly uplifted mountains. Much of these plains were covered by swampy equatorial forests (Figure 5.2a), where thick peat deposits accumulated. These became buried by younger sediments and eventually formed the coal measures upon which Britain's Industrial Revolution was founded.

After the Carboniferous, during the Permian, there were dunes comparable to those now seen in the Sahara (Figure 5.2b). Towards the end of this period, seawater flooded much of what is now north-west Europe, including the region that is presently the North Sea. Because of the hot and dry climatic conditions, evaporation rates were high, and the shallow shelf seas became highly saline, but nonetheless supported large and diverse populations of fishes, now abundantly preserved as fossils (Figure 5.3). Thick **evaporites** – salt deposits (in this case chiefly gypsum, $CaSO_4.2H_2O$, and halite, $NaCl$) produced by repeated evaporation of enclosed bodies of water – provide additional evidence of the dry conditions of this period.

Plate tectonics, carbon and climate

Figure 5.2
Artist's reconstruction of two 'British' landscapes at different times in the past.
(a) The Carboniferous. This was a time of equatorial forests and swamps, where thick peat deposits accumulated, later becoming coal.
(b) The Permian. What is now the Vale of Eden, with the predecessor to the Pennines in the distance. Sands similar to those forming the dunes in the valley centre now form the reservoir rocks of the gas fields of the southern North Sea.

(a)

(b)

Figure 5.3
Fossil of a fish that lived in shallow waters off 'Britain' during Permian times.

Coals, desert sandstones and evaporite deposits are all examples of **paleoclimatic indicators**. Their presence in the British rock record was known as long ago as the 19th century, long before paleomagnetic evidence of moving continents became available (that did not happen until the 1950s). Such indicators occur all over the world, and it was their distribution that provided much of the original supporting evidence for the theory of continental drift – an idea that only became accepted when independent confirmation became available in the form of paleomagnetic data (cf. Figure 5.1).

However, paleoclimatic indicators may not always provide unambiguous evidence of past climates. Coal, which forms from the remains of vegetation (particularly the woody parts) preserved in anoxic (oxygen-poor) environments such as bogs or swamps, provides a good example of this. The first coals to be studied were clearly the remains of tropical rainforests, and coals found subsequently were similarly interpreted.

- Why do you think we should be wary of using coal as a paleoclimatic indicator of tropical conditions?

- Because coal-forming environments may occur in middle and high latitudes as well as low latitudes. Coal deposits could be the remains of a high-latitude peat bog, or of a temperate forest, growing perhaps along a delta, as well as of an equatorial or tropical rainforest. To be certain of the kind of conditions that led to the formation of a coal deposit, we would need to know the type of vegetation that grew there.

You may have been realizing another problem with paleoclimatic indicators, namely that climatically determined vegetational 'zones' (as seen in Figures 1.4 and 3.4) do not in fact lie strictly east–west. Instead, their distribution is strongly influenced by topography and the sizes and shapes of continents and oceans, and the extent to which these influence – and are in turn affected by – winds and currents. Nevertheless, it is interesting to see whether, and to what extent, the distribution of paleoclimatic indicators over the continents for particular times in the past exhibit a zonal pattern resembling today's climatic belts.

5.1.1 Rearranging the continents

The maps in Figure 5.4b,c show the distribution of another paleoclimatic indicator: **tillites**. Tillites are the fossil equivalent of tills (also referred to as boulder clay) left behind by glaciers as they melt, and so are a fairly unequivocal indicator of cold conditions. They consist of clays in which 'float' a jumble of pebbles and boulders, of all shapes and sizes (Figure 5.4a, below).

Figure 5.4
(a) Photograph of a till, also known as a boulder clay. It consists of pebbles and boulders (in a mixture of sizes, or 'ill-sorted'), suspended in a very fine-grained clay, and has since been eroded to form a cliff. This particular till was deposited by the ice-sheet that covered Britain about 400 000 years ago.

Plate tectonics, carbon and climate

145

(b) ☐ tillites

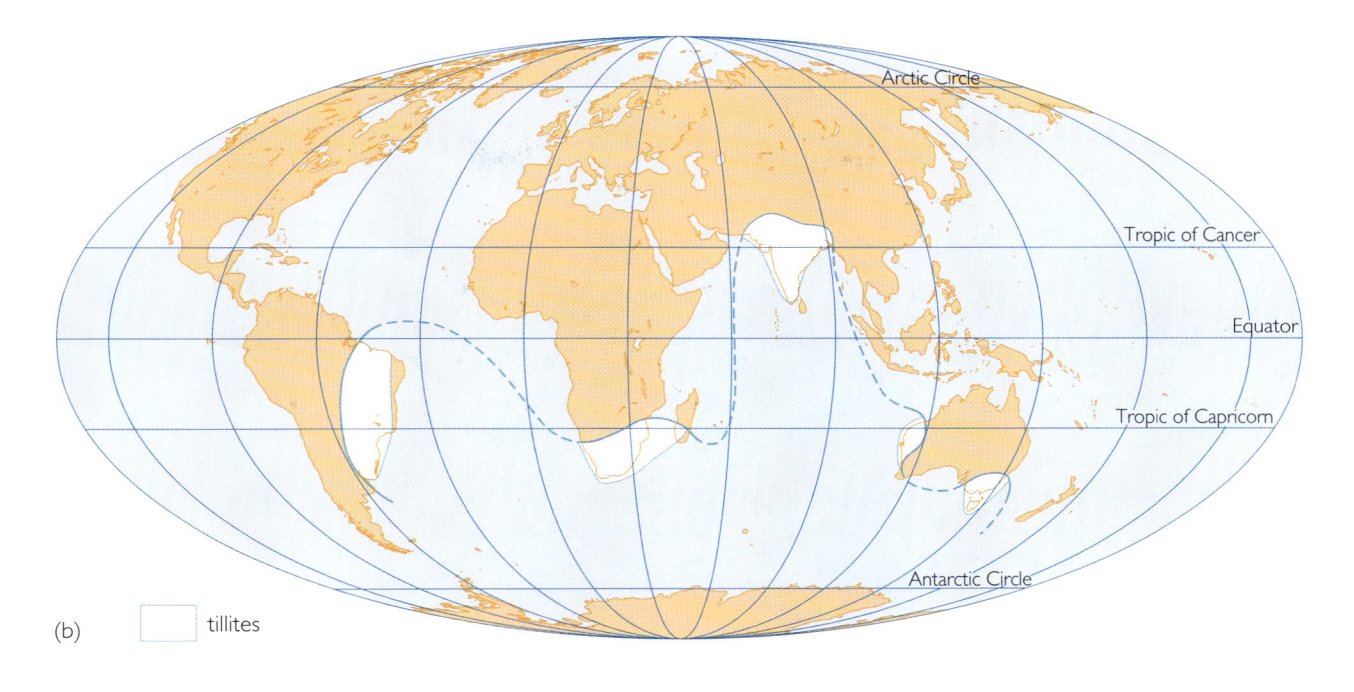

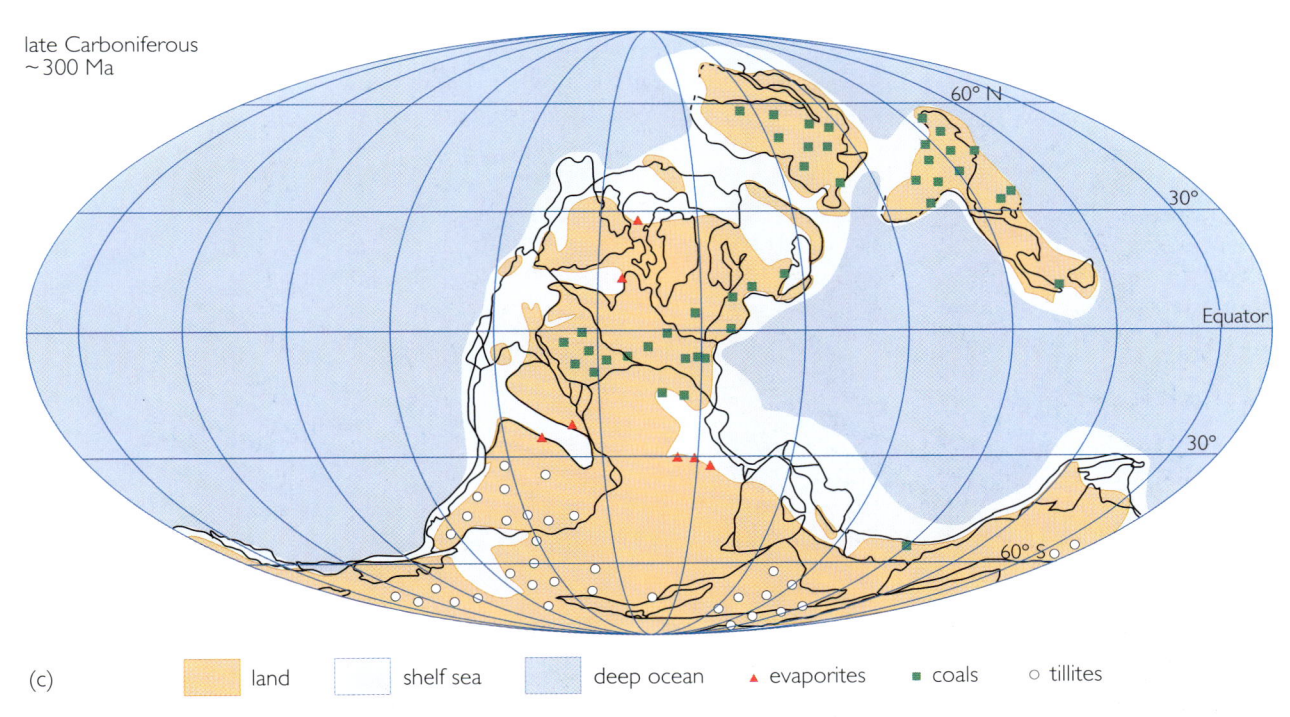

late Carboniferous
~300 Ma

(c) ☐ land ☐ shelf sea ☐ deep ocean ▲ evaporites ■ coals ○ tillites

Figure 5.4
(b) The global distribution of tillites, deposited as tills about 300 million years ago, during the late Carboniferous, plotted on a present-day map of the world. The blue dashed line indicates the extent of an ice-sheet that could have led to such a distribution. (c) The global distribution of tillites and two other paleoclimatic indicators (coals and evaporites) plotted on a map showing the continents in the positions they are believed to have occupied 300 million years ago (as determined by paleomagnetic studies), along with inferred positions of ancient coastlines and shelf seas. *Note:* In the map projections employed in (b) and (c) (often used for showing paleogeographic reconstructions), lines of latitude remain parallel to the Equator.

Figure 5.4b shows the distribution of tillites deposited 300 million years ago, plotted on a *present-day* map of the world. This distribution seems to suggest that there was a huge southern ice-cap, extending northwards across the Equator to India – an indication of the problems faced by geologists before continental drift was an acceptable concept! By re-assembling the continents to their likely positions 300 million years ago (Figure 5.4c), the size of the ice-sheet becomes comparable to ice-sheets which existed in the Northern Hemisphere during the past two million years. Moreover, at low latitudes there was a zone of equatorial/tropical rainforests (indicated by coals), more or less parallel to the Equator and flanked by two hot arid zones (indicated by evaporites).

However, given what you have read about the various influences on the climate system, it will probably not surprise you to learn that there is evidence to suggest that in the past, distributions of climatic belts were sometimes significantly different from what we see today. As an example, Figure 5.5 is a map similar to Figure 5.4c, but for the very late Permian, about 250 million years ago. The distribution of land and sea is not significantly different in the two maps, but the same cannot be said of the paleoclimatic indicators. There are no tillites, suggesting that by this time there were no ice-sheets, even at high southern latitudes, while the distribution of coals is consistent with forests (i.e. high rainfall) at high latitudes in both hemispheres, as well as in the vicinity of the Equator. In contrast, evaporites indicate extensive arid regions in the northern subtropics. Evaporites in the vicinity of the Equator and in southern subtropical regions may also have been extensive, but we cannot know for certain because there is little evidence in the rock record. It must always be borne in mind that paleoclimatic reconstructions like those in Figures 5.4c and 5.5 are necessarily incomplete.

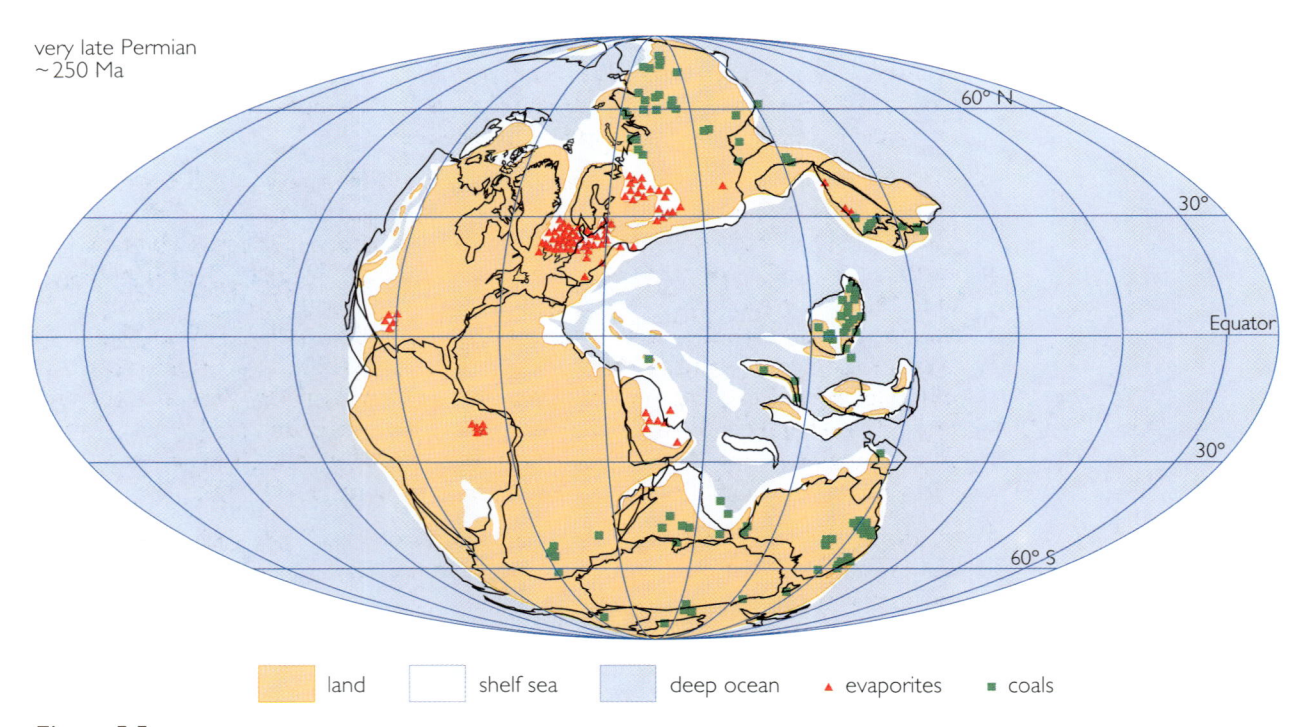

Figure 5.5
The global distribution of the evaporites and coals (cf. Figure 5.4c), plotted on a map showing distribution of continental areas and shelf seas in the very late Permian, ~ 250 million years ago. There are no tillites of this age.

Plate tectonics, carbon and climate

Much of the foregoing has implicitly assumed that the poles will always be colder than low latitudes, and that there were always rain belts in the vicinity of the Equator, dry subtropical regions, temperate zones, and so on. However, it's important to be aware that in relating climatic conditions to latitude in this way we are making an even more fundamental assumption.

Question 5.1
Thinking back to Box 1.1, can you suggest what this assumption might be?

We can expand briefly on the answer to Question 5.1. Because there is evidence of ancient glaciations at low latitudes, some scientists have suggested that in the past the angle of tilt of the Earth's axis has been much greater than it is now.

- Look at Figure 1.8. What would be the implications of the angle of tilt exceeding 45°?

- If the angle of tilt exceeded 45°, the lines of latitude corresponding to the 'Tropics' would lie at higher latitudes than the 'polar circles'. For example, if the angle of tilt was 50°, the 'Tropics' would be at 50° of latitude and the 'polar circles' would be at 90° − 50° = 40° of latitude.

Under these circumstances, seasonal changes in temperature would be much more extreme, especially at the poles where summers of permanent daylight would be much hotter than summers at the Equator. It has been argued that for a tilt greater than 54° climatic zones would effectively be reversed, as would the directions of zonal winds. This is an extreme example, but it is obvious that if the angle of tilt of the Earth's axis were well outside its range of recent millennia (21.8°–24.4°, Section 1.2), the relative widths of climatic belts, and hence the distribution of ecosystems, would be considerably different from those at the present day. Many geoscientists are sceptical about the idea that the Earth's angle of tilt has changed over time, but this possibility cannot be discounted.

So we cannot assume that the distribution of climatic belts observed today has been the same throughout geological time. In particular, the extent – or indeed existence – of ancient ice-sheets cannot be taken for granted. The periods when the Earth has extensive ice-sheets are referred to as Ice Ages. We are presently within an Ice Age, known as the Quaternary Ice Age: there is a large ice-cap over the landmass of Antarctica, another over Greenland, and permanent ice cover over much of the Arctic Ocean. If the present Ice Age is typical, we can assume that within Ice Ages, ice-caps grow and retreat many times, as glacial periods alternate with interglacial periods. At the present time, the Earth's ice-caps are by no means at their maximum extents and the climate even at fairly high latitudes is moderate – in other words, we are currently experiencing an interglacial period.

Over the history of the Earth there have been five or possibly six Ice Ages, three of them in the last 500 million years, and they occur at intervals of hundreds of millions of years.

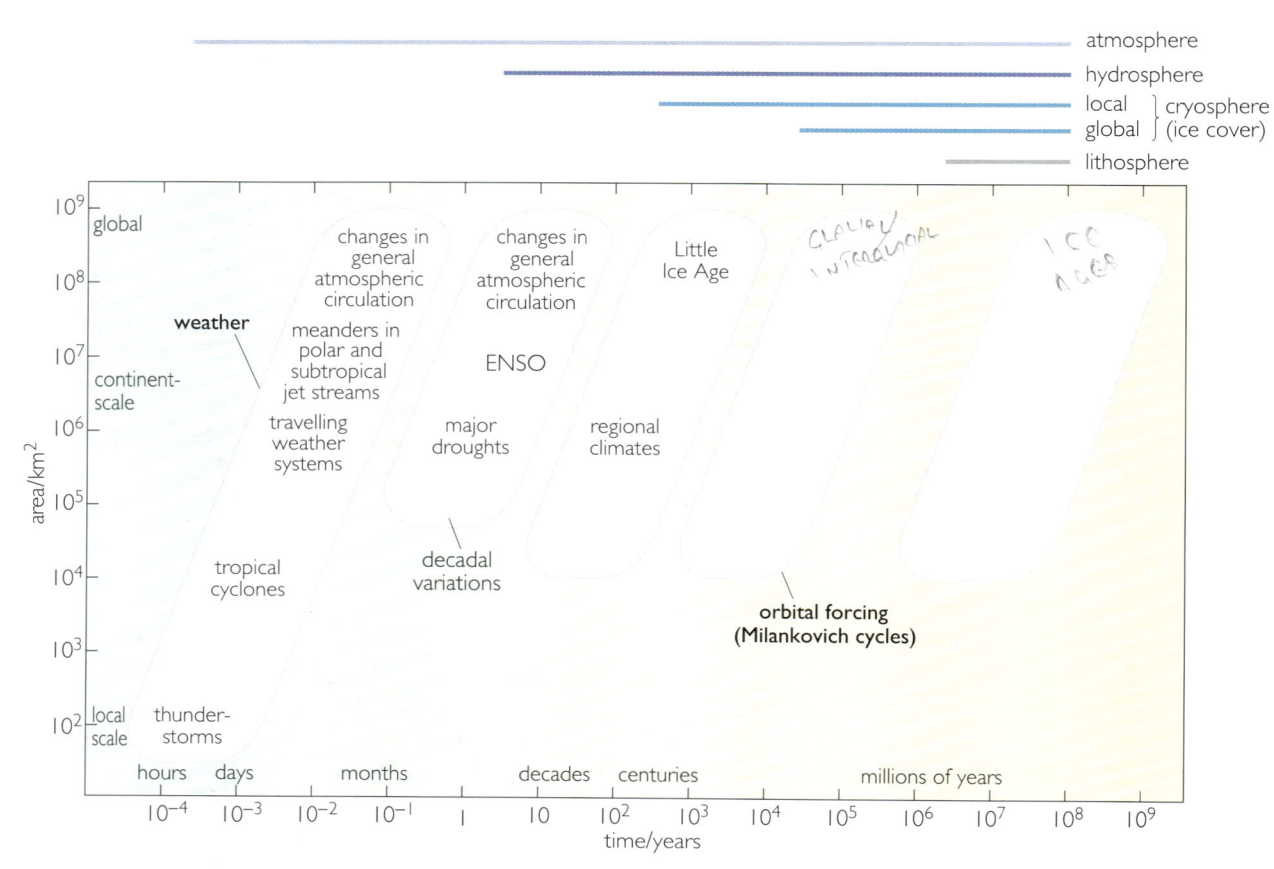

Figure 5.6
Schematic diagram (partly completed) showing the time- and space-scales of various phenomena affecting conditions on Earth. (The continual increase in solar luminosity since the formation of the Solar System is not included.) The bars along the top are to indicate the general principle that while short-term variations are mostly limited to the atmosphere, longer-term variations involve progressively more components of the Earth system.

Question 5.2

(a) Figure 5.6 is a version of Figure 2.41 as completed so far (cf. Figure AA1). Given their frequency of occurrence, where would Ice Ages plot?

(b) Why might this suggest that large-scale changes in the positions of the continents could be at least a contributory cause of the initiation and ending of Ice Ages?

In other words, plate motions would seem to be an obvious candidate for a major forcing factor implicated in the coming and going of Ice Ages, because they act over a similar time-scale. However, plate motions are most unlikely to be the only factor – of themselves they cannot bring about climate change. What they can do, though, is change the distribution and relative positions of continents and oceans in ways that enable other influences to come into play. We can begin our exploration of how the distribution of land and sea might affect climate by looking at models which use greatly simplified situations to simulate conditions that might have obtained in the past, when distributions of continents and oceans were different. Now read Box 5.1.

Plate tectonics, carbon and climate

Box 5.1 Model worlds and their climates

Although contrasting global climates appear to be associated with different continental distributions, there is no way we can prove absolutely that changing continental configurations change global climate. Possible relationships between the two can be investigated through computer simulations, but before we look at that it is worth glancing at the imaginative speculations of Charles Lyell (the 'Father of Geology'). In his book *Principles of Geology*, first published in 1837, Lyell redrew the map of the world to show the present-day continents distributed either in an equatorial/tropical 'ring' (Figure 5.7, top) or in two polar 'caps' (Figure 5.7, bottom). This was long before geologists had begun to think about continental drift, but Lyell's vision was sufficiently far-sighted for him to recognize the consequences of shifting continents relative to climatic belts. It was a prodigious intellectual leap, and it is further evidence of Lyell's genius that he also proposed the corollary: that shifting continents might themselves be an agent of global climate change.

Today, when the complexity of a problem overwhelms us, or our imagination is inadequate, we can turn to computer modelling. Idealized extremes of continent–ocean distribution have been used to investigate the effects of continental configurations on climate by means of a model that simulates the more familiar features of the Earth's climate system. These include seasonal and latitudinal variations of incoming solar radiation, of surface temperatures, of cloud cover, of precipitation and evaporation, and snow and ice cover.

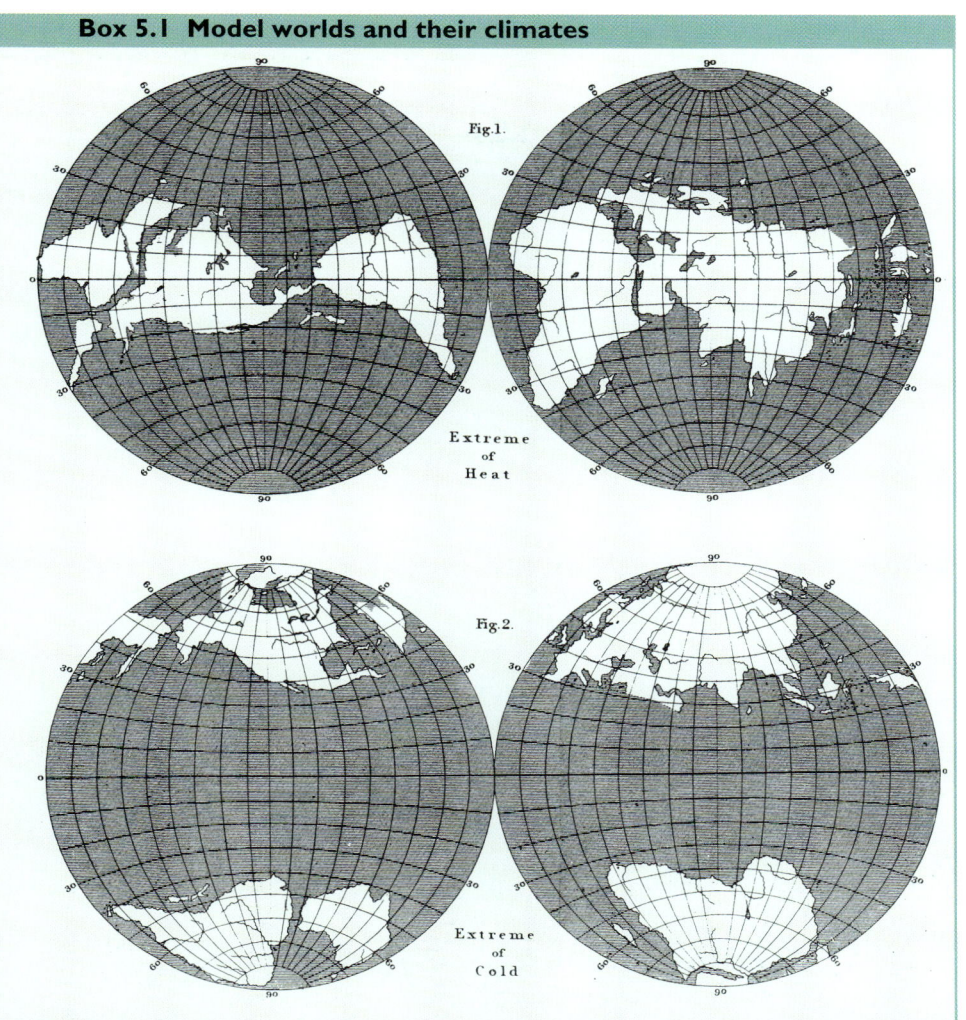

Figure 5.7
Facsimiles of Charles Lyell's maps of present-day continents redistributed to form (top) a 'ring world', with the continents concentrated in the tropics, and (bottom) a 'cap world' with the continents gathered together around the two poles.

The model was run using two idealized continental geometries based on the present-day total land area: two 'caps' of land extending from the poles to 45° of latitude (both with and without ice-caps); and a tropical 'ring' of land extending 17° north and south of the Equator (Figure 5.8a). These idealized continental distributions are nothing like those of today, but the geography of the Earth may have approximated to a tropical 'ring world' 700–600 million years ago, and could have approximated very roughly to a 'cap world' with one polar ice-cap in the late Carboniferous, about 300 million years ago (Figure 5.4c).

(continued overleaf)

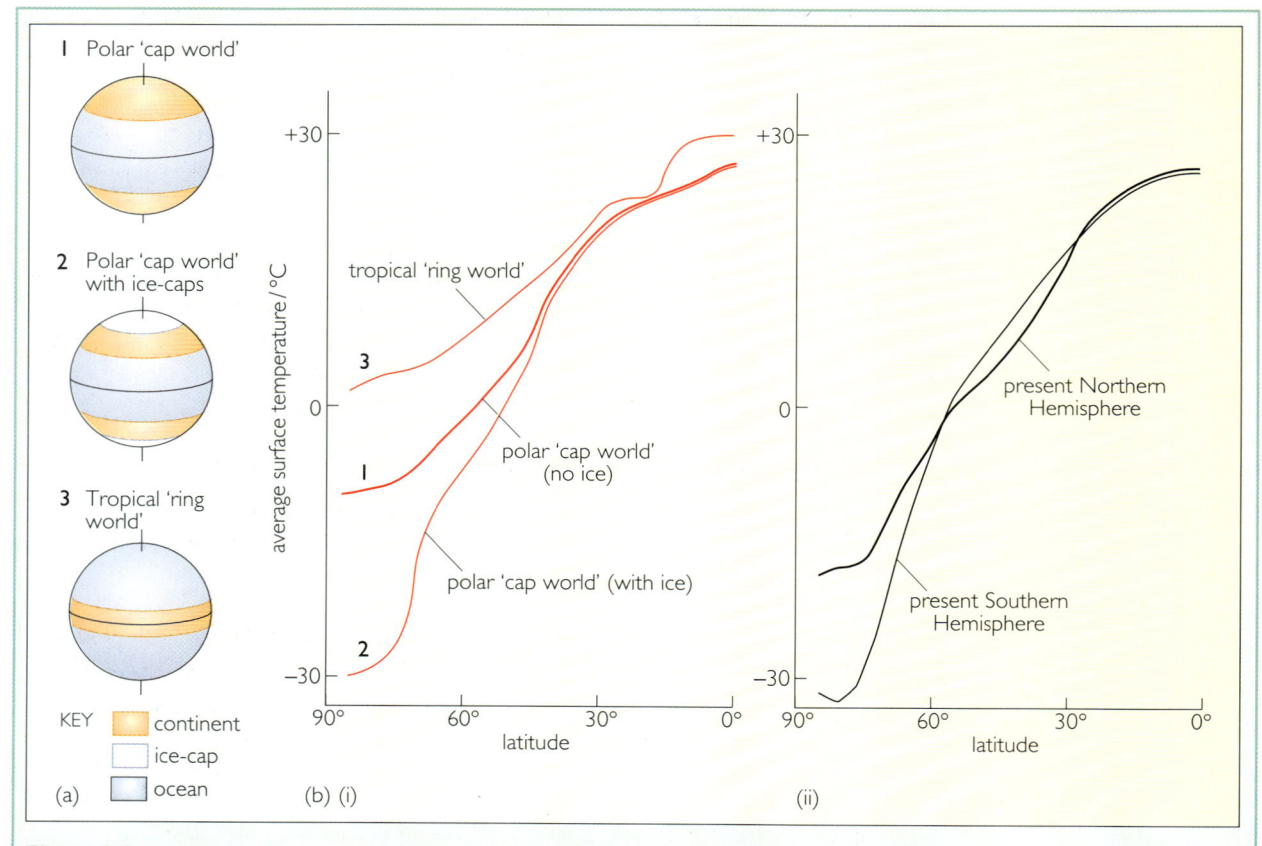

Figure 5.8
(a) Simplified models of the Earth with present-day total land area redistributed to form (1) a polar 'cap world' without ice-caps, (2) a polar 'cap world' with ice-caps extending equatorwards to latitude 70°, and (3) a tropical 'ring world'. (b) (i) Mean annual surface temperatures (zonally averaged), simulated for the cap and ring worlds shown in (a). (ii) Mean annual surface temperatures (zonally averaged), for each hemisphere, simulated for present-day geography.

As you might expect, for a polar cap world with ice-caps (Figure 5.8a, 2) the meridional variation in temperature (Figure 5.8b(i)) looks rather similar to a simulation of the temperature distribution for the present-day Southern Hemisphere (Figure 5.8b(ii)), which is not so different from reality; and both of the cap worlds give average equatorial temperatures close to those of the present day. By contrast, the tropical ring world (Figure 5.8a, 3) is significantly warmer than either of the cap worlds.

- In terms of radiation balance it is easy enough to understand why addition of ice-caps would lead to a cooler world. Why is that?

- Because ice has a much higher albedo than exposed or vegetated continental crust (Table 1.1), and so an 'ice-cap world' reflects more solar radiation back into space than an ice-free cap world.

In summary then, simple models of the kind shown in Figure 5.8a confirm our original supposition that the relative distribution of continents and oceans can influence global climate. Other configurations have also been modelled, including a so-called 'slice world' (Figure 5.9), in which the continents are arranged meridionally, a distribution approximating to that on the present-day Earth. We shall not consider these particular models further because they ignore important aspects of the Earth's climate system.

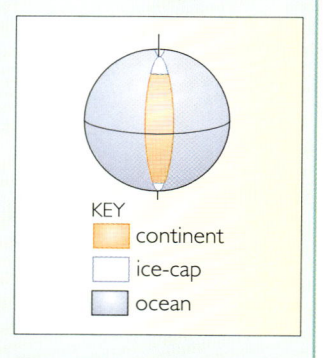

Figure 5.9
A 'slice world', showing the longitudinal configuration of one of the continental areas and parts of two polar ice-caps.

As demonstrated in Box 5.1, computer models can be useful in helping us to isolate the effects of different variables (in this case, the distributions of continents, ice-caps and oceans), while keeping other variables (e.g. the total continental area, total incoming solar radiation) constant. However, it is important to remember that (1) model results consistent with reality do not *prove* that the modelled changes are the causes of the situation observed in reality; and (2) computer models are only as good as the information put into them.

In this case, two important components of the climate system *not* incorporated into the model are: (1) heat transport by surface currents and by the deep thermohaline circulation (Sections 2.2.2 and 2.2.4); and (2) changes in the concentrations of greenhouse gases in the atmosphere. We will come to the second of these factors later in the chapter, but first let's examine how ocean currents might be affected by changes in the global distribution of continents and oceans.

5.1.2 Continental drift, ocean currents and climate change

Cap, ring, and slice worlds (Figures 5.8a and 5.9) are very simplistic representations of the Earth, but they do help us to see how continental configurations might influence current patterns and hence climate.

- Which configuration – cap world, ring world or slice world – would affect the surface current pattern so as to intensify cold conditions in polar regions?

- Cap world, where currents carrying heat from lower latitudes would be unable to penetrate to very high latitudes, with the result that there would be a strong temperature contrast between equatorial and polar regions.

In a cap world, the polar regions would be thermally isolated, making the development of polar ice-caps more likely: strong eastward currents, comparable with today's Antarctic Circumpolar Current (Figure 2.21), could flow around the polar continents under the influence of westerly winds, further isolating the polar continents from warm currents flowing from low latitudes.

- Look carefully at Figure(s) 2.21 and/or 2.43. In today's world, are northern polar regions similarly isolated?

- No, but they are nevertheless largely cut off from warm currents flowing from lower latitudes. The Arctic Ocean is almost completely surrounded by land, with the result that the only warm water penetrating the region is the North Atlantic Drift (the downstream extension of the Gulf Stream) which flows northwards through the Norwegian and Greenland Seas.

By contrast, in a tropical ring world (Figure 5.8a, 3), ocean circulation between tropical and polar regions would be unimpeded, and so heat could be transferred from mid-latitudes to high latitudes, resulting in lower meridional temperature contrasts.

Figure 5.10 shows three more configurations of oceans and continents, still very simple but slightly more realistic than those shown in Figures 5.8 and 5.9. In the past, the configuration of oceans and continents has been very different, so it is worth using these simple models to aid our thinking about how current patterns may affect the distribution of temperature over the Earth's surface. Try Question 5.3, and begin by reminding yourself about some fundamental features of surface current patterns.

Question 5.3

(a) On the basis of what you know about winds and currents on the real Earth (Figures 2.7, 2.21 and 2.43), you could yourself have sketched surface current patterns similar to those shown in Figure 5.10. Identify *at least three* features of the circulatory pattern, visible in all three configurations, that are unavoidable consequences of fundamental characteristics of the rotating Earth and its fluid envelopes.

(b) Would you expect the temperature distributions over the oceans in Figure 5.10 to be symmetrical, with temperatures along the western boundary the same as those along the eastern boundary at similar latitudes?

(c) In which of the three configurations do you think the ocean would be warmest overall?

Of course, discussing the transport of heat around the Earth by means of currents alone is unrealistic – we are ignoring the redistribution of heat by winds, including the effects of evaporation, transport and condensation of water vapour. Nevertheless, the high specific heat of water (Box 2.1) means that heat transport in the ocean is an extremely important influence on climate.

Implicit in the discussion of Figure 5.10 is the effect of *gateways* on the pattern of surface and deep ocean currents. Gateways are gaps between the continents that permit significant longitudinal or latitudinal connections to be made between oceans. Their opening and closing, and the resulting changes in heat transport, can cause climate for a particular landmass to change much more rapidly than the slow drift of the continent across climatic belts.

Mention of deep currents should remind you that we have not yet considered this extremely important aspect of the oceanic heat transport system. The effect of wind-driven surface currents in transferring heat from low to high latitudes is reinforced by the transport of cold water away from polar regions in the density-driven thermohaline circulation (Sections 2.2.2 and 2.2.4). It may be facilitated by the opening of gateways in the deep ocean, or impeded by the development of topographic barriers there. At the present day, mid-ocean ridges interfere with the flow of deep and bottom waters, but barriers and gateways may also result from crustal uplift or volcanic activity on the one hand, and subsidence related to lithospheric plate movements on the other.

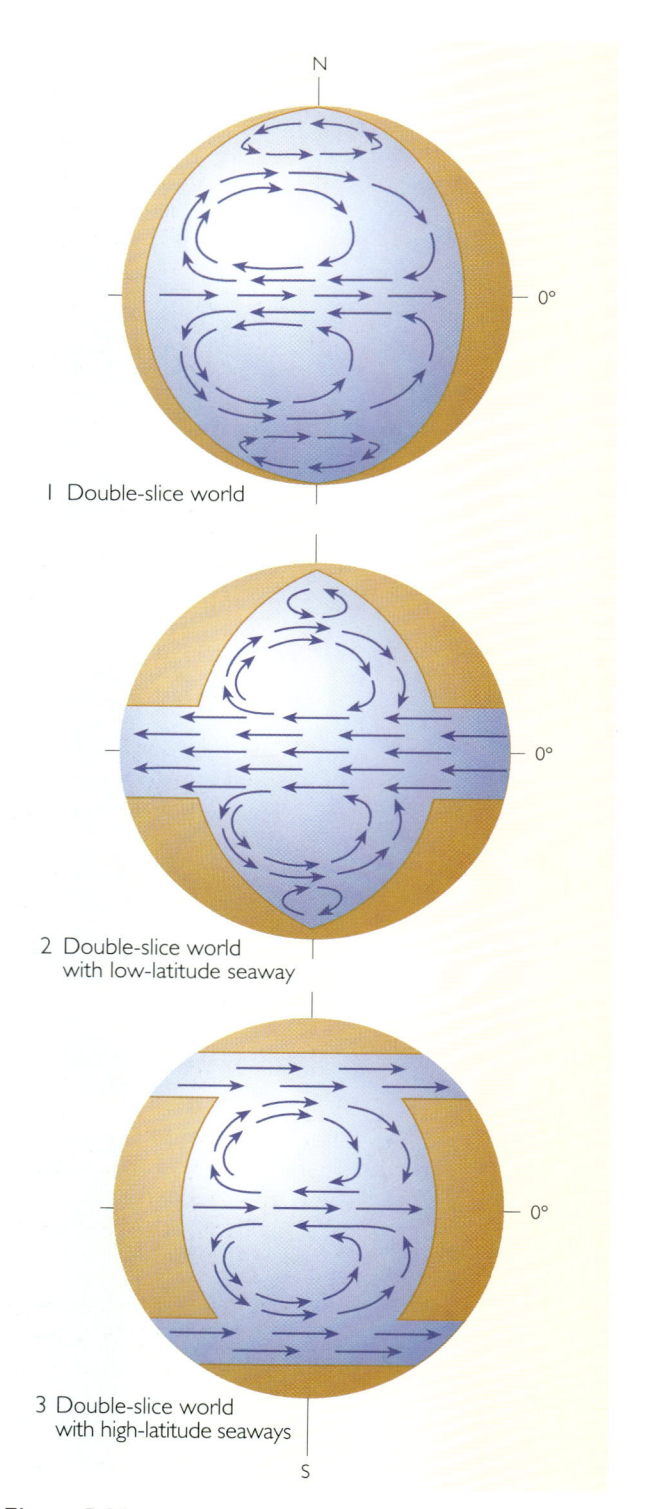

1 Double-slice world

2 Double-slice world with low-latitude seaway

3 Double-slice world with high-latitude seaways

Figure 5.10
Three simple configurations of oceans and continents, with appropriate generalized surface current patterns: (1) a 'double-slice' world, with two continents and two oceans (only one visible here); (2) as (1), but with open ocean around the globe at low latitudes; (3) as (1) but with open ocean around the globe at relatively high latitudes.

Plate tectonics, carbon and climate

The break-up of Pangea

Having used our simple models to consider some possible effects of different continental configurations on ocean circulation and global climate, let us now consider how events in the break-up of the supercontinent known as Pangea (Greek for 'all land') might have influenced climate change. Pangea formed some 250 million years ago, when what is now North America, most of Europe, and Asia came together with the pre-existing southern supercontinent (comprising South America, Africa, India, Antarctica and Australia) which we now refer to as Gondwanaland. Gondwanaland had been extensively glaciated about 50 million years previously (i.e. 300 million years ago, Figure 5.4c), but by 250 million years ago the Earth was probably already beginning to warm up (Figure 5.5), and by 100 million years ago it was a 'greenhouse' planet some 10 °C warmer (on average) than today. Subsequently, it began to cool again, albeit gradually, to arrive at its present 'icehouse' state.

Before Pangea began to break up, the supercontinent must have been accompanied by a superocean. This superocean – which has been named Panthalassa ('all ocean') – extended from the North Pole to high southern latitudes (about 50°–60° S), and must have extended for some four-fifths of the Earth's circumference around the Equator. We have no reason to suppose that the factors determining the global wind and surface current systems were any different from what they are today, and can therefore propose some likely wind and current patterns. If present-day oceanic current patterns are any guide, there could have been a number of linked subtropical gyres in each hemisphere (the present-day Pacific Ocean has a more complex gyral system than the narrower Atlantic, Figure 2.21), and presumably also a number of subpolar gyres.

- Is there any evidence in Figure 5.5 to suggest that at low latitudes the western and eastern sides of Panthalassa had contrasting climates?

- Not a great deal, but Figure 5.5 suggests that near the Equator, evaporites – implying arid conditions – tended to be found on the eastern side of the ocean (i.e. the western side of the continent), and coals – implying warm humid conditions – tended to be found in the west (in a location analogous to that of Indonesia in the Pacific at the present day).

This is not unlike the situation in present-day low latitudes, where the Trade Winds blow across the tropical oceans, carrying moisture-laden air westwards. Around the Pacific, in particular, rainforests are relatively more abundant in the west, while the American coastline to the east tends on the whole to be drier (cf. Figures 2.42 and 3.4).

Figure 5.4c suggests that the vast extent of Gondwanaland in the south could well have isolated high southern latitudes from the influence of warm currents – and we know that a south polar ice-cap existed during the late Carboniferous. Although there was little to obstruct heat transport to northern high latitudes (Figure 5.4c), there could have been a seasonally ice-covered Arctic Ocean at that time also. As discussed in Section 5.1, the presence of late Carboniferous coals at high northern paleolatitudes does not necessarily mean that the vegetation from which they formed was tropical.

Figure 5.11 shows a series of maps which illustrate successive stages in the break-up of Pangea over the past 175 million years. On the basis of what we know about global winds, and the effects of landmasses on winds and currents, we can not only infer surface current patterns but also propose regions where deep and bottom water masses might have formed.

154　　　　　　　　　　　　　　　　　　　　　　　　　　　　　　　　　**The Dynamic Earth**

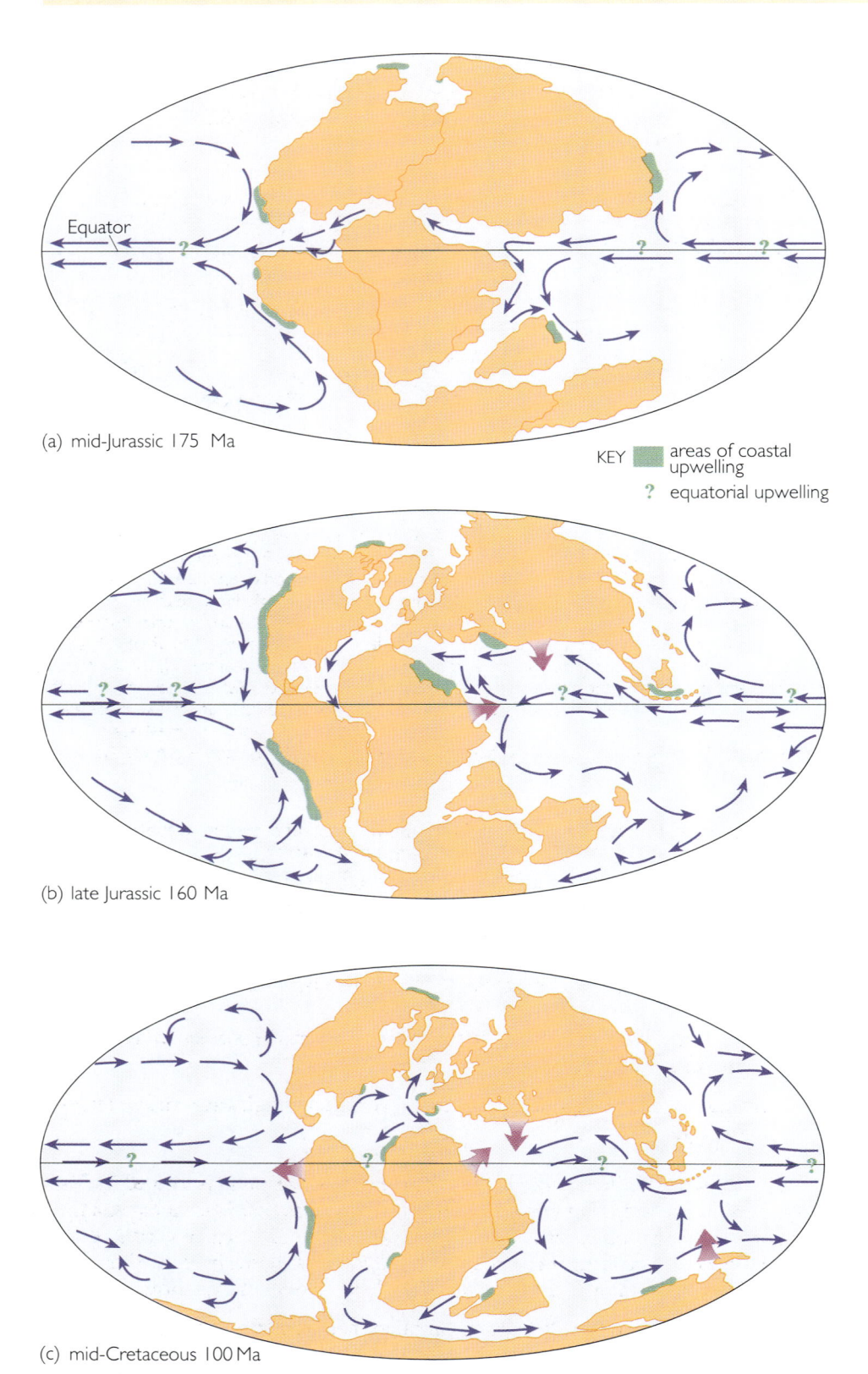

(a) mid-Jurassic 175 Ma

KEY: ▇ areas of coastal upwelling　　? equatorial upwelling

(b) late Jurassic 160 Ma

(c) mid-Cretaceous 100 Ma

Plate tectonics, carbon and climate

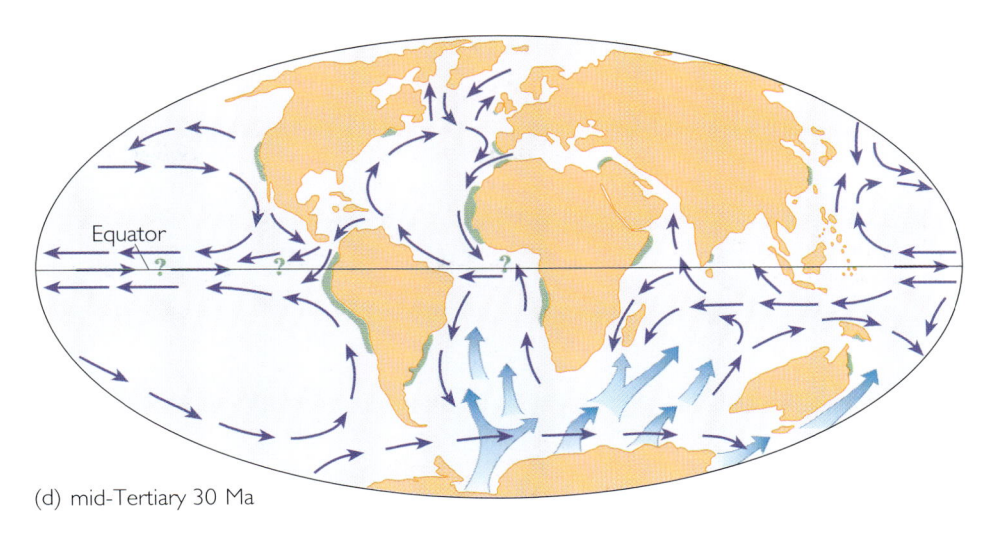

(d) mid-Tertiary 30 Ma

Figure 5.11
Maps showing the changing distribution of the continents and changing surface current patterns during the break-up of Pangea (no attempt has been made to show actual coastlines or shelf seas). Likely areas of upwelling (discussed later) are shown in green. In (b) and (c), pink arrows indicate locations where warm, saline water masses may have flowed down into the ocean from shallow evaporating basins. (a) 175 million years ago, there was no Atlantic Ocean, only the great ocean Panthalassa. (b) By 160 Ma, the North Atlantic was a long narrow ocean but the South Atlantic had barely begun to open. (c) By 100 Ma, the opening of the 'Straits of Gibraltar' and the submergence of 'central America' had provided a low-latitude seaway. (d) By 30 Ma, the oceans were approaching their present configuration. Gyres were now established in the North and South Atlantic, and currents could flow unimpeded around Antarctica under the influence of westerly winds. The climate was by now cooler, and cold bottom waters formed near Antarctica and flowed north into all the ocean basins – the equivalent of today's Antarctic Bottom Water.

The break-up of Pangea began about 200 million years ago, and an equatorial gateway started to open within the supercontinent. By 175 Ma, in Jurassic times (Figure 5.11a), circum-equatorial flow was blocked only by a relatively narrow isthmus at what is now Gibraltar, and the ancestral Mediterranean (known as the Tethys Ocean) had formed. Surface flow is shown as westwards on both sides of this isthmus in the reconstruction. It is possible that to the east of 'Gibraltar' the return flow to the east was at depth, in the form of a dense water mass.

- In a relatively warm Earth, at low latitudes, how might surface water be made sufficiently dense to sink?
- By high rates of evaporation, which would remove freshwater and so increase the salinity.

In fact, in today's Mediterranean a warm saline deep water mass is formed in this way, aided by some cooling in winter. It is thought that by 160 Ma, similar saline but relatively warm water masses could have been forming in shallow coastal basins, subject to high rates of evaporation, in the areas shown in Figure 5.11b,c. At this time, there was a land bridge between North and South America.

By the early Cretaceous, the 'Straits of Gibraltar' had opened and rising sea-levels had caused 'central America' to become submerged, providing a shallow-water gateway. Circum-equatorial currents could now flow unimpeded round the Earth, and by the mid-Cretaceous (100 million years ago, Figure 5.11c), this seaway was quite wide.

■ Bearing in mind the discussion related to Figure 5.10, how could the presence of this low-latitude seaway have affected the average temperature of the Earth?

■ Surface currents could flow uninterrupted around the globe at low latitudes, being warmed by the high levels of solar radiation. This very warm water would eventually be transported to higher latitudes by the subtropical gyral systems. As a result, the whole ocean, and hence the globe, would, on average, be warmer than it would otherwise have been.

Ocean circulation patterns became more complicated as the Atlantic and Indian Oceans opened, but it would seem that by the mid-Tertiary (30 million years ago, Figure 5.11d) much of the surface circulation resembled the pattern we see today. Closure of the Tethys Ocean was well advanced, shutting off the equatorial seaway. Submergence of the Tasman Ridge had opened a gateway between Australia and Antarctica about 40 million years ago; this had reduced flow between the Pacific and Indian Oceans but allowed surface currents to flow around Antarctica. Opening of the deep-water channel between the southern end of South America and Antarctica (today's Drake Passage) at around 25 Ma, resulted in the circumpolar circulation we know today. As mentioned above, the Antarctic Circumpolar Current isolates Antarctica from poleward transport of warm water; its initiation is generally considered to have led to significant cooling of Antarctica, and the growth of the southern polar ice-cap.

The equatorial gateway between the Atlantic and Pacific Oceans closed relatively recently, only about 3 million years ago, when central America emerged above sea-level (so providing a land bridge linking the long-isolated mammalian fauna of North and South America). Equatorial flow would no longer have been continuous from the eastern Atlantic to the western Indian Ocean; closure may well have contributed to cooling in the Northern Hemisphere and the initiation of Arctic glaciation. In fact, cooling of Antarctic waters began as long ago as 65–60 Ma or so (cf. Figure 5.12), possibly in response to closure of the 'Tethyan gateway' at Gibraltar. Temperatures roughly stabilized from then until the early Miocene, when the Drake Passage opened and Antarctica became isolated.

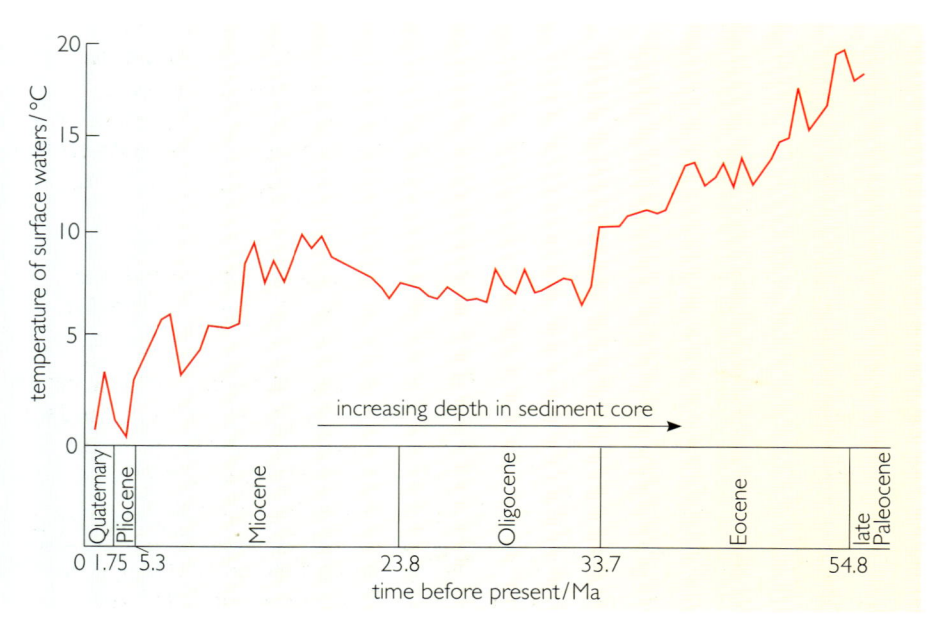

Figure 5.12
Surface water temperatures in the vicinity of Antarctica during the past ~ 60 Ma, as determined from oxygen isotope ratios of the remains of planktonic foraminiferans.

Plate tectonics, carbon and climate

In our study of Figures 5.11 and 5.12, we have reached the Ice Age affecting the Earth at the present time. Remember that the reason why northern Europe and America are not buried beneath giant ice-caps is that we are currently enjoying the relative warmth of an interglacial period. Before turning to the alternation of glacials and interglacials, however, we should make a brief digression to remind ourselves that changing climate is only one of the factors affecting life on Earth.

Plate tectonics and life

Plate-tectonic processes can have more direct effects than simply those of moving continents relative to climatic belts, transporting species across the globe or causing them to become isolated. Over relatively short periods of time they can also alter the relative proportions of *types* of environment available as living space.

In considering how ocean circulation has been affected by changing continental configurations, we omitted an aspect of great importance for life, namely upwelling.

- Why is upwelling of particular importance for life? And how might we be able to deduce where upwellings occurred in the distant past?

- Upwelling is important for life as it brings nutrient-rich water up into the photic zone where it can support high levels of primary production – i.e. large phytoplankton populations – and hence, directly and indirectly, other organisms. We could deduce where it might have occurred in the past from our understanding of how the positions of the continents affect the wind field (Figure 2.18), and from the distribution of organic remains in marine sediments. High productivity results in large fluxes of organic debris to the sea-bed, some of which may be preserved in the sedimentary sequence.

Areas where coastal upwelling may well have occurred in the Jurassic and mid-Cretaceous are shown in green on Figure 5.11a–c. Those along the western sides of the 'American' continents would have resulted from the same equatorward winds (the equivalent of the present-day Trade Winds) that drove the currents along the eastern side of the gyres (cf. Figure 2.27a).

Now let's move from the open ocean to the continental shelf. Look at Figure 5.13, which shows how the numbers of families in nine phyla of invertebrate shelf-dwelling animals have varied since the beginning of the Cambrian, ~ 545 million years ago. Compare the shape of the curve with the diagrammatic representation of changes in the numbers of continents over the same period. Note that stages D to F correspond to the coming together and break-up of Pangea, shown in Figures 5.4c, 5.5 and 5.11.

- Is the information in Figure 5.13 consistent with the hypothesis that the global diversity of shelf-dwelling animals is related at least in part to the availability of shores and shallow seas?

- Yes. In general, the greater the number of continental fragments, the greater the total length of shoreline available for marine animals to colonize, and the greater the number of separate, isolated environments in which endemic coastal faunas could evolve. The number of families increased from Cambrian to Ordovician times as an ancient supercontinent broke up (A, B), remained more or less constant until Pangea began to be assembled (C, D, E), and decreased dramatically during the Permian, when the total length of shoreline was at a minimum. Thereafter, as Pangea fragmented (F, G), diversity rose sharply once more.

Figure 5.13
Changes in numbers of families in nine phyla of benthic shelf-dwelling invertebrate animals with hard parts, and changes in number of continental fragments (shown diagrammatically), from the Cambrian (~ 545 million years ago) to the present. Positions of letters indicate approximate times when the continents had the numbers of fragments shown. Stages D to F correspond to the coming together and break-up of Pangea, discussed earlier.
This diagram dates from 1970; a similar diagram published today would show even more dramatic changes with time, particularly as far as the increase in numbers of families over the past 250 Ma is concerned.

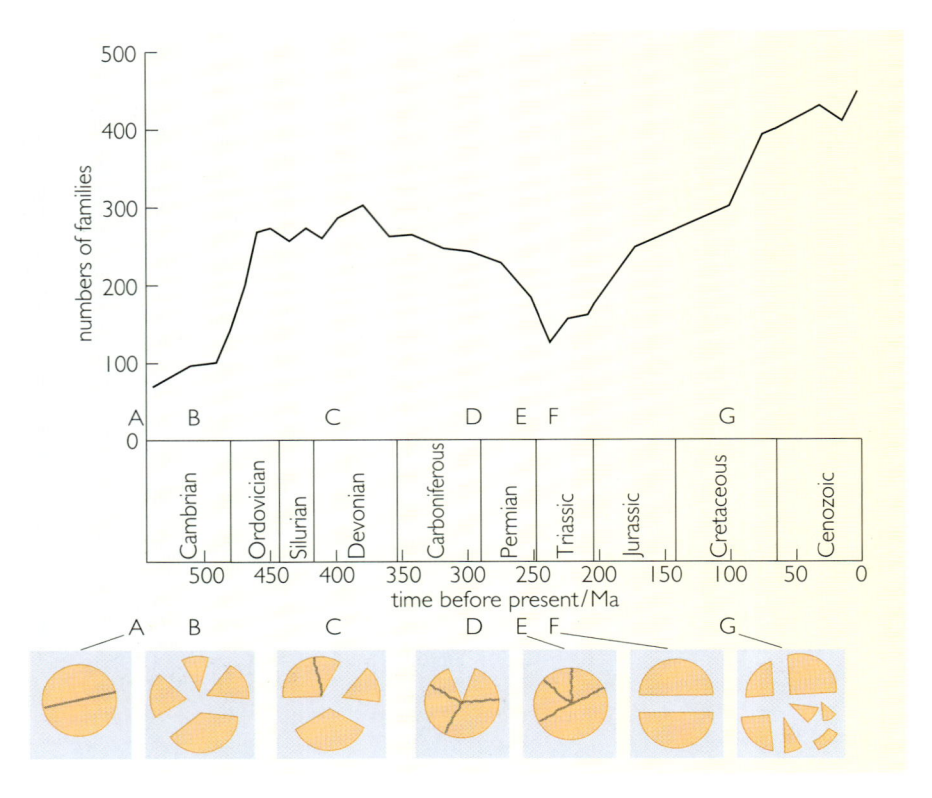

Clearly, the correlation is a fairly general one and shoreline length is only one possible factor in determining diversity of shelf-dwelling organisms. It would be ludicrous, for example, to attribute the great Permian extinction of species (responsible for the dip at F) mainly to this cause, though some scientists think it could have been a contributory factor.

5.2 Glacials and interglacials

Figure 5.14a shows how the volume of ice in Northern Hemisphere ice-caps and glaciers has varied over the past 600 000 years (left-hand axis), which gives a good indication of the variation in sea-level over the same period (right-hand axis).

■ Why is sea-level related to ice volume?

■ Water evaporated from the Earth's surface and transported polewards may fall as snow at high latitudes; it is the accumulation and compaction of this snow that leads to the growth of ice-caps. When ice-caps are growing, more and more water is being removed from the oceans, and is prevented from returning to the oceans until the ice melts.

The figure shows that glacial periods last for tens of thousands of years, at most about 100 000 years, during which time the water remains locked in ice-caps and sea-level is lower.

The ice volumes (and sea-levels) shown in Figure 5.14a were determined indirectly from the ratios of oxygen isotopes in the shells of marine organisms that have accumulated in deep-sea sediments. The marine organisms concerned are usually foraminiferans (cf. Figure 3.16c), and are obtained from drill cores of deep ocean floor sediments. To see why oxygen-isotope ratios can be used in climate studies, see Box 5.2.

Plate tectonics, carbon and climate

Figure 5.14
(a) The variation in the amount of ice in Northern Hemisphere ice-caps and glaciers, and accompanying changes in global sea-level, over the past 600 000 years. As ice-sheets grow and retreat, so ice is removed from or added to the ocean. As explained in the text, the curve is actually the variation in $\delta^{18}O$, a measure of the oxygen-isotope ratio.
(b) Computed curve showing the variation in incoming solar radiation in summer at northern latitudes (65°N) over the past 600 000 years. (This is the same plot as shown in Figure 1.14.)

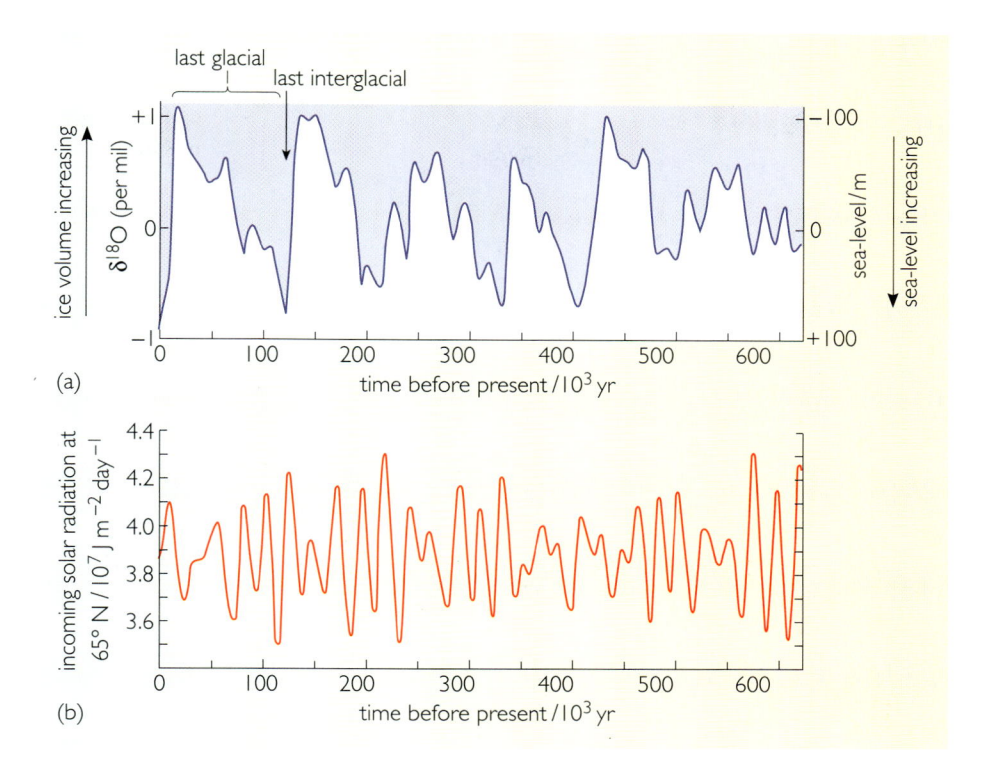

Box 5.2 Oxygen isotopes and the climate record

Oxygen has three stable isotopes with relative atomic masses 16, 17 and 18. Over 99% of natural oxygen is made up of ^{16}O, with most of the balance being ^{18}O. Water that evaporates from the ocean eventually condenses as cloud and falls as rain or snow (Figure 2.34). When seawater evaporates from the ocean, water molecules with the lighter oxygen isotope ($H_2{}^{16}O$) evaporate more readily, so atmospheric water vapour is relatively enriched in the lighter isotope. When water vapour condenses and is precipitated back into the ocean, the heavier isotope ($H_2{}^{18}O$) condenses preferentially. Both processes deplete water vapour in the atmosphere in $H_2{}^{18}O$ relative to $H_2{}^{16}O$. When ^{18}O-depleted water vapour is precipitated as snow in polar regions, the snow will also be depleted in ^{18}O relative to the oceans. The larger the ice-caps, the higher the relative proportion of ^{18}O in seawater and the lower the relative proportion of ^{18}O in ice-caps.

Marine organisms which form hard parts (shells or skeletons) of calcium carbonate incorporate different proportions of ^{16}O and ^{18}O according to the temperature: the lower the temperature, the greater the $^{18}O:{}^{16}O$ ratio in the calcium carbonate secreted. Although all organisms secreting calcium carbonate have higher $^{18}O:{}^{16}O$ in cold than warm water, the actual ratio for a particular temperature depends on the species concerned and the $^{18}O:{}^{16}O$ ratio of the water it is living in.

The fossils in sediment cores used for oxygen-isotope analysis are usually micro-organisms with calcium carbonate shells, often foraminiferans (cf. Figure 3.16c). The amount of ^{18}O in their shells is very small, but it can be measured accurately by mass spectrometry. The result is not given as a simple ratio, but as a delta (δ) value, which is determined by comparison of the sample with a standard, and results in a value expressed in parts per thousand (‰ or 'per mil'):

$$\delta^{18}O = \frac{(^{18}O/^{16}O)_{sample} - (^{18}O/^{16}O)_{standard}}{(^{18}O/^{16}O)_{standard}} \times 1000$$

(Equation 5.1)

Generally, the standard used nowadays is seawater which has a $\delta^{18}O$ value close to zero. Snowfall in polar regions has $\delta^{18}O$ values of −30‰ to −50‰, the negative values indicating depletion of ^{18}O. The higher (or less negative) the measured $\delta^{18}O$ value in marine fossils, the greater the enrichment of ^{18}O in seawater and the larger the ice-caps on land at the time the organisms were alive.

$\delta^{18}O$ values may be determined for the hard parts of both planktonic and benthic species of foraminiferans.

▪ Bearing in mind that oxygen-isotope ratios are affected by the temperature of the water in which an organism lived, which would give the most reliable estimates of sea-level changes over the course of glacials and interglacials — planktonic foraminiferans or benthic foraminiferans?

> - Benthic foraminiferans, because the temperature variation of the cold bottom waters (Figure 2.29) is less than that of surface waters.
>
> To summarize, the $\delta^{18}O$ value of remains of foraminiferans, particularly of benthic species living in cold bottom waters, can be taken as a measure of the amount of water held in ice-sheets at any given time, and hence as an indicator of global sea-level. $\delta^{18}O$ values of benthic foraminiferans are thus the 'imprints' of ancient water masses, referred to at the end of Section 2.2.4. As far as sea-level is concerned, it has been calculated that a change in $\delta^{18}O$ of 0.1‰ (0.1 per mil) is equivalent to a change in sea-level of ~10 m.
>
> However, useful information about past climate can also be obtained from planktonic foraminiferans, particularly those living at very high latitudes where seasonal temperature variations are quite small. Their $\delta^{18}O$ values reflect changes in ice volume *and* global surface temperature; and, of course, when global temperatures are lower, ice-caps are larger.
>
> - What change of sea-level would have been accompanied by a *rise* in $\delta^{18}O$ of 0.25‰?
> - A fall in sea-level of $2.5 \times 10 = 25$ m.

If you found Box 5.2 difficult to follow, the essential points to remember are these: larger $\delta^{18}O$ ratios in marine fossils indicate more enrichment of the ^{18}O isotope in seawater, which in turn indicates *larger* ice-caps, *lower* temperatures and *lower* sea-levels.

Study Figure 5.14a. Can you see that the successive glacial 'peaks' (or interglacial 'troughs') occur every 100 000 years or so?

Question 5.4
Plot 'glacial/interglacial' onto Figure 5.6, assuming that a large part of the globe is affected by such climatic fluctuations. What does the result of your plot tell you? Can you be even more precise about the particular forcing factor involved?

Figure 5.14b is a repeat of Figure 1.14, showing how the intensity of summer sunshine has varied over the past 600 000 years (cf. answer to Question 5.4). There is a clear correlation between high levels of incoming solar radiation in Figure 5.14b and high sea-level (low ice volume) in Figure 5.14a. What's more, the graphs indicate that the decline in the ice-sheets every 100 000 years or so occurs when northern summer sunshine levels are rising, despite the fact that winter sunshine levels must be falling. This indicates that the growth and decay of ice-caps is determined by how warm it gets in summer, rather than how cold it gets in winter (cf. Croll's contrasting view, discussed in the answer to Question 1.6a(ii)).

Figure 5.15 extends the fluctuations of global sea-level back to about 1.6 Ma, and provides dramatic evidence that the interval between successive warm and cold phases (not necessarily glacials) has not been constant during that time.

- According to Figure 5.15, how long ago did the eccentricity cycle become the dominant control on the frequency of glaciations?
- About 600 000–700 000 years ago. Before that, the frequency seems to have been greater – and the amount by which sea-level fluctuated was rather less.

The record in Figure 5.15 suggests that between 1.5 and 1 Ma ago the 40 000-year tilt cycle was the dominant factor determining climate change; then, between one million and ~600 000 years ago there was a transition to the 110 000-year eccentricity cycle, which has dominated since. The record is more equivocal before 1.5 Ma. Clearly, there is still much to be learned about the astronomical forcing factors that have controlled the Earth's climate during our current Ice Age.

Plate tectonics, carbon and climate

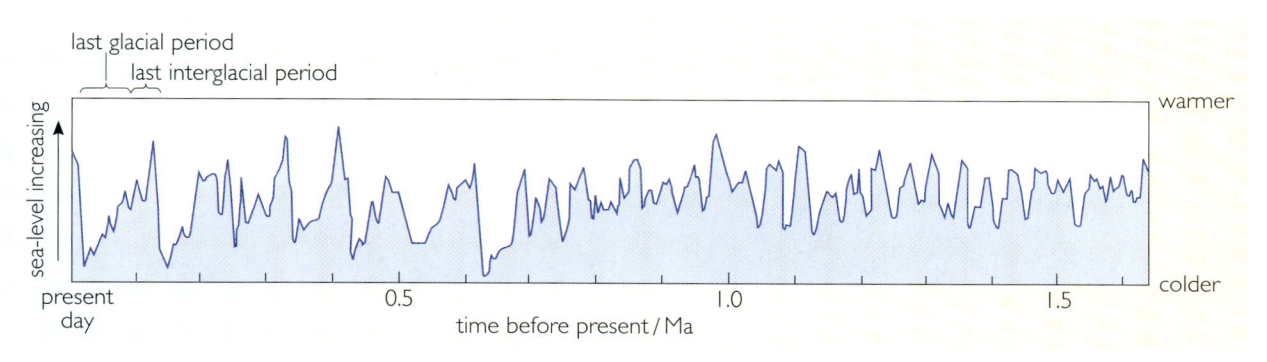

Figure 5.15
The global sea-level record extended back to about 1.6 Ma, as determined from oxygen-isotope ratios of marine microfossils in sediment cores. Note that the left-hand part corresponds to the plot in Figure 5.14, but with the axes plotted so that sea-level (rather than ice volume) increases upwards. The shape of the plot is not identical, however, as oxygen isotope measurements from different sediment cores give slightly different results.

Sea-level fluctuations between glacial and interglacial periods are a consequence of changes in global ice volume. The fluctuations have been on the order of several tens up to about 150 metres, with – as we have seen – a time-scale that has ranged from 40 000 to 100 000 years. There have been much greater sea-level changes in the more distant geological past, however, on much longer time-scales – millions to tens of millions of years. Such variations cannot possibly be caused by growth and decay of ice-caps, not least because they occur during periods of geological time when the poles were virtually if not entirely ice-free. So what causes them? The next section addresses this question.

5.3 Sea-level change: causes and consequences

Nowadays, it is general knowledge that sea-levels are rising world-wide because of global warming, but it is often forgotten that – even in the absence of 'greenhouse' warming – this would be occurring anyway as a result of climatic warming during the present interglacial. The warming is causing sea-level to rise for two reasons: it is causing continued melting of glaciers and ice-caps, and it is causing thermal expansion of the upper layers of the ocean (i.e. above the thermocline, Figure 2.29). However, the evidence of present-day sea-level change is often contradictory, and at first sight inconsistent with our general knowledge that sea-level is rising all over the world.

Around the coasts of Britain, there is abundant evidence indicating past sea-level changes (Figure 5.16). Submerged forests found at low tide, and the drowned river valleys so characteristic of south-west England, indicate that the sea has advanced over the land. Raised beaches, on the other hand, indicate that the sea once lapped higher against our landscape. The recent geological evidence thus indicates both sea-level rises and sea-level falls around the British Isles. Similar apparent contradictions can be found in historical sea-level records from different parts of the world. We should begin our consideration of sea-level change by resolving this apparent paradox.

Figure 5.16
Evidence for sea-level change around Britain.
(a) A river valley in South Devon drowned by sea-level rise after the last glacial.
(b) Fossilized remains of a forest at Marros Sands, Pembrokeshire, South Wales. The trees, which grew during a period of lower sea-level, can now only be seen at low spring tides.
(c) A raised beach in South Devon; the sandy sediments were deposited when the sea-level was higher than today. How this raised beach can be found in close proximity to the drowned valley in (a) will become clear in Section 5.3.1.

Plate tectonics, carbon and climate

5.3.1 Contributions to sea-level change

Figure 5.17 shows changes in annual mean sea-level as determined from selected tide-gauge records from the Atlantic and Pacific coasts of the USA. The data only cover the period 1890–1970, but the trends continue to the present day. It appears that mean sea-level is nearly constant at one locality, but rising or falling at others. How can this be? The answer lies in the tectonic setting of the measuring stations.

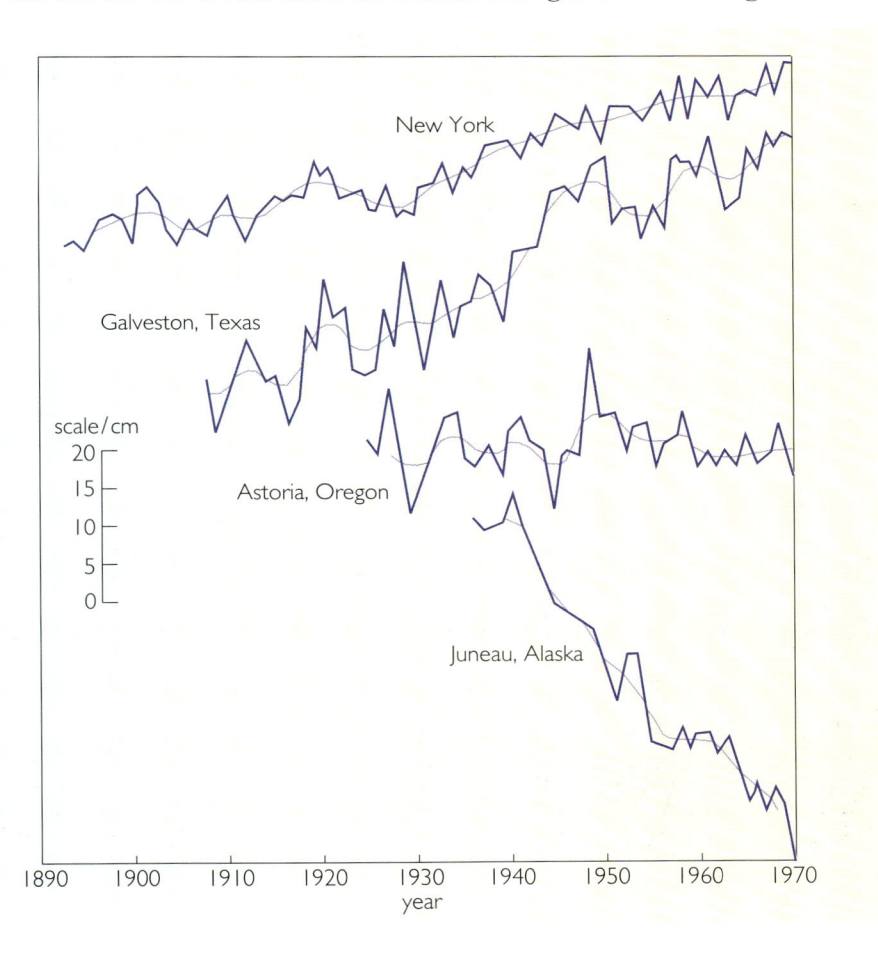

Figure 5.17
Tide-gauge records of annual mean sea-level from selected coastal localities in the USA. The wide annual variations in sea-level are caused by meteorological changes (e.g. low pressure causes higher sea-levels, and winds can drive water up against coasts). The smoother light blue curves (five-year averages) give a closer approximation to long-term changes in sea-level.

The curve for New York corresponds, more or less, to the global average rise in sea-level. New York is on an area of stable continental crust. In contrast, the more or less constant sea-level at Astoria, Oregon, results from tectonic uplift cancelling out or reversing the global sea-level rise, because this station is close to a convergent plate boundary (along the western side of North America, see Figure 4.1). The same is true of Juneau in Alaska, where, in addition, the crust is rebounding, following removal of the weight of overlying ice (see Box 5.3). The high rate of sea-level rise at Galveston, however, is interpreted as being due to the subsidence of the land because of abstraction of oil and/or groundwater.

By now, it should be clear that the sea-level changes indicated in plots like Figure 5.17 are made up of two components:

1   **Eustatic changes of sea-level** are world-wide changes that affect all oceans and have the potential to cause global climatic changes, as we shall see. Over the past 2 million years or so, such changes have been caused

largely by ocean water becoming frozen into, or melted from, continental ice-caps (Section 5.2). Over longer time-scales, as hinted above, eustatic changes of sea-level are caused by other mechanisms, which we discuss presently.

2   **Isostatic (or epeirogenic) changes of sea-level** are caused by vertical movements of the crust. Such movements may be caused by changes in the thickness and/or density of the lithosphere, and by loading or unloading with ice or sediments. As described in Box 5.3, such changes cause lithosphere to ride higher or lower on the underlying asthenosphere, rather as blocks of wood may float higher or lower in water.

Box 5.3 Isostasy

Consider a block of wood floating in water (Figure 5.18). The thicker the wooden block, the greater the thickness of wood that emerges above the water. Similarly with an iceberg: the larger it is, the more of it can be seen above the sea-surface. The tendency for the Earth's lithosphere to behave in a similar manner with respect to the underlying asthenosphere is known as **isostasy**.

Continental crust (i.e. the upper part of continental lithosphere; Box 4.1) is mostly granitic in composition. Its average thickness is about 35 km, but beneath mountain ranges it can be as much as 90 km thick. Oceanic crust is mostly basaltic in composition with an average thickness of about 7–8 km, and it is denser than continental crust. By analogy with the blocks of wood and icebergs, therefore, at isostatic equilibrium, continental lithosphere 'rides' or 'floats' higher on the underlying asthenosphere than oceanic lithosphere – which is why the ocean floors are below sea-level.

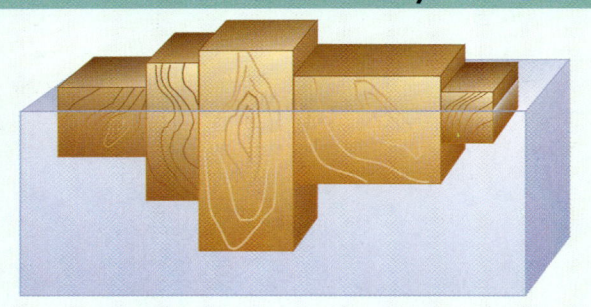

Figure 5.18
Wooden blocks floating in water; an approximate analogy for continental lithosphere of different thicknesses, illustrating why mountains have deep 'roots'.

The analogy with wooden blocks is a simplification because the real lithosphere increases in density with depth, but for the purposes of this discussion you can imagine that the density of the blocks is equivalent to the average density of the lithosphere.

Crustal thickening occurs during mountain-building (through magmatic intrusions and/or convergence of plates at subduction zones); this is why mountains are high. Crustal thinning occurs by stretching during continental break-up, and results in subsidence of the crust that may continue over long periods of time, allowing thick sequences of sediment to accumulate. Loading the lithosphere (by infilling a basin with sediment or by piling ice on top of the crust) will cause its surface to be lowered, and unloading it (by eroding a mountain belt or by melting an ice-sheet) will cause it to rise. In the case of the north-western American tide-gauge stations in Figure 5.17, the nearby mountains are being eroded and have only relatively recently lost their ice-caps from the last glaciation.

Continental shelves and shelf seas

As outlined above, continental crust is thinned by stretching during the break-up of continents, and it therefore subsides. This explains why many continental margins are low-lying, with coastal plains bordered by shallow waters (about 200 m deep on average) overlying continental shelves, which in many cases can be extensive enough to be called shelf seas – the North Sea and the Baltic are good examples. Beyond the edge of the continental shelf, where the thinned continental crust ends and oceanic crust begins (cf. left-hand side of Figure 2.37), water depths increase into the ocean basins proper, which are a few kilometres deep. Substantial thicknesses of sediment can accumulate on continental shelves, deposited there from rivers running off the adjacent land. This additional load upon the crust causes it to continue subsiding isostatically, but average water depths remain of the order of a couple of hundred metres.

Continental shelves are important in the context of sea-level change: because they typically have a rather flat topography, relatively small rises or falls can flood or expose substantial areas of shelf.

Plate tectonics, carbon and climate

It is easy enough to decide whether features such as those in Figure 5.16 or records such as those in Figure 5.17 can be attributed to eustatic or isostatic (epeirogenic) causes. For the more distant past, the distinction is much harder to make, and it is better to use the term *relative sea-level change*, indicating a combination of eustatic and isostatic effects.

Figure 5.19 compares changes in sea-level and (estimated) global temperature over the past 540 million years. We have seen that during past glacial periods, sea-level was as much as 150 m below its present level, while Figure 5.19 shows that in the Cretaceous it was considerably higher than today (though for reasons discussed in Box 5.4 overleaf we must be careful not to read too much into such sea-level curves). The relationship between global sea-level and climate over the past 540 million years (i.e. during the Phanerozoic) is not as clear as it is over the past couple of million of years or so (Figure 5.15). The plots of sea-level and global temperature change (warming and cooling) correlate reasonably well during the Jurassic, Cretaceous and Tertiary. Earlier in Earth history, the correlation between the two curves is less good, though cooling during the Carboniferous is initially matched by falling sea-level.

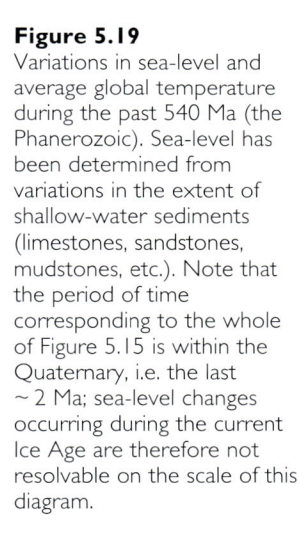

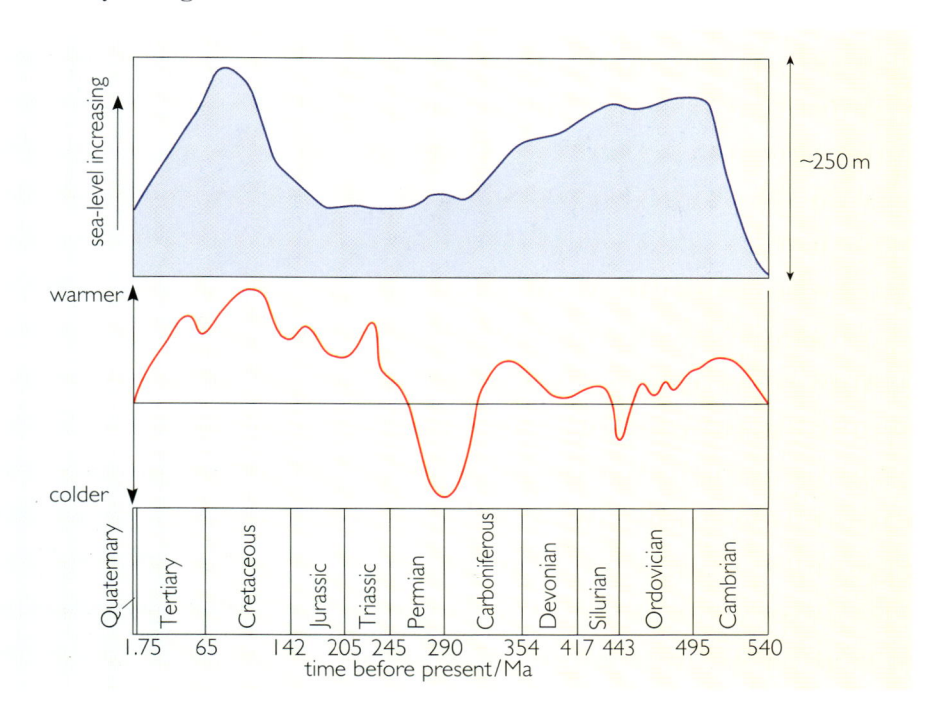

Figure 5.19
Variations in sea-level and average global temperature during the past 540 Ma (the Phanerozoic). Sea-level has been determined from variations in the extent of shallow-water sediments (limestones, sandstones, mudstones, etc.). Note that the period of time corresponding to the whole of Figure 5.15 is within the Quaternary, i.e. the last ~ 2 Ma; sea-level changes occurring during the current Ice Age are therefore not resolvable on the scale of this diagram.

When examining Figure 5.19, you may have wondered whether climate controlled sea-level during the Phanerozoic, or sea-level controlled climate. It is fairly clear that climate has controlled sea-level during the last couple of million years of the Phanerozoic, i.e. during the current Ice Age. Global cooling triggered the growth of polar ice-sheets which removed large amounts of water from the oceans to the land, resulting in a drop in sea-level. Global warming led to melting of the ice, and sea-level rose again. In addition, changes in the global mean temperature of the oceans increased or decreased the water volume by thermal expansion or contraction. Expansion and contraction cause much smaller sea-level changes than the formation or melting of ice-sheets: an increase of 10 °C throughout the water column in all oceans would cause a eustatic sea-level rise of about 10 m.

An important aspect of sea-level change, however, is that it has a significant effect on the total area of 'emergent' continental crust, i.e. of continental crust forming land above sea-level. Today, 30% is emergent, but during glacial periods this increased to as much as about 35%, whereas in the Cretaceous (cf. Figure 5.19) it was probably as little as about 25%. The reason for such large changes in area above or below sea-level is of course the extensive areas of continental shelf and coastal plains (Box 5.3), of which large areas can be flooded or exposed by even relatively small rises or falls of sea-level.

Question 5.5
Might such changes in relative areas of land and sea affect the climate so as to either reinforce or counteract changes caused by other forcing factors? Think back to discussions in previous sections concerning (i) the opening and closing of oceanic gateways, and (ii) the Earth's albedo, and explain how climate might be affected in each case.

Implicit in the answers to Question 5.5 is the existence of feedback mechanisms linking sea-level change and climate. The corollary of this is that if sea-level were raised or lowered by some process *independent* of climate, then the changing sea-level might itself initiate climate change. To explore this aspect, we need to consider other processes that can affect sea-level. Before doing so, however, we should take another quick look at measurement of sea-level change (see Box 5.4).

Box 5.4 Measuring sea-level changes in the distant past

Changes in sea-level before the current Ice Age are more difficult to measure than those that occurred during it because, in the absence of polar ice-caps, oxygen-isotope data can give information only about water temperatures (Box 5.2). Recourse must be made to examination of sequences of sedimentary rocks which enables relative changes in water depth through time to be determined. It is not possible to determine absolute depths; all that can be said is that a particular sequence indicates either shallowing or deepening trends through time, and it is not possible to separate the eustatic and isostatic contributions. So how can quantitative estimates of eustatic changes in sea-level in the geological past be made? The only direct method of measuring past sea-level change involves determining the areal extent of continental crust covered by shallow seas at successive intervals through geological time (by plotting the extents covered by various shallow-water sediments, cf. Figure 5.19), and constructing paleogeographic maps.

This method inevitably has its own shortcomings because the record becomes progressively blurred further back in time, as greater and greater proportions of older sediments are removed by erosion, buried, or altered by heat and pressure (metamorphosed). Continental and shoreline deposits are more likely to be removed by erosion than those deposited deeper in the sea, and so the data available will be biased towards the marine realm (i.e. towards flooding).

Also, paleomagnetic data cannot tell us about changes in position in an east–west direction, so paleogeographic maps often cannot take account of the fact that past plate movements may have telescoped parts of the crust together. This is a minor problem over old, stable areas of the Earth's crust, but is obviously a major one in former mountain belts where it could lead to an under-estimate of areas flooded. Remember that we do not have a fixed datum against which to measure changes in sea-level over geological time. So, while relative rises and falls in sea-level on diagrams like Figure 5.19 are likely to be fairly reliable, it is not advisable to use them to estimate absolute sea-levels. For example, according to Figure 5.19, sea-level has not been below its present level since Cambrian times. In reality, sea-level may have been below its present level much more recently, and you may see sea-level curves indicating such a situation at the beginning of the Triassic.

5.3.2 Causes of eustatic sea-level change

There are two basic processes that change global sea-level. The first involves changing the volume of water filling the ocean basins, the second results from changes in shape and size of the ocean basins themselves.

Changes in ocean water volume

The principal cause of these changes is formation and melting of ice-caps, which was discussed in Section 5.2 and requires little further discussion here. Suffice it to say that if all of the present-day Antarctic ice-sheet melted, global sea-level would rise by 60–75 m (cf. Question 2.6). Disappearance of the Greenland ice-sheet would add about another 5 m. However, the load of the extra water in the oceans would depress the ocean crust and increase the depth of the ocean basins, so the overall rise would be in the range of 40–50 m. As already mentioned, an increase of 10 °C in the average temperature of water in the oceans would further raise sea-level by about 10 m, through expansion of seawater.

Changes in shape and size of ocean basins

It has been proposed that a major cause of rising sea-level could be an increase in the rate of formation of oceanic crust at spreading axes, which would increase the volume of ocean ridges, thus displacing water onto the continents. This suggestion needs to be treated with some caution, for the following reason. Sea-floor spreading is believed to be the major means of heat loss from the Earth's convecting interior: the rate of heat loss has declined since the Earth formed 4.6 billion years ago, and is unlikely to have *increased* in the past 200 million years. However, many scientists do consider the expansion of ridges associated with increased rates of spreading to be a major cause of sea-level rise and we should not dismiss this possibility out of hand.

Another explanation for rising sea-levels – and a much simpler one – involves simply displacing water from a contracting deep basin into a number of shallow ones. Consider the break-up of Pangea (Figure 5.11), with new oceans opening between the dispersing continental fragments, and Panthalassa contracting as the continental fragments disperse. A shrinking deep basin would not become significantly deeper, so water in it would have to be displaced elsewhere. If it could only be displaced into a rather small number of other basins – newly forming and still shallow – then the net result would be a global rise of sea-level. Eventually, deepening of the new ocean basins would have become more rapid than shrinkage of the original Panthalassa, which was now approaching the dimensions of the present-day Pacific; this would have occurred in the mid- to late Cretaceous (at about 100 Ma), when sea-level was at a maximum (cf. Figures 5.11c and 5.19).

Distributions of shallow-water marine sediments suggest that at that time global sea-level could have been ~ 200 m higher than at present (Figure 5.19), and some estimates are much higher than that. It seems likely that such high sea-levels might have been at least partly caused by production of new ocean floor, despite the reservations discussed above concerning spreading ridges.

- ■ Thinking back to Section 4.5, can you recall another way in which material is added to oceanic crust?

- ■ The eruption of enormous amounts of basaltic lavas to form oceanic plateaux, as a result of mantle superplumes.

Significant volumes of rock were added to the oceanic crust in this way during the Cretaceous (Figures 4.21 and 4.22). However, if you look carefully at Figure 4.22, you can see that between 120 and 80 Ma ago part of the increase in oceanic crust production cannot be attributed to the superplume events and must therefore be a result of increased production of oceanic crust at spreading ridges. (Incidentally, it is thought that superplume events represent the release of a temporary build-up of heat within the Earth, not an increase in the rate of heat production.)

- What side-effect of the increase of volcanism during the Cretaceous might have had an *indirect* effect on sea-level?

- The release into the atmosphere of large amounts of carbon dioxide, leading to 'greenhouse' warming.

The rise in sea-level in response to global warming would have been modest, because it would have been due only to thermal expansion of ocean waters. There would have been no melting of ice-caps, as there were none at that time.

Note that although CO_2 would be *directly* supplied to the atmosphere only by those eruptions above sea-level (those forming volcanic islands and subduction zone volcanoes) – on time-scales up to 1000 years or so, the CO_2 released into the ocean at mid-ocean ridges (via hydrothermal vents and volcanoes) would begin to enter the atmosphere (Section 4.5). In other words, on geological time-scales, it does not matter whether the volcanoes are above or below sea-level. We will come back to the greenhouse effect of atmospheric CO_2 in the next section.

Eustatic sea-level rise can also be caused by the deposition in the ocean of sediments transported from continental areas, because that will effectively decrease the size of the ocean basins by displacing water (though this effect may be offset by isostatic depression of the crust by the weight of the sediment). Sea-level *falls*, on the other hand, could occur during the aggregation of supercontinents (the reverse of the break-up of Pangea, discussed above), when continental collisions result in thickened continental crust and the formation of mountains, which would be isostatically elevated (Box 5.3).

5.3.3 Sea-level, climate and atmospheric CO_2

In Question 5.5, you considered the possibility that changes in sea-level might lead to secondary effects which would in turn affect climate. Both the effects considered – changes in albedo and changes in ocean current patterns by opening or closing of gateways – are essentially physical in nature. What happens if we extend this idea to include chemical and biological consequences?

Question 5.6

(a) Think back to Section 3.3.3 (on long-term changes in the carbon cycle) and suggest possible ways in which changing the relative areas of land and sea might indirectly influence climate through resulting changes in:
(i) continental weathering and its effect on atmospheric CO_2; (ii) the areal extent of environments found in low-lying coastal regions; and (iii) the areal extent of shallow seas.

(b) In answering (a), you will have effectively been discussing a number of feedback mechanisms. Explain whether each of the feedback mechanisms you identified in (i) to (iii) is positive or negative, and hence what general effect it might be expected to have on the climate system. What is the situation for the feedback mechanisms you identified in Question 5.5?

Plate tectonics, carbon and climate

Of course, in reality no feedback effect resulting indirectly from sea-level change would be acting on its own, and the real situation would be much more complicated than we have been suggesting. Nevertheless, it is interesting that of the five aspects we considered in Questions 5.5 and 5.6, the two that bring about negative (i.e. stabilizing) feedbacks are those that involve the activities – lives and deaths – of organisms, through their effects on the global carbon cycle. This finding is consistent with a 'Gaian' view of the world, where the activities of organisms contribute to keeping conditions – in this case temperatures – within a range suitable for life.

We should now look more closely at some aspects of the factors you considered in (i)–(iii) of Question 5.6, beginning with continental weathering.

- Bearing in mind what you know about the factors which affect chemical weathering, do you think that global average weathering rates would be greater for an Earth where the distribution of continents approximated to a 'cap world' or one where they approximated to a 'ring world' with land concentrated in low latitudes (Figure 5.8)?

- At low latitudes, conditions are likely to be warm and wet, conducive to chemical weathering. Therefore, the continents of a tropical 'ring world' would be subject to more chemical weathering than would those where the continents were concentrated at higher latitudes.

Thus, weathering of silicates would occur more readily on a tropical ring world than on a polar cap world, and – at first sight at least – CO_2 could be more rapidly removed from the atmosphere, leading to global cooling.

- Very briefly, what happens to the carbon removed from the atmosphere during weathering of silicates (i) on the short term, (ii) on the intermediate term (up to 1000 years), and (iii) on the long term (millions of years)?

- (i) It is carried to the sea in rivers, and enters the marine carbon cycle where (ii) it takes part in biological and chemical processes for periods up to ~ 1000 years. However, some of this carbon ends up being preserved in sediments as organic remains or carbonate debris (shells, skeletons, etc.); (iii) these sediments eventually become lithified, thus being removed from the atmosphere for millions of years.

So carbon that is removed from the atmosphere by weathering of silicates *stays* removed if it is first 'fixed' through the activity of marine organisms and then buried and preserved.

Finally in this section, we consider the circumstances in which deposition of carbon in the oceans may have prevented excessive global warming (often referred to as a 'runaway greenhouse effect') during the Cretaceous, when large amounts of mantle-derived CO_2 were being released via eruptions of enormous volumes of lava.

Preservation of organic carbon in marine sediments during the Cretaceous

As discussed in Section 5.1.2, it is likely that during the Cretaceous continental configurations were such that oceanic gateways allowed both circum-equatorial flow and polewards transport of heat in surface currents (Figure 5.11c). Together with higher atmospheric CO_2 levels, this would have resulted in a very warm, virtually ice-free planet, with relatively low meridional (north–south) temperature gradients.

■ Remembering Figure 5.11 (which shows changes in current patterns during the break-up of Pangea), what can you say about the oceanic thermohaline circulation on an Earth so much warmer than today?

■ As shown in Figure 5.11b and c, the only significant deep water masses forming at that time were probably warm and saline, rather like today's Mediterranean water.

It is possible that there was some deep-water formation at higher latitudes, but the lack of a large temperature difference between the Equator and the poles, and the absence of polar ice-caps, would probably mean that the thermohaline circulation was very sluggish compared with that observed today.

■ Would ocean bottom waters during the Cretaceous have contained more, or less, oxygen than present-day bottom waters?

■ They would have contained less oxygen. Being warm rather than cold, much less oxygen would have dissolved in the water masses when they were at the surface in contact with the atmosphere (Section 2.3.2). Also, a sluggish thermohaline circulation would have meant that water oxygenated at the surface would take much longer to reach the sea-bed, during which time the oxygen would be removed by respiring organisms and (especially) decomposing organic debris.

The lack of oxygen being carried into the deep ocean during the Cretaceous led to much of the water column and extensive areas of sea-bed being anoxic. Furthermore, because of the warm conditions, the Cretaceous saw a massive increase in the growth of land plants, and this in turn resulted in large amounts of terrestrial organic matter being transported into shelf seas and oceans, joining the remains raining down from high planktonic productivity. The combination of an abundant supply of organic debris and an already low supply of dissolved oxygen meant that compared with today, a much larger proportion of organic debris reached the sea-floor without being consumed or decomposed (cf. Figure 3.15). As a result, large amounts of organic carbon survived the journey to the deep sea-bed and were buried and eventually lithified (cf. Figure 5.20a). They were therefore effectively removed from the global carbon cycle (and hence from the atmosphere) for millions of years, so helping to prevent a runaway greenhouse effect.

'Carbonate factories': the accumulation of inorganic carbon during the Cretaceous

Carbonate rocks are by far the largest reservoir of carbon on Earth – as discussed in Section 3.3.3, they contain about four times as much carbon as is preserved in carbonaceous rocks (Figure 3.20). In our discussion of the marine carbon cycle, we concentrated mainly on precipitation of calcium carbonate by organisms in the deep oceans (Section 3.3.2), but the greater proportion of carbonate rocks are, in fact, made up of shallow-water carbonates. In shallow waters around coastlines and over continental shelves, especially in warm tropical and subtropical conditions, corals, bivalves and certain types of benthic algae all precipitate calcium carbonate from seawater (Figure 5.21).

■ How can changes in global sea-level affect the volume of carbonate rocks produced in shallow-water carbonate factories?

Plate tectonics, carbon and climate

(a)

(b)

Figure 5.20
Crustal reservoirs of carbon. (a) Kimmeridge Clay cliffs, composed of organic-rich sediments (these were in fact laid down during the Jurassic Period). (b) Chalk cliffs at Beachy Head (southern England), built mainly of the remains of coccolithophores that lived in shallow seas during the Cretaceous.

(a)

(b)

(c)

(d)

Figure 5.21
Some organisms that contribute to shallow-water carbonate factories, all typical of warm climates. (a) Corals in ~ 8 m of water in the Florida Keys. A variety of corals can be seen, but the most prominent are the rounded *Diploria* corals, and the branching *Acropora palmata*. Reef-building corals make calcium carbonate for their skeletons with the help of algae living symbiotically within them. (b) Close up of a carbonate-precipitating ('coralline') alga. This is a specimen of the red alga *Goniolithon* whose branches are encrusted with other carbonate-precipitating organisms. (c) Giant stromatolites near the Exuma Islands in the Bahamas. Stromatolites are built up of mats of photosynthesizing cyanobacteria, which trap sediments and become calcified; they also become encrusted with carbonate-precipitating organisms. (d) Accumulation of shells of cerithid gastropods – a mollusc found in high-salinity environments such as marine-fed ponds. Such shells are commonly found between layers of calcareous algal peat.

■ As mentioned earlier, flooding of continental shelves during periods of high sea-level increases the area of shallow water available for occupation by carbonate-secreting organisms. During periods of low sea-level, continental shelves are largely emergent, and shallow-water carbonate factories are drastically reduced in area.

During the Cretaceous, vast areas of continental crust were flooded, and extensive shallow-water carbonate factories produced thick limestone sequences. The formation of these deposits must have removed a significant proportion of the 'extra' CO_2 introduced into the atmosphere as a result of the superplume events, so providing a negative feedback which helped prevent a runaway greenhouse effect (cf. Question 5.6b).

What of deep-water carbonate factories? For most of geological time, these have had a much less important role in the removal of carbon into sediments than have shallow-water carbonates. This is because the principal organisms that form calcium carbonate sediments in the deep oceans, the coccolithophores and the foraminiferans (Figure 3.16), did not begin to become abundant until about 100 Ma ago, during Cretaceous times.

■ Given that animal biomass is roughly a tenth of plant biomass, which would you expect to make the greater contribution in the deep-water carbonate factories – coccolithophores or foraminiferans?

■ Coccolithophores are algae, i.e. plants / primary producers, and so they must make the greater contribution.

Incidentally, although they are pelagic organisms inhabiting the water column of the open ocean, coccolithophores can also be abundant in shelf seas, and most of the chalk deposits of north-west Europe (including the chalk cliffs of southern England, Figure 5.20b) are formed of accumulations of coccoliths, not so different from those shown in Figure 3.16b.

■ From what you read in Section 3.3.2, would you say that there must be carbonate sediments everywhere on the deep ocean floor at the present day?

■ No. Below the carbonate compensation depth (CCD), seawater is undersaturated with respect to calcium carbonate, and so calcareous remains dissolve.

Although calcium carbonate skeletal material is being produced throughout the surface oceans, only that which reaches the sea-bed above the CCD is likely to be preserved. The ocean floor is shallowest near to spreading axes (Figure 4.3) and subsides isostatically with increasing distance from ocean ridges, eventually sinking below the CCD, where calcareous remains reaching the sea-bed would be dissolved. It is important to stress that this dissolution occurs only *at the sea-bed*, so the substantial thicknesses of carbonate sediments which accumulate when the sea-floor is above the CCD are preserved, to be buried beneath the land-derived sediments which cover the deepest parts of the ocean floor.

Plate tectonics, carbon and climate

Question 5.7

(a) (i) Would you expect the CCD to be deeper or shallower beneath areas of high productivity?

(ii) What would you expect to happen to the CCD when large amounts of extra CO_2 are released into the oceans at times of increased submarine volcanism?

(b) Is your answer to (a) consistent with Figure 5.22, a generalized diagram showing how the level of the CCD is thought to have varied since early in the Cretaceous?

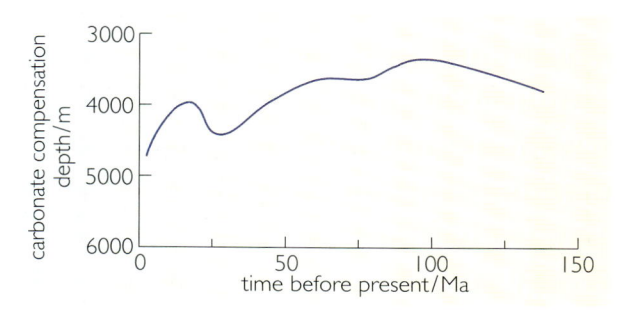

Figure 5.22 Schematic and highly generalized curve to show variations in the depth of the carbonate compensation depth over the past 150 Ma.

At the present time, the deep-water carbonate factory is an important mechanism for removing excess atmospheric CO_2 that humans have released by combustion of fossil fuels and deforestation (Figure 3.24). In Figure 5.23, the conditions accompanying the relatively low sea-levels of the present day are contrasted with those accompanying the high sea-level of the Cretaceous.

Figure 5.23 Highly schematic diagram comparing the preservation of carbon in the oceans during (a) the Cretaceous, a period of high global sea-level and (b) the present day, a period of relatively low sea-level. During the Cretaceous, continental shelves were flooded, enabling carbonate-secreting organisms to flourish in extensive tropical/subtropical seas. Deep ocean currents were warm and sluggish, leading to anoxic conditions at the sea-bed and the preservation of large amounts of organic carbon. By contrast, at the present time, the area of shallow-water carbonate factories has been drastically reduced and deep oceanic circulation is more vigorous, preventing widespread anoxic conditions on the deep sea-floor.

(a) early Cretaceous

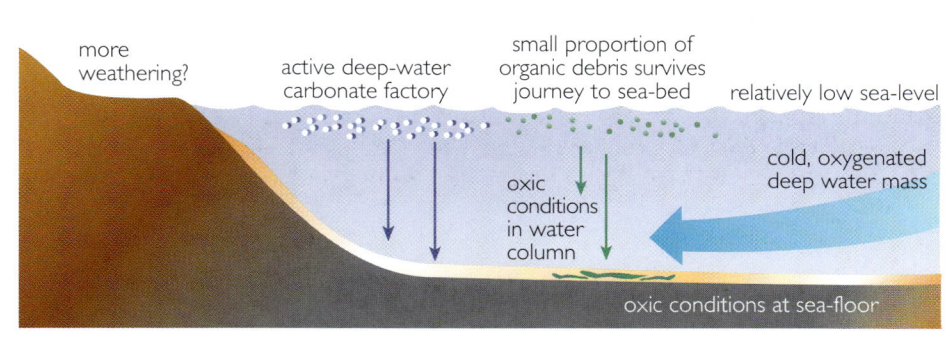

(b) present day

- In what important way do accumulations of carbon preserved in deep marine sediments during the early Cretaceous differ from those accumulating at the deep sea-bed at the present day?

- Before the deep-water carbonate factory became properly established at ~ 100 Ma, carbon was accumulating in the deep sea as organic remains (which would eventually have become carbonaceous rocks). At the present day, carbon is accumulating in the deep sea mainly as shells and skeletons (inorganic carbon, which will eventually become carbonate rocks).

Deductions about the roles that organisms have played in the Earth system in the geological past can only be based on a combination of evidence from sediments and sedimentary rocks, and what we observe occurring on the Earth today. Given that we are always discovering new things about how organisms interact with the Earth at the present time, there must be many aspects of past interactions that will remain at best uncertain. We have already seen how DMS production by algal blooms can result in increased cloud cover, increasing the local albedo (Section 2.3.2). Another example of a surprising recent discovery is the large effect that coccolithophore blooms can have on the local albedo: coccolithophores are examples of algae that 'self-shade' – the highly reflective platelets (the coccoliths), which are continually being shed, initially float on the surface, keeping conditions at shallow depths optimum for coccolithophore growth. Coccolithophore blooms can cover enormous areas (Figure 5.24) – who can say to what extent their appearance in the oceans might have changed the local physical environment (or, indeed, the chemical environment) for other organisms?

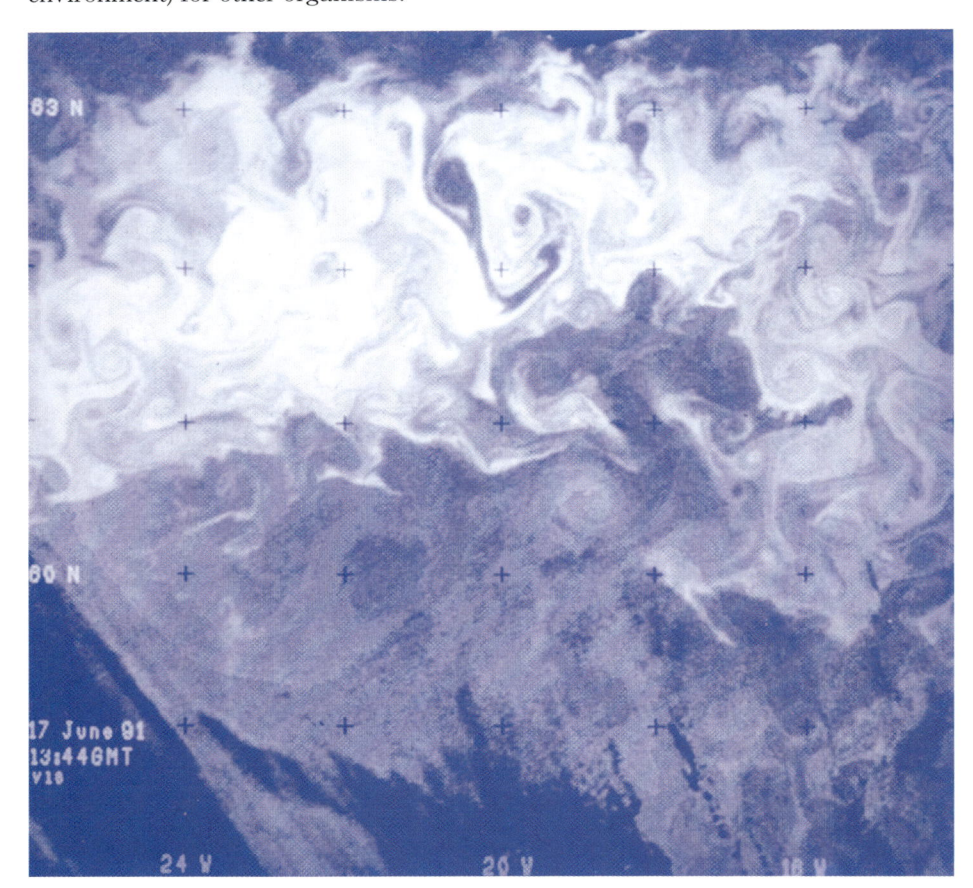

Figure 5.24
Satellite image showing a coccolithophore bloom to the south of Iceland. The bloom is made visible by its high reflectance, and broken up into swirls by complex current patterns in surface waters. The north–south distance across the densest part of the bloom is ~ 300 km.

5.4 Plate tectonics, mountain-building and climate

In the previous section, we have been mainly concerned with processes occurring in the oceans. Now let's turn our attention to that part of the Earth above sea-level, in particular to mountains.

5.4.1 Mountain-building

As you know (Box 4.1), ocean floor is eventually subducted at a destructive margin (Figure 4.3). An important consequence of subduction is that the mantle above the subducted lithosphere can melt. Some of the resulting magma rises to the surface and is erupted as lava from volcanoes, while the remainder forms intrusions within the crust of the overlying plate. A large proportion of continental crust is made up of low density igneous rocks, and therefore continental lithosphere (lithosphere with continental crust as the upper layer) cannot be subducted and provides an irreversible addition to the Earth's outer layer.

- Bearing in mind what you read in Box 5.3 about the tendency for the lithosphere to 'float' in isostatic equilibrium, can you suggest what the consequences of thickening the crust would be?

- A thicker crust means a thicker lithosphere with a lower average density. Buoyancy forces in the mantle will cause such a lithosphere to rise, leading to land surface at high elevations.

The strong correlation between high surface elevation and a large thickness of lithosphere – as seen, for example in the case of the Andes or Himalayas – is a clear indication of the effect of increased buoyancy forces on thickened crust.

Descent of the ocean crust into subduction zones (Figure 5.25a) eventually draws the continents either side of the ocean closer together, ultimately closing the ocean basin completely (Figure 5.25b). Inevitably, the result is a collision between two masses of low-density continental crust (Figure 5.25c). Neither plate can be subducted into the much denser asthenosphere, so the system appears to be 'locked'. At this stage, further convergence of the plates must be taken up by thickening of one or both of the plates of continental lithosphere. In the upper part of the crust, the sedimentary sequences along the original continental margins are thickened, as sheets of rock are thrust under one another along low-angle faults (Figure 5.25d). At deeper levels in the crust, the rocks are hotter and more ductile and likely to thicken by folding, and – as mentioned earlier – the crust is further thickened by the intrusion of low-density magmatic rocks.

The high topography that results – mountains – can affect climate through both physical and chemical mechanisms. We will look at the physical mechanisms first.

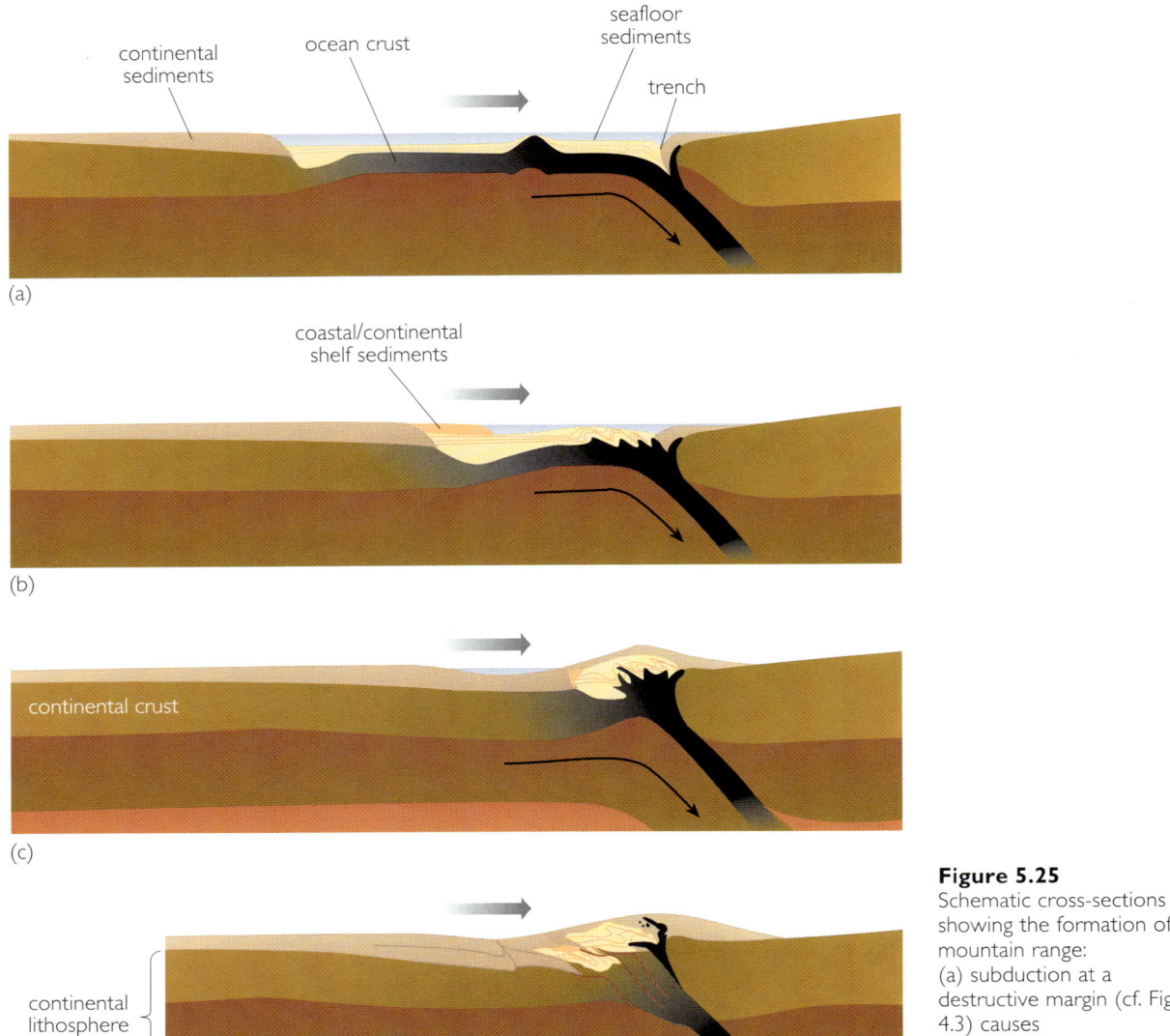

Figure 5.25
Schematic cross-sections showing the formation of a mountain range:
(a) subduction at a destructive margin (cf. Figure 4.3) causes
(b) contraction of the ocean basin which
(c) leads to collision, and
(d) thickening of the continental lithosphere.

5.4.2 Mountain climate

The climate over mountains is distinctive. It is important to remember this in the context of the global Earth system, because *local* climate can affect *global* climate.

Even at low latitudes, high mountains may be capped with snow (cf. Figure 1.3).

- Remembering Figure 2.4, can you suggest why this is?
- At the altitude of the tops of high mountains, the air is cold (red curve in Figure 2.4). Furthermore, air forced to rise by the presence of a mountain will expand in response to the decreasing atmospheric pressure (black curve in

Plate tectonics, carbon and climate

Figure 2.4) and cool adiabatically. Eventually, any water vapour in the air will condense, forming clouds and precipitation, which at high altitudes will fall as snow.

- So, how can the extent of mountainous terrain at low latitudes affect global climate?

- Through its effect on the global albedo. A large area of highly reflective snow at low latitudes, where a large proportion of incoming solar energy reaches the Earth, could lead to a significant increase in the albedo of the Earth as a whole.

It is obviously extremely difficult to estimate what effect low-latitude mountains may have had on the Earth in the distant past, not least because we don't know for sure where the mountains were or how high they were. Nevertheless, it is something that we should bear in mind, particularly when considering what the climate might have been like when a large proportion of the global land area was at low latitudes, as may have been the case during the Precambrian (Box 5.1).

So much for how mountains may affect climate through their very existence. Now let us look at how the plate-tectonic *processes*, including mountain-building, can affect climate by influencing the concentration of carbon dioxide in the atmosphere.

5.4.3 Subduction, mountains and atmospheric CO_2

Volcanic eruptions at constructive and destructive margins, eruptions of flood basalts and eruptions of sea-floor volcanoes (seamounts) all result in addition of CO_2 to the atmosphere (even those below sea-level, see Section 5.3.3). As already discussed, the global flux of CO_2 from volcanism (currently equivalent to about 0.02×10^{12} kgC yr^{-1}) has varied through time: we have seen that it is large during times when great volumes of flood basalts are being erupted, and it will also be affected by sea-floor spreading rates because higher rates of sea-floor spreading mean not only more volcanism at spreading ridges, but also more at subduction zones, where the sea-floor generated at the ridges is eventually destroyed (Figure 4.3).

- What is the source of the CO_2 released at mid-ocean ridges (dissolved in hydrothermal fluids or magma), as well as that released during the formation of flood basalts, seamounts and volcanic islands?

- Its source is the mantle.

On the other hand, subduction zone volcanoes – of which there are at present an enormous number, particularly around the Pacific Ocean (Figure 4.1) – have an extra source of CO_2. When subducted ocean crust melts, generating magma which rises to produce volcanic activity in the overlying crust (Figure 4.3), the components of the erupting magma derive both from the basaltic oceanic crust *and* its covering of sediments, as well as the lithosphere through which the rising magma passes.

If calcium carbonate and silica are heated together, at around 400 °C a **decarbonation reaction** occurs, which releases CO_2:

$$\underbrace{SiO_2(s)}_{\text{silica}} + \underbrace{CaCO_3(s)}_{\text{calcium carbonate}} \longrightarrow \underbrace{CaSiO_3(s)}_{\text{calcium silicate}} + \underbrace{CO_2(g)}_{\text{carbon dioxide}} \quad \text{(Equation 5.2)}$$

- Where have the silica and calcium carbonate on the descending slab of ocean floor come from? And why would decarbonation reactions not have occurred in these circumstances before about 100 Ma?

- Both the silica and the calcium carbonate are in the remains of planktonic organisms (notably diatoms and radiolarians on the one hand, and coccolithophores and foraminiferans, on the other). There were no calcium carbonate remains in the deep sea before about 100 Ma because planktonic organisms that precipitate calcium carbonate had not yet evolved (Section 5.3.3).

Eventually, an ocean basin with a subduction zone will shrink to nothing and two continents will collide (Figure 5.25c,d). All the time that subduction has been occurring, deep-sea sediments will have been scraped off the surface of the subducting oceanic plate, building up great wedges of sediment at the side of the subduction zone trench (as shown schematically at the right-hand side of Figure 4.3). As the continents finally come together, these sediments – including calcareous and siliceous remains – will be subjected to great heat and pressure (cf. Figure 5.25), and will become altered. As on subducting ocean floor, the decarbonation reaction is just one of many that can occur, but is the most important from our point of view because it adds CO_2 to the atmosphere.

However, in these circumstances, deep-sea sediments are not the only source of calcium carbonate reacting in the decarbonation reactions. At least some products of the shallow-water carbonate factory – carbonate-secreting algae, corals, bivalve remains, etc. – will also become squeezed between the colliding continents. The silica in shallow-water sediments is, however, not biogenic (i.e. produced by organisms) but is mainly in the form of quartz sands which have weathered from the land and accumulated on the continental shelf. It is important to realize that the decarbonation reaction will only occur when the silica and carbonate are mixed together. While this is the usual situation for planktonic remains in deep-sea sediments, it is not so common for shallow-water sediments where (say) coral-reef debris is not so likely to be intimately mixed with sand.

The CO_2 produced by decarbonation reactions occurring as continents collide is not usually released into the atmosphere via volcanism, as volcanoes are not generally found in such tectonic settings (Figure 4.1). Instead, it seeps out via faults and fractures.

Of course, the subduction of sediments of *pure* calcium carbonate, unmixed with silica, will mean that the carbon they contain will not find its way back into the atmosphere for millions of years. Similarly, the 'piling up' of carbonate rocks which occurs during continental collision can mean that large volumes of carbon-containing rocks are buried deep in the crust, and prevented from participating in the global carbon cycle for a very long time.

Mountain-building may therefore result in the deep burial of carbon within the continents. At the same time, however, it provides large amounts of exposed rock at high altitudes, and the breaking down of this rock may *remove* CO_2 from the atmosphere.

- By reference to Figure 3.21, can you suggest how this could occur?
- Through the weathering of silicate minerals.

Plate tectonics, carbon and climate

As shown diagrammatically in Figure 3.21, for every two carbon atoms taken from the atmosphere during the weathering of silicate minerals, one has the potential for being removed to sea-bed sediments in carbonate remains. By contrast, weathering of carbonate minerals followed by precipitation in the ocean does *not* result in net removal of CO_2 from the atmosphere.

Mountains are sites of vigorous erosion. At the highest altitudes, **physical weathering** occurs: rocks are shattered by repeated freezing and thawing of water that seeps into cracks (a consequence of the fact that ice occupies more space than the water from which it forms, Box 2.1), and crushed and ground into small particles by the action of glaciers. On the other hand, at lower, warmer altitudes, the wet windward slopes of mountains (Figure 2.20a) tend to be regions of strong chemical weathering, which continues as rock fragments carried down in fast mountain streams accumulate and become chemically altered in reactions with rain and surface waters (Figure 5.26).

One illustration of the importance of mountains in enhancing weathering rates is provided by the River Amazon. Even at its mouth, over three-quarters of the dissolved material it carries has been derived from the eastern slopes of the Andes, more than 3000 km away – only one-quarter comes from weathering of the vast tracts of bedrock that underlie the Amazon Basin between the source and the mouth of the river.

Having looked briefly at the role that mountains play in the climate generally, in the final chapter of this book we consider a particularly high part of the Eurasian continent – the Himalayas and the Tibetan Plateau. We will look both at the role they may have played in the climate system during the last few tens of millions of years, and at the effect they have had – and continue to have – on the overlying atmosphere and the Arabian Sea.

Figure 5.26
Fragments of rocks forming a vast fan at the outlet of a river eroding the Karakoram Mountains to the west of the Himalayas.

5.5 Summary of Chapter 5

1. Climatic conditions for a particular continental area change simply as a result of that landmass moving over the globe through different climatic belts. In the case of the landmass that is now southern Britain, paleomagnetic measurements and paleoclimatic indicators in the rock record show that over the past 450 million years or so, this particular area of continental crust moved northwards from southern mid-latitudes to northern mid-latitudes, crossing a number of global climatic belts (dry subtropical, wet tropical/equatorial, dry subtropical again) in the process. The global distribution of paleoclimatic indicators for the Carboniferous and Permian provide evidence consistent with the existence of a supercontinent (Pangea), which had an ice-cap around the South Pole and tropical forests in the vicinity of the Equator. There may have been high-latitude forests later in the Permian. Reconstructions of continental configurations using paleoclimatic indicators must be approached with caution because: (1) paleoclimatic indicators may be ambiguous; (2) climatic belts do not, in reality, run simply east–west; (3) we cannot know for sure whether the tilt of the Earth's axis, and hence the seasonally varying distribution of solar energy over the Earth's surface, was similar in the past to what it is today.

2. Climate models using highly simplified continental configurations suggest that a tropical 'ring world' would be significantly warmer than polar 'cap worlds'. It is thought that the geography of the Earth may have approximated to the former 700–600 million years ago, and to the latter (with one polar ice-cap) during the late Carboniferous.

3. It is thought that the changing distribution of continents in response to plate-tectonic processes affects global climate on the million-year time-scale through its effect on the radiation budget, and hence indirectly on the hydrological cycle and weathering (which affects the CO_2 concentration of the atmosphere). Furthermore, the closing and opening of oceanic gateways as continents change their relative positions has a significant effect on shallow and deep oceanic circulation, and can contribute to global warming and cooling on significantly shorter time-scales.

4. The break-up of Pangea began at about 200 Ma. The opening of a low-latitude seaway may have contributed to global warming in the Cretaceous, at about 100 Ma. Around this time, deep water masses were probably warm and very saline, having formed at low latitudes. It is thought that thermal isolation resulting from the initiation of the Antarctic Circumpolar Current at 25 Ma accelerated cooling of Antarctica and the growth of the south polar ice-cap.

5. Ice Ages are periods when the Earth has large areas of ice-cover. *Within* Ice Ages, there are glacials and interglacials when ice-sheets alternately grow and retreat (and sea-levels fall and rise). Between 1.5 and 1 million years ago, these alternations occurred with a cyclicity of ~40 000 years (corresponding to the tilt component of the Milankovich cycles) but by 600 000 years ago the 110 000-year eccentricity cycle had become the most dominant. Our knowledge of changes in ice volume (and hence sea-level), and of global temperature, largely derives from $\delta^{18}O$ values obtained from the remains of foraminiferans (notably planktonic species).

6. Relative sea-level change results from a combination of eustatic and isostatic (epeirogenic) sea-level change. Eustatic changes are global in extent, whereas isostatic changes result from local or regional uplift or subsidence of the lithosphere. Eustatic sea-level changes are due either to changes in the volume of ocean waters (resulting from the formation and melting of ice-sheets), or to changes in the size and shape of the ocean basins (resulting either from the

Plate tectonics, carbon and climate

formation of new oceanic crust, notably as submarine plateaux, or from the replacement of a few large ocean basins by a number of smaller ones). During periods of global warming, some sea-level rise is attributable to expansion of the ocean water; however, an increase of 10 °C throughout the water column in all oceans would lead to a sea-level rise of only ~ 10 m.

7 Over the past 540 million years, there have been two periods of high global sea-level – in the Ordovician–Silurian and the Cretaceous – while sea-levels were generally low in the Permo-Carboniferous and from the late Tertiary to the present time. The distribution of shallow-water sediments, and paleogeographic maps of the proportion of continents flooded at different periods in the Earth's history, can be used to determine past changes in sea-level. During glacial periods over the past 2.5 million years, sea-level has been as much as 150 m below its present-day level, and in the Cretaceous (about 100 million years ago), it was probably 250 m higher than it is today (and some researchers believe that it was even more than this). Because of the extensive areas of flat coastal plains and continental shelves, fairly small changes in sea-level can dramatically alter the relative areas of land and shallow coastal seas, with important implications for the operation of the carbon cycle, and hence for climate.

8 Global warming during the Cretaceous was probably the result of the addition to the atmosphere of huge amounts of CO_2 as a result of volcanism (as flood basalts and as a consequence of increased rates of sea-floor spreading). However, the warming was counteracted by the removal of atmospheric CO_2, through deposition and preservation of carbon in the ocean. Large amounts of organic carbon were preserved and buried in the warm anoxic oceans, and large amounts of inorganic carbon were precipitated in shallow coastal seas (and, to some extent, via the newly established deep-water 'carbonate factory').

9 Snow-covered mountains (particularly those at low latitudes) affect the global climate through their effect on the Earth's albedo. Over geological time-scales, the uplift of mountains as a result of continental collision can affect the operation of the global carbon cycle, and hence the amount of CO_2 in the atmosphere, in various ways. Subduction of ocean-floor sediments and collision of continental masses both result in decarbonation, whereby silica and calcium carbonate react together to produce CO_2 which eventually escapes to the atmosphere. However, mountains are sites of vigorous weathering, physical at high altitudes and chemical lower down; chemical weathering (followed by accumulation and burial of carbon in the ocean) results in a *loss* of CO_2 from the atmosphere. (In addition, continental collision may result in large volumes of carbon in rocks being removed from contact with the atmosphere for many millions of years.)

Now try the following question to consolidate your understanding of this chapter.

Question 5.8

As mentioned in Box 5.1, 700–600 million years ago, during the late Precambrian, the distribution of the continents may have approximated to a tropical 'ring world'. Rocks of this age often contain tillites and remains of shallow-water carbonates, including stromatolites (Figure 5.21c), in close proximity.

(a) Why might the close association of these deposits be a cause of puzzlement?

(b) One possible explanation for this situation relates to the tilt of the Earth's axis. Thinking back to Question 5.1 and the associated text, can you suggest what this might be?

Chapter 6
Tibet, the Himalayas and the Arabian Sea

As discussed in Section 5.4, large mountain belts can have a strong influence on climate, both regionally and globally. The larger the mountains, the more widespread their effects are likely to be. Of all the uplifted regions of the world, Tibet is unique in terms of height and extent. This final chapter begins by considering how climatic changes occurring on time-scales of millions of years might be related to the uplift of Tibet, and concludes with a brief look at how the sedimentary record in the nearby Arabian Sea provides important clues about changes in climate and primary productivity over time-scales of thousands of years. Intriguingly, changes on both these time-scales have been illuminated through consideration of changes that occur on a seasonal time-scale – namely, the seasonally reversing winds over southern Asia, the monsoons.

6.1 Raising the roof of the world

Tibet is a plateau of two million square kilometres, about the size of western Europe (Figure 6.1). It is on average 5 km above sea-level and includes over 80% of the world's land surface higher than 4 km. The Himalayan and Karakoram mountains, which define its southern and western margins, include the only peaks reaching more than 8 km above sea-level.

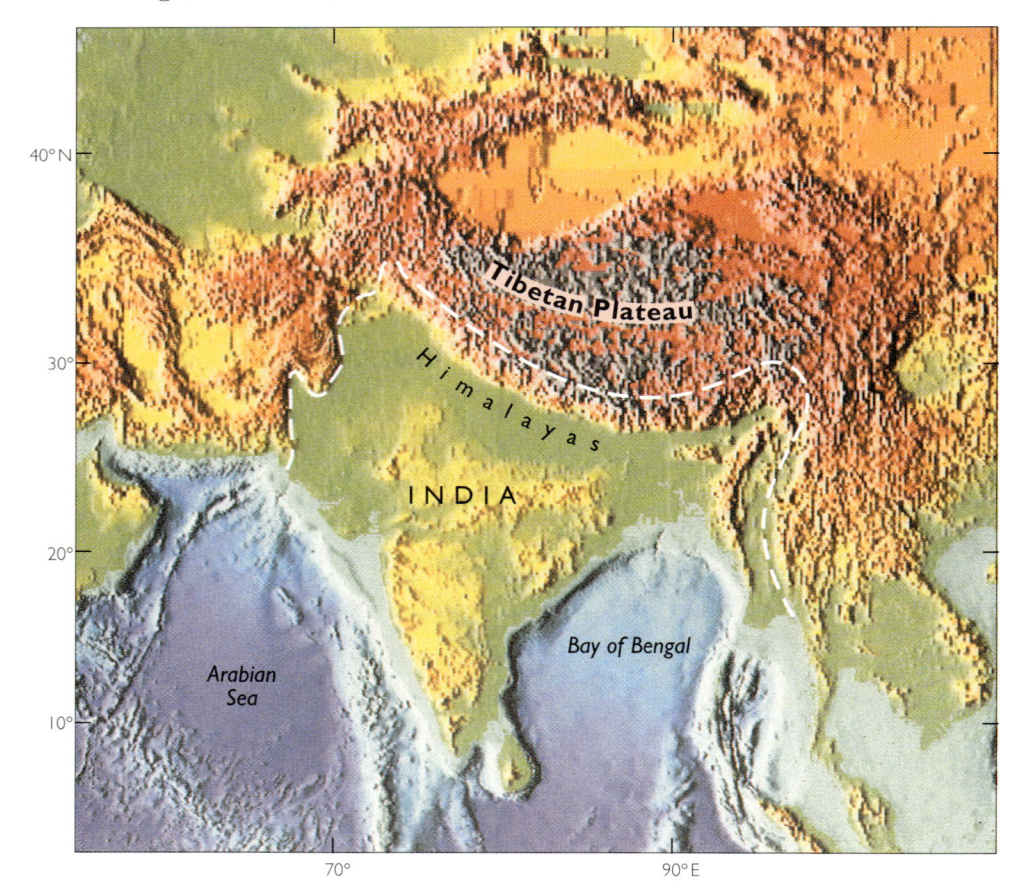

Figure 6.1 Computer-generated map showing the topography of India and Tibet. Grey land areas indicate altitudes > 5000 m, red 1000–5000 m, yellow 500–1000 m, and green < 500 m. The dashed line indicates the boundary between the Indian Plate and Eurasian Plate.

Tibet, the Himalayas and the Arabian Sea

Before examining the climatic effects of these mountains and the Tibetan Plateau, we need to know how they came to be so high in the first place.

6.1.1 Why are the Himalayas and Tibet so high?

One hundred million years ago, India lay well south of the Equator (Figure 5.11c). The northern part of the Indian Plate was oceanic crust lying beneath the Tethys Ocean. Subduction of this oceanic lithosphere resulted from the northerly migration of the Indian continental landmass on a collision course with Eurasia (Figures 6.2 and 6.3). As a result of the subduction, magmas were generated beneath the southern margin of Eurasia, intruding and thickening the crust in the same way that the Andes are being thickened today (right-hand side of Figure 4.3).

Figure 6.2
Diagram showing the northward movement of India between 100 and 50 million years ago, causing shrinkage of the Tethys Ocean as India and Eurasia converged.

By about 50 Ma, i.e. by the mid-Eocene, that part of the Tethys Ocean separating the Indian and Eurasian plates had completely closed. The two continents then collided, with the result that the continental crust on both sides of the collision zone was thickened and the surface elevation rapidly increased (Figures 5.25 and 6.3). Today, the join between the Indian and Eurasian plates can be traced along the northern edge of the Himalayan chain in southern Tibet (Figure 6.1) by a line of outcrops comprising rocks characteristic of oceanic crust. These rocks were once part of the oceanic lithosphere underlying the great Tethys Ocean.

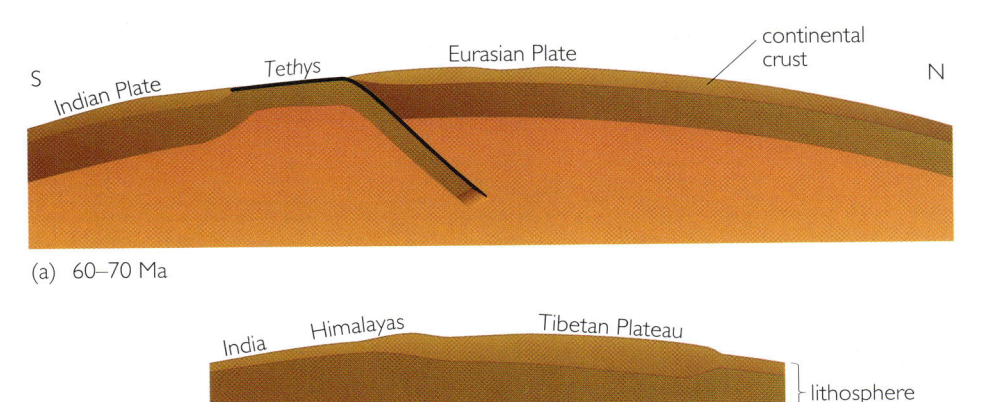

Figure 6.3
The India–Asia collision in cross-section (the scale is realistic, i.e. there is no vertical exaggeration).
(a) At about 60–70 Ma.
(b) Today.

The Himalayas are by no means the only mountains on Earth formed by continental collision. The Alps are the result of a collision that began 120 million years ago, between Africa and Eurasia; and during earlier geological times, collisions initiated the building of many of the world's lesser mountain belts, including the Urals and the Scottish Highlands. Why then did the Himalayas, and their hinterland, the Tibetan Plateau, become so high?

The uplift of the Himalayas and of the Tibetan Plateau are a response to considerable crustal thickening. Some geologists maintain that the continental edge of Eurasia was unusually thick *before* collision, due to magmatic thickening above the subduction zone. In any event, further thickening occurred after collision, by both folding and faulting of the rocks caught near the leading edges of the colliding plates (Figure 5.25d). The great forces that led to such dramatic crustal thickening resulted from two unusual characteristics of the collision. First, it was 'head-on' rather than oblique, with a high closure velocity of ~ 20 cm yr^{-1}. Secondly, the convergence persisted long after the intervening ocean had closed; India has migrated nearly 2000 km northwards since then, and a good deal of the convergence has had to be taken up by squashing and uplifting the rocks of the Himalayas.

The simple model for uplift outlined above works pretty well for rather narrow mountain ranges like the Himalayas, but has Tibet been uplifted in the same way? It is not immediately clear why an area as broad as the Tibetan Plateau should be uplifted so high following collision. However, in recent years, geophysicists have suggested an additional mechanism for rapid uplift.

Up until now, we have concentrated on the thickening of the *crust* during collision. But the crust is just a small part of the lithosphere, and following collision the entire lithosphere will be thickened (Figure 5.25d), creating a lithospheric 'root' which protrudes down into the asthenosphere. Since the early 1980s, geophysicists have been using computer modelling to study the thermal properties of the lithospheric root and the surrounding asthenosphere. These studies suggest that, following lithospheric thickening, the deep, cold root will obstruct convection in the asthenosphere, become 'eroded' by melting, and eventually break off and sink into the asthenosphere. This process, illustrated in Figure 6.4b, is known as *convective thinning* of the lithosphere.

Now let's return to the simple analogy in which floating wooden blocks represent the lithosphere (Figure 5.18), and imagine that the taller blocks are weighed down by a high-density lowermost layer.

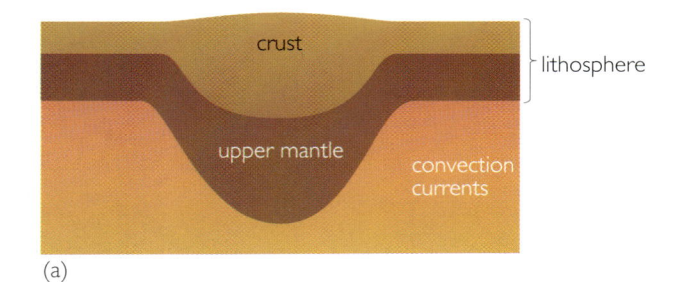

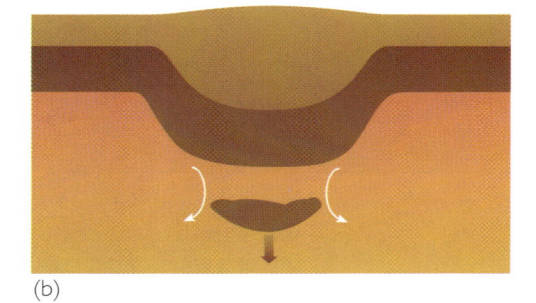

Figure 6.4
Schematic diagram to illustrate how convective thinning of the lithosphere may have contributed to the uplift of the Tibetan Plateau. (a) The lithosphere thickened as a result of collision. (b) Melting and erosion of the dense lithospheric root by convection currents in the mantle. (c) Uplift of the plateau in response to the removal of the base of the lithosphere.

Tibet, the Himalayas and the Arabian Sea

- What would be the result of removing the dense base from the floating blocks?

- Clearly, the blocks will bob upwards to a new position, higher in the water.

It is thought that, in a similar way, the loss of the lithospheric root caused the rapid elevation of the Tibetan Plateau (Figure 6.4c).

So, to summarize: the high impact velocity between India and Eurasia, followed by continuing compression over tens of millions of years, led to an unusual degree of lithospheric thickening beneath Tibet, and the convective thinning that resulted allowed the lithosphere to 'bob up'. This last may be the principal reason why Tibet is so much larger and higher than other plateaux on the Earth today.

6.1.2 When was the Tibetan Plateau uplifted?

It is very difficult to trace the uplift of a plateau through time because, if there *is* such a thing as a reliable paleo-altimeter, no-one has managed to discover it yet. If we were to attempt to plot the change of altitude of the Tibetan Plateau against time, we would have two – and only two – firm points. First, we know the present-day altitude which is, on average, 5 km above sea-level. Secondly, we know that rocks now 5 km high were somewhat *below* sea-level 70 million years ago, because limestones with remains of marine organisms originally deposited in shallow seas at about that time have been found in southern Tibet. However, details of the uplift during the intervening 70 million years are highly uncertain.

One approach to determining the elevation history of the plateau is to exploit the changes in surface temperature resulting from its elevation. The morphology of successful plant species is quite different in cold and warm climates, and because temperatures decrease with increasing altitude (Figure 2.4), the higher the altitude the more that cold-climate species are favoured. Paleobotanists have attempted to use this fact to infer altitude changes from fossil flora collected from sedimentary rocks deposited over the past 50 million years. Work along these lines has led to the publication of several different altitude–time paths for the Tibetan Plateau (Figure 6.5).

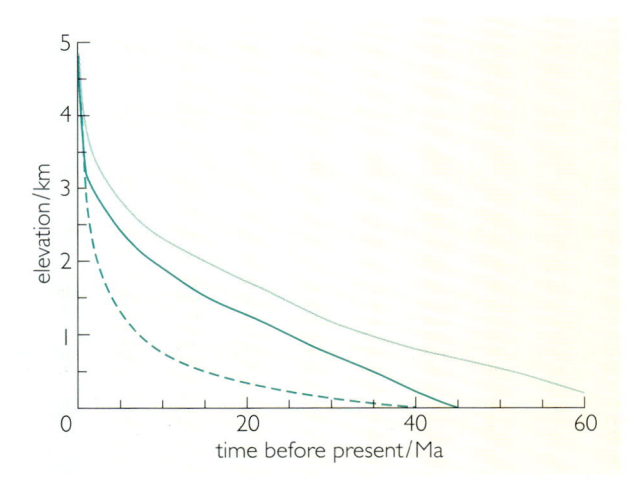

Figure 6.5
The history of uplift of the Tibetan Plateau as determined in studies of three different fossil plant assemblages, using the 'nearest living relative' approach.

Unfortunately, there are a couple of flaws in this approach. First, even if the region corresponding to the Tibetan Plateau had not been uplifted at all, its climate would have changed considerably over the course of 70 million years, not least because of the closure of Tethys. Furthermore, within the plateau itself there will always have been small-scale climatic variations resulting from local topography, with some areas being more sheltered, some areas wetter, etc. To some extent, these latter complications can be allowed for, but there is another problem which is more fundamental. As the fossil species found in sediments of Eocene age (~ 50 million years old) are usually extinct, paleobotanists have adopted the 'nearest living relative' approach. In practice, this involves identifying the nearest living relative to the fossil species concerned, and then assuming that the climatic conditions under which the ancient plants lived were similar to those of their relatives living today.

- Can you suggest why such an approach might give misleading results?
- Species evolve, and adapt to changing environmental conditions, so the extinct species might have lived under somewhat different conditions from its nearest living relative.

Imagine, for example, that yaks are extinct, but that their fossilized remains are well-documented by paleontologists who have concluded that cattle are the yak's nearest living relative. The inferred habitat of yaks would then be characterized by a temperate or warm climate, similar to where cattle are found. The truth is, of course, that yaks are adapted to high-altitude, cold conditions (Figure 6.6). In other words, the 'nearest living relative' approach ignores evolutionary adaptation.

Clearly, this approach has inherent problems, and the resulting estimates of uplift rates need to be treated with care. An alternative approach is to examine and classify the shapes of plant leaves which are adapted to different environments (Figure 6.7). This technique has not yet been applied to Tibetan flora but could potentially provide the most accurate constraints on the uplift history of Tibet.

Another indirect way of measuring the uplift history of the Tibetan Plateau depends on dating movement along a particular kind of fault. These faults, sometimes referred to as **normal faults**, result from stretching of the crust and can be seen at the surface as steep escarpments (Figure 6.8). A series of such normal faults, which have resulted in local collapse of fault-bounded blocks, have now been found across Tibet (Figure 6.9).

- Does Figure 6.9 suggest any relationship between the distribution of normal faults and their elevation above sea-level?
- The normal faults seem to occur on the highest parts of the Tibetan Plateau.

Some of these faults remain active today, causing earthquakes. Detailed studies of the signals of these earthquakes confirm that they only occur on that part of the plateau which reaches an altitude greater than about 3.5 km. It seems that movement on such faults has allowed some of the highest regions of the plateau to subside, and that the faults have been initiated only when the plateau reached a critical altitude, at which point it began to collapse.

- Why should this be?

Figure 6.6
Adult yaks on the Tibetan Plateau: an example of adaptation to high-altitude, cold conditions.

Figure 6.7
Fossil leaf from Tertiary sediments on the Tibetan Plateau. By examining the leaf's shape, size and margin characteristics (leaf physiognomy), it is possible to determine the climatic regime under which the plant was growing when it was alive. Thus, as plants live at the Earth's surface, fossil leaves, along with an independent (e.g. isotopic) dating method, can be used to estimate the height of a land surface at any time in the past.

Tibet, the Himalayas and the Arabian Sea

Figure 6.8
The steep scarp of a normal fault, with accumulations of scree, form the backdrop to the Tibetan horse and rider. The flat-bottomed valley is the result of subsidence due to movement along the fault.

Earlier, to help explain mountain uplift, we drew on the analogy of blocks of wood floating in water (Figure 5.18). However, rocks do not always behave as rigid blocks. As a plateau is elevated, the pressure on the underlying rocks causes them to have an increasing tendency to flow. In the case of the Tibetan Plateau, a critical altitude was reached at which – although the plateau continued to rise upwards – the crustal rocks could no longer be supported by the compressional stresses around them and the rocks at depth began to deform and flow outwards from the margins of the plateau. However, rocks at the surface (which would be under less pressure) would be cold and brittle. As the rocks beneath them began to flow, the uppermost rocks dropped downwards, fracturing along fault lines. Today, these normal faults provide evidence for the collapse of the 'roof of the world'.

These conclusions are important because they provide a means of dating the final stages of the uplift of Tibet. If the initial movement on the faults could be dated, it would give us an idea of when the plateau approached its maximum elevation before beginning to collapse. Happily, it is possible (though not easy) to date fault movement. During movement along a fault, rocks that are caught up in the fault

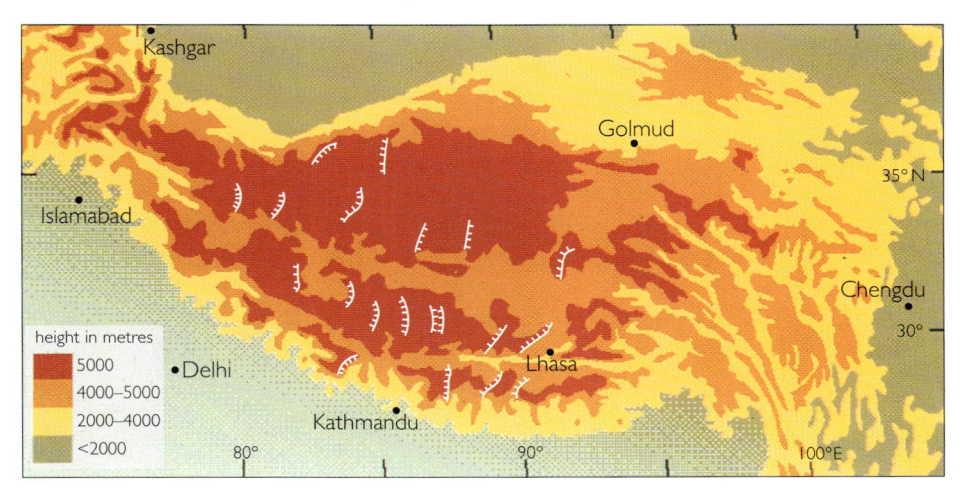

Figure 6.9
The distribution of normal faults on the Tibetan Plateau. The short bars along the fault lines indicate the side of the fault that has collapsed.

zone are deformed and the minerals that make up these rocks recrystallize. Such recrystallized minerals can be recognized because they regrow in parallel alignment to the direction of movement along the fault. Once identified, these minerals can be dated using radioactive isotopes.

At the time of writing (1997), only two normal faults, in different parts of Tibet, have been dated. Results from these two studies suggest early movement at 14 ± 3 Ma and movement again at 8 ± 1 Ma. No-one knows how representative these dates are for the plateau as a whole but we can tentatively conclude that although parts of the Tibetan Plateau reached a maximum elevation by 14 million years ago, other regions did not reach their maximum elevation until 8 million years ago. Subsequent collapse resulted in the plateau as we know it today.

> **Question 6.1**
> Do the results from dating normal faults discussed above agree with the uplift history inferred from paleobotanical studies based on the 'nearest living relative' approach (see Figure 6.5)?

Your answer to this question should confirm that the uplift history of the Tibetan Plateau is not yet well understood. Given that different dating methods predict different rates of uplift for the Tibetan Plateau and the Himalayas, it is interesting to see the extent to which any of them might fit in with the timing of climate change in the region.

6.2 The uplift of Tibet and climate change

As mentioned at the very start of this chapter, our understanding of how the climate of southern Asia has changed over time depends very largely on assumptions concerning changes in the monsoons. It makes sense, therefore, to begin our discussion of climatic change over long time-scales with a brief look at climate in this region at the present day.

6.2.1 Present-day climate

To a large extent, the climate of southern Asia is dominated by the monsoons. As discussed in Section 2.2.1, and illustrated in Figures 2.18b and 2.19a, in the northern winter, the region is affected by cold, dry north-easterly winds blowing out from the intense high pressure region over the continent. By contrast, in the northern summer, when there is a strong low pressure region over the Eurasian continent (Figures 2.18a and 2.19b), the area receives moisture-laden air in the South-West Monsoon.

The whole of southern Asia benefits from the monsoon rains. In particular, the areas of tropical rainforest along south-west India, Burma and Sri Lanka (Figure 3.4) are the results of the south-westerlies releasing much of their moisture over high land. On approaching the southern slopes of the Himalayas, the still-moist air mass is driven upwards further, causing summer rainfall over northern India. As a result, annual rainfall in Darjeeling on the southern slopes of the Himalayas is over 3000 mm, of which 88% falls between June and September, i.e. during the South-West Monsoon. Less than 100 km away, to the north of the Himalayan mountains, the Tibetan town of Gyangtse receives an annual rainfall of only 270 mm. The contrasting climates result in very different floras and faunas: the southern slopes of the Himalayas are covered by rainforests supporting a population density of 200 people per square kilometre; Tibet – effectively in the rain shadow of the Himalayas (cf. Figure 2.20a) – is characterized by semi-arid steppe, and in places is a rocky desert, hardly supporting one person per square kilometre (Figure 6.10b).

Tibet, the Himalayas and the Arabian Sea

Figure 6.10
Two locations only 50 km apart, but on different sides of the Himalayan watershed.
(a) On the southern slopes of the Himalayas, rhododendron forests flourish in abundant rainfall.
(b) Rocky desert conditions on the northern side of the Himalayas.

As Figure 2.18 shows, the seasonal shift in the ITCZ means that seasonally changing winds – monsoons – affect large parts of the globe at low latitudes. However, the extreme change in pressure over a large part of central southern Eurasia from intense high pressure in winter (Figure 2.19b) to very low pressure in summer (Figure 2.19a) means that seasonal changes in the vicinity of southern Asia and the Arabian Sea are by far the most dramatic. It seems that the reason for this may lie in the existence of the Tibetan Plateau itself.

In 1989, two American climatologists, Maureen Raymo and Bill Ruddiman, published the results of a series of experiments using a sophisticated computer model of the global climate (a 'general circulation model' or GCM) designed to investigate the effect on climate of such an extensive high-altitude plateau. Starting with a simulation of present-day climate, they changed just one variable: the topography of present landmasses. When the Tibetan Plateau was 'removed', the heavy summer rainfall in northern India all but disappeared. An even larger, higher plateau in central Asia greatly *increased* the area of summer monsoon rainfall throughout extensive regions south of the plateau, caused desert conditions over vast areas to the north of the plateau, and decreased summer precipitation much further west in the Mediterranean region.

Of course, such experiments had their critics, many of whom emphasized that no model can take into account all the possible variables. After all, if we think about climate change over the past 60 million years, uplift of the Tibetan Plateau is not the only change to have taken place within the Earth system. Not only did the Indian landmass move northwards across the globe, traversing climatic belts, but the global climate was itself changing. Nevertheless, the results of the modelling clearly suggested that the uplift of Tibet could have had a dramatic effect on atmospheric circulation and precipitation throughout much of the Northern Hemisphere, and may well have affected the strength of the monsoon winds – particularly those of the South-West Monsoon over southern Asia.

The strength of the South-West Monsoon is determined by the pressure difference between the high over the tropical Indian Ocean and the low over the southern part of the continent (Figure 2.18a). At the end of winter, the large rocky mass of the Tibetan Plateau heats up fast once its high-albedo covering of snow has melted; the overlying air is warmed, and the pressure over the continent falls. However, the warming of the continent is not the whole story.

Question 6.2
Because they have blown over the Indian Ocean and the Arabian Sea, the winds of the South-West Monsoon are initially laden with moisture. Referring to Section 2.2.1 if necessary, can you suggest why the release of this moisture as the Himalayan monsoon rains helps to intensify the circulatory pattern shown in Figure 2.19b?

The warming of the air over the Himalayas and Tibet has a particularly large effect because the heat source is provided at a critical point in the circulating air. Furthermore, at such altitudes the air is thin and its temperature is therefore more sensitive to changes in heat content.

As discussed in the answer to Question 6.2, the monsoon rains themselves indirectly help to warm the air over Tibet. The condensation of moisture to form rain over the southern Himalayas releases latent heat, and so the summer winds driving from the south into Tibet are not only dry but warm (cf. Figure 2.20a). At this latitude, in the absence of a plateau the air temperature at 5 km above sea-level would be around −20 °C (Figure 2.4). As it is, during the summer months temperatures rarely drop below freezing. Nevertheless, the dramatic summer storms that produce rain over the Himalayas may produce hail over Tibet (Figure 6.11).

(a)

(b)

Figure 6.11
(a) The Royal Crest of the Himalayan Kingdom of Bhutan. The crossed 'thunderbolts' are a common symbol in Tibetan culture, reflecting the power of summer storms in the Himalayas.
(b) Dark clouds presaging a violent hailstorm on a Tibetan summer afternoon. High plateaux like Tibet are notorious for their vicious hailstorms, caused by sudden updrafts within clouds under cold conditions.

■ By reference to Figure 2.17, can you suggest another way in which a high, extensive plateau at the latitude of Tibet (~ 30°–40° N) could affect the circulation of the atmosphere?

■ A large, high-altitude plateau could interfere with flow in the upper troposphere, and Figure 2.17 shows that the path of the northern subtropical jet stream could well be affected.

Indeed, the Tibetan Plateau is so high that the subtropical jet stream passes either to the north or south of it. The implications of this for climate (and weather) are not well understood, but it is clear that before the uplift of Tibet the jet stream would not have been diverted in this way.

The overall effect of the Tibetan Plateau and the Himalayas on atmospheric circulation is therefore determined by both their high elevation and their geographical position. Uplift must have caused major changes in atmospheric circulation across the Northern Hemisphere. As far as the South-West Monsoon was concerned, because summer heating of the atmosphere over Tibet would have increased as the plateau rose, it is possible that at some stage during its elevation a threshold altitude was reached, above which the monsoon winds were greatly strengthened. Partly for this reason, scientists have been looking for evidence of climate change in southern Asia that can be linked to uplift of Tibet. To find such a link, we need to know something about the timing of climate change in southern Asia over the relevant period.

6.2.2 Evidence for climate change

As discussed in Section 6.1.2, the Tibetan Plateau must have begun to rise some time after 70 million years ago (probably after 50 million years ago), and stopped rising between 14 and 8 million years ago (or more recently, according to studies using the 'nearest living relative' approach). In this section, we will first look briefly at some evidence for how climate in southern Asia might have changed over the past 10 million years or so, and then set this in the context of global climatic change during the Tertiary Period (i.e. since 65 Ma).

Climate change in southern Asia

Evidence for climate change in southern Asia over the past 10 million years is drawn from a number of different lines of study. We will begin by looking at two techniques which utilize the remains of living organisms: zooplankton which lived in the surface waters of the Arabian Sea; and terrestrial plants, along with the mammals and rodents that fed on them.

We will begin with the planktonic organisms, but before we go into the detail, look at Figure 6.12 (overleaf) which shows the concentration of phytoplankton in surface waters of the Indian Ocean (a) during the inter-monsoon period, when winds are generally light, and (b) during the summer monsoon, when winds over the northern Indian Ocean and Arabian Sea are strong and from the south-west.

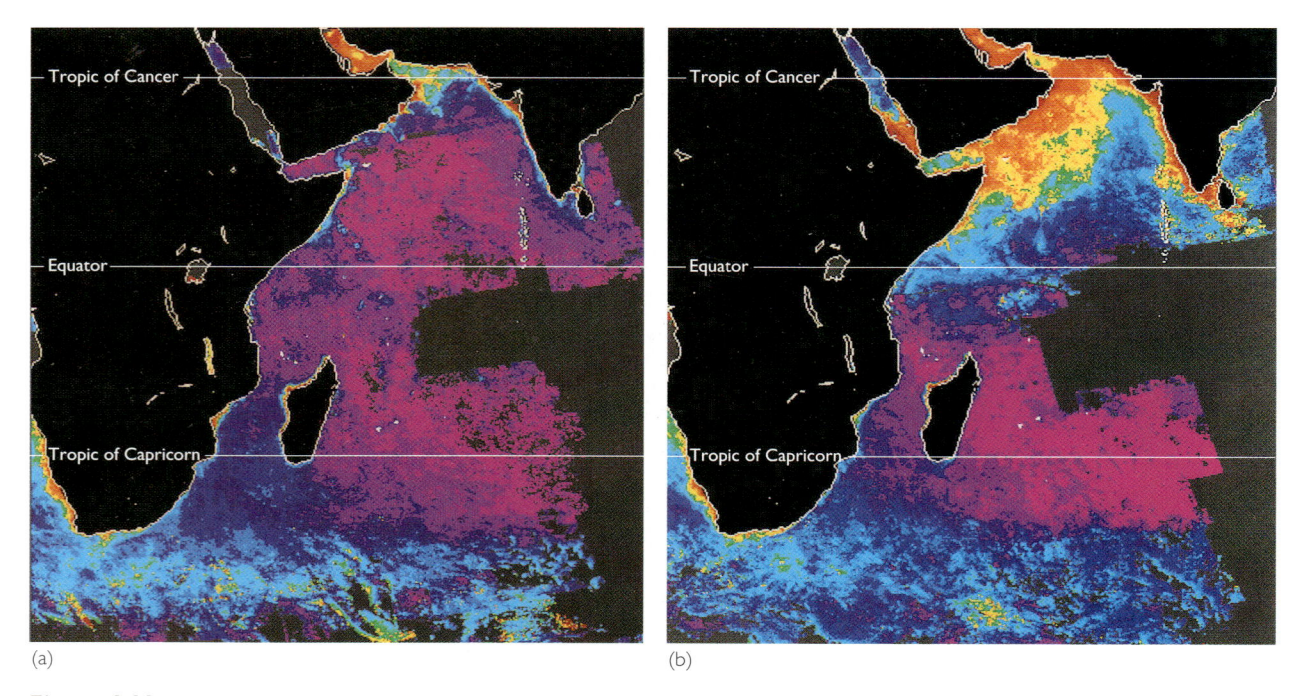

Figure 6.12
Seasonal variations in phytoplankton concentrations, on the basis of chlorophyll pigment recorded by the satellite-borne Coastal Zone Color Scanner. Highest concentrations are shown in red, lowest in pinkish purple, and no data in black. (a) A composite image for May–June, a period of light winds. (b) A composite for September–October, during the South-West Monsoon.

■   Without going into details, to what could you attribute the different levels of primary productivity in the northern Indian Ocean/Arabian Sea, shown in the two images?

■   To the fact that upwelling is stronger and more widespread in this region during the South-West Monsoon.

As discussed in earlier chapters, upwelling brings nutrient-rich subsurface water into the photic zone, supporting populations of phytoplankton, on which feed zooplankton and larger organisms. If you compare Figure 6.12 with the wind patterns shown in Figure 2.18a, you will see that the nearshore upwelling occurring here cannot result from longshore equatorward winds (Figure 2.27a), as it does in the tropical regions of the other oceans. Instead, it mainly occurs where surface waters diverge in cyclonic gyres (Figure 2.27b(ii)), and at places where surface currents diverge from the coast.

Now back to our discussion of climatic change. The zooplanktonic organism that we are going to consider is a species of foraminiferan known as *Globigerina bulloides* (cf. Figure 6.13a). Its abundance at various times in the past 14 million years, relative to other species of microplankton (those between 20 and 200 μm across), has been estimated on the basis of its fossil remains in sediment cores drilled from the floor of the Arabian Sea (cf. Figure 6.13b). *G. bulloides* is presently abundant in nutrient-rich tropical waters, and in the Arabian Sea its abundance (expressed as a proportion of the total microplankton population) increases by three orders of magnitude during periods of upwelling.

Tibet, the Himalayas and the Arabian Sea

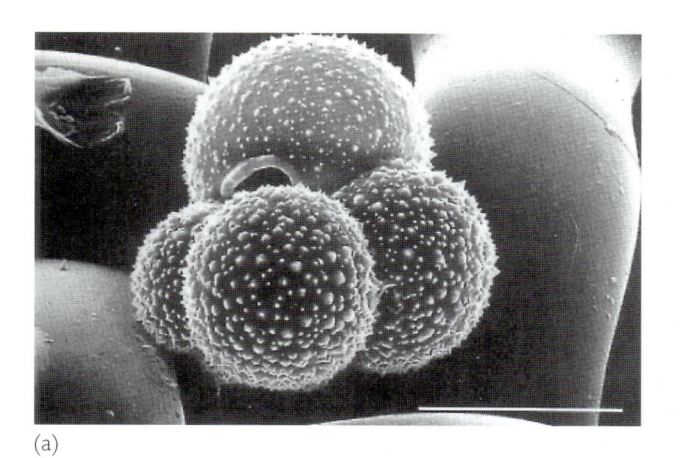

Figure 6.13
(a) The remains of a specimen of *Globigerina* sp. The background is the mesh of the sampling net. Scale bar = 50 µm.
(b) Variation in the proportion of the microplankton population made up of *G. bulloides* over the past 14 million years, from a sea-bed drill core from the Arabian Sea.

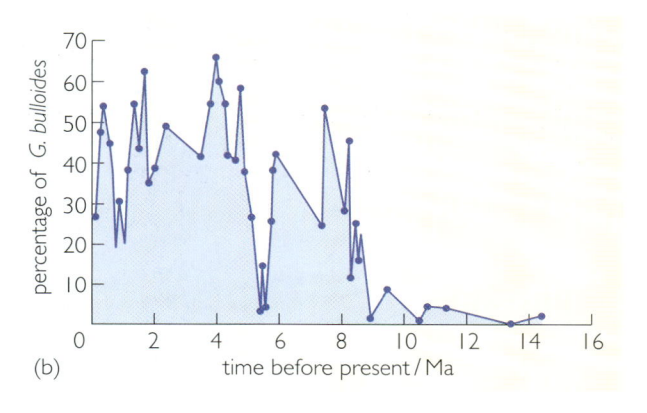

Question 6.3

(a) In general terms, how would you describe the variation in the abundance of *Globigerina bulloides* over the past 14 million years?

(b) Bearing in mind the high levels of primary productivity shown in Figure 6.12b, does the plot in Figure 6.13b tell us anything about changes in the strength of the South-West Monsoon over the time in question?

So, the patterns of relative abundance of *G. bulloides* suggest that the South-West Monsoon became stronger at about the time that the Tibetan Plateau was attaining its maximum elevation.

For the second clue to past climate, we turn to organic debris in sediments eroded from the Himalayas. As the Himalayas rose, so the rate of erosion of the steepening slopes increased. Great rivers flowing southwards deposited much of the eroded material into a large subsiding basin, with the result that these sediments from the Himalayas are now exposed in northern Pakistan and India. Their use as climatic indicators lies in the proportion of the different isotopes of carbon that they contain, expressed in terms of the ratio $\delta^{13}C$; for more about $\delta^{13}C$, see Box 6.1.

> **Box 6.1 Carbon isotopes and $\delta^{13}C$**
>
> Carbon occurs in nature as two stable isotopes: ^{12}C and the much rarer ^{13}C. During photosynthesis, fixation of the lighter $^{12}CO_2$ is favoured over that of the heavier $^{13}CO_2$ because $^{12}CO_2$ diffuses into cells more rapidly and more readily takes part in chemical reactions. As a result of this fractionation of isotopes, organic matter produced by photosynthesis is enriched in ^{12}C and depleted in ^{13}C relative to the inorganic carbon in the atmosphere and hydrosphere (CO_2 gas, plus bicarbonate and carbonate ions in solution). Enrichment or depletion of ^{13}C is expressed using the ratio $\delta^{13}C$, which is calculated in an analogous way to $\delta^{18}O$ (Box 5.2) to give a value in parts per thousand or 'per mil', often written as ‰:
>
> $$\delta^{13}C = \left(\frac{(^{13}C/^{12}C)_{sample}}{(^{13}C/^{12}C)_{standard}} - 1 \right) \times 1000$$
>
> (Equation 6.1)
>
> $(^{13}C/^{12}C)_{standard}$ is the ratio calculated for a standard carbonate sample, and is 1/88.99. If $^{13}C/^{12}C$ is greater in the sample than in the standard, the ratio of ratios will be greater than one and the expression in the large brackets, and hence $\delta^{13}C$, will be positive; if $^{13}C/^{12}C$ is less in the sample than in the standard, the expression in the large brackets, and hence $\delta^{13}C$, will be negative. Higher values of $\delta^{13}C$ (or less negative ones) correspond to a higher proportion of ^{13}C.
>
> $\delta^{13}C$ for atmospheric CO_2 is −8‰, and because of fractionation, plant tissue generally contains a lower proportion of $^{13}CO_2$ by about 2% (= 20‰); as a result, plant tissue has an average $\delta^{13}C$ value of about −(8 + 20) = −28‰.
>
> ■ So does organic material (plant or animal tissue) have a higher or lower value than inorganic carbon?
>
> ■ It is always lower (more negative).
>
> The $\delta^{13}C$ of a sample tells us more than just whether it is plant-derived, however. Because of the different mechanisms of photosynthesis that they use, plants and shrubs that flourish under warm conditions tend to incorporate more of the isotope ^{13}C than those that thrive in colder conditions. As a result of plant respiration through the roots and the accumulation of plant debris, soil acquires a $\delta^{13}C$ 'signature' similar to that of the plants that grew in it. In particular, carbonates that precipitate in soil have a $\delta^{13}C$ value somewhere between that of atmospheric CO_2 and that of the living plants and plant debris in the soil. Measurement of $\delta^{13}C$ for carbon (as organic carbon or calcium carbonate) in soil of different ages can therefore provide an indicator of the temperatures that prevailed during the growing season.

The information that you need to take away from Box 6.1 is simply this: the $\delta^{13}C$ of plant (and hence animal) tissue is always negative, but is higher for the kinds of plants that flourish under warm conditions.

The sediments from the Himalayas, referred to above, eventually formed soil, within which carbonates were precipitated. As discussed in Box 6.1, the $\delta^{13}C$ of these carbonates would reflect the $\delta^{13}C$ of organic debris in the soil. Figure 6.14 is a plot of $\delta^{13}C$ against time for these soil carbonates.

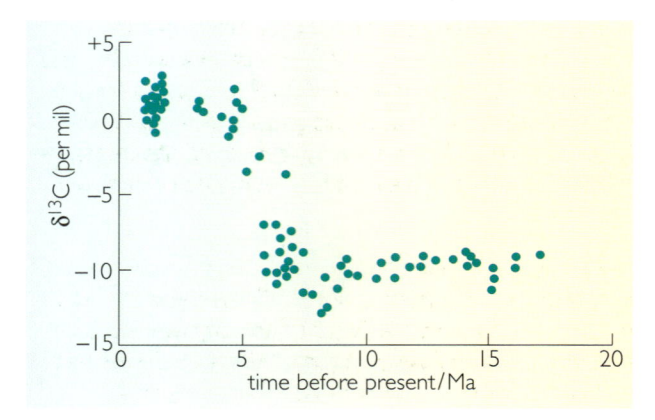

Figure 6.14 Variation of $\delta^{13}C$ with time in carbonates in soil formed from debris eroded from the Himalayas of northern Pakistan.

■ According to Figure 6.14, have there been any marked changes in the $\delta^{13}C$ values of organic debris eroded from the Himalayas during the past 20 million years?

■ Yes. There was a sharp increase in $\delta^{13}C$ values from −10‰ to values of around 0, between 8–7 and 5 million years ago.

The increase in organic material incorporating more ^{13}C has been interpreted by some paleoclimatologists as an explosion of plant production during the growing season, resulting from increased summer rainfall because of a strengthening of the South-West Monsoon 8–7 million years ago. However, because this increase in $\delta^{13}C$ values has been recognized in plant remains from other continents and is not a peculiarity of southern Asia, other workers in the field have argued that it marks a spread of plants using a slightly different mechanism for photosynthesis, which evolved within the last 14 million years. Some modern plant groups rely on this newer mechanism exclusively, and are referred to as C4 plants; the majority, using the original mechanism, are referred to as C3 plants. Figure 6.14 might therefore chart a transition of plant type which could be part of a global response to climate change that may or may not be directly linked to the strengthening of the South-West Monsoon.

Interestingly, the remains of fossilized mammal teeth now preserved in the Himalayan sediments suggest a change in the vegetation in the region about 8 million years ago. The shapes of the teeth suggest that at about this time there was a marked change from browsers (which feed on trees and shrubs) to grazers (grass-eaters); furthermore, some identifiable forest-based mammals (e.g. orang-utans) disappeared from the region at that time. Trees and shrubs are C3 plants, grasses are C4 plants.

Finally, the *type of sediment* originating from the Himalayas may be used as an indicator of the temperature at which weathering occurred. Before about 7 million years ago, the deposits carried down from the Himalayas were predominantly sands and silts; sediments like these suggest strong erosional forces – i.e. physical weathering, by freeze–thaw or glaciers, acting on exposed rock surfaces lacking soil cover. In contrast, sediments younger than 7 million years old include plenty of muds and clays, suggesting chemical weathering, producing thick layers of soil in the source areas of the rivers. Like the $\delta^{13}C$ data, the size-distribution of sediments in the sedimentary record points towards a sudden increase in summer rainfall in southern Asia around 7 million years ago.

We can summarize as follows: Although it is not possible to be precise about the timing of *either* the uplift of the Tibetan Plateau *or* climate change in southern Asia, there is evidence that summer rainfall associated with the South-West Monsoon increased between 9 and 6 million years ago. It is not known whether the Tibetan Plateau suddenly increased in elevation at about this time, but the dating of normal faults from the highest parts of the plateau suggests that it did (Section 6.1.2). The apparent simultaneity of rapid uplift and strengthening of the monsoon is certainly consistent with (but not conclusive proof of) the causal link suggested by computer-modelling experiments.

Global climate change during the Tertiary

So far, we have investigated a possible link between uplift of the Tibetan Plateau and a strengthening of the South-West Monsoon. Although this may well have affected a large part of the globe, it was essentially a regional effect on the climate. It is also possible that uplift of a high plateau could have an effect on *global* climate. If so, we might expect the change to have taken place not over the past 10 million years as the monsoon was strengthening, but over the past 50 million years, i.e. since uplift of the plateau was initiated by continental collision. Before we look more closely at possible global consequences, we should first establish how global climate has changed since the collision between India and Eurasia.

The evidence for past temperature changes rests largely on the use of oxygen isotopes (Box 5.2). The variation in global average temperature over the past 120 million years, deduced from oxygen-isotope studies of the remains of deep-sea benthic foraminiferans, is shown in Figure 6.15. Despite fluctuations, there is a clear downward trend, with a net cooling of nearly 20 °C over the period concerned. There is strong independent geological support for this inference. For example, during the past 50 million years, the distribution of sediments deposited by glaciers has generally been increasing; such sediments were being deposited further and further from the polar regions and reached their maximum extent during the peak of the present (Quaternary) Ice Age.

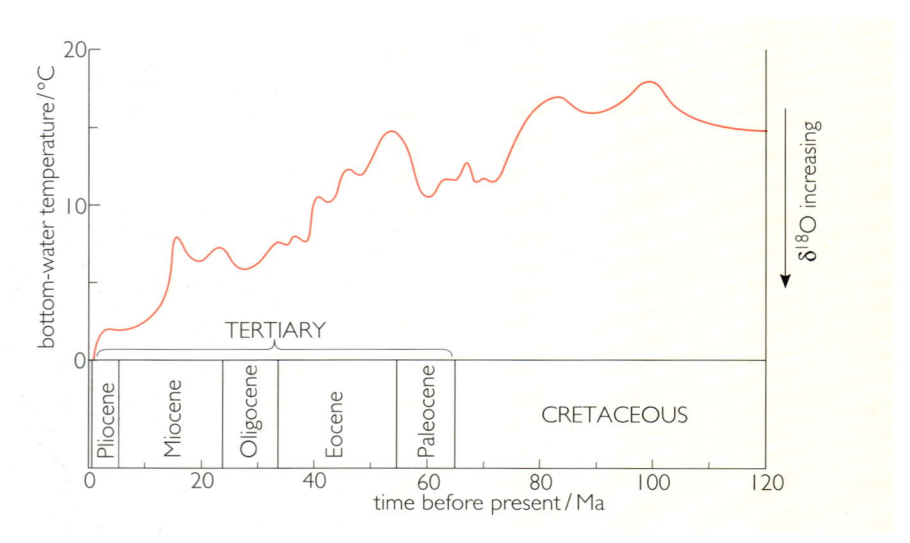

Figure 6.15
Variations in the temperature of ocean bottom waters over the past 120 million years, estimated using oxygen-isotope ratios ($\delta^{18}O$) from deep-sea benthic foraminiferans. (This is an expanded version of the left-hand side of the temperature plot in Figure 5.19; the unlabelled division on the left corresponds to the Quaternary Period.)

What caused such a decrease in global temperatures? Perhaps the rearrangement of the continents played a key role, as discussed in the previous chapter. But however the continents are re-arranged, computer-based climate models cannot reproduce the long-term cooling pattern suggested by Figure 6.15 and similar plots. What's more, the climate-modelling experiments that investigated the effects of the Tibetan Plateau on global climate failed to show that the uplifting of Tibet could, by itself, cause long-term cooling. It seems that something else is required to explain it.

- Bearing in mind our discussion of climate during the Cretaceous (Section 5.3.4), can you suggest what other factor might have been involved?

- A change in the composition of the atmosphere – in particular, a long-term *decrease* in the concentration of CO_2 (as opposed to an increase, which occurred during the Cretaceous).

For the sake of argument, let us assume that the temperature variations shown in Figure 6.15 resulted entirely from fluctuations in atmospheric CO_2 concentration. If this were so, to account for the overall fall in temperature, the CO_2 concentration would have to have declined from a value about eight times that of the present day. Many geoscientists believe that the building of large mountain belts in central Asia could well have played a role in bringing about such a dramatic change in atmospheric composition. This particular aspect of mountain-building, discussed briefly in Section 5.4.3 and in more detail in the following section, is thought by some to be an important mechanism for changing atmospheric CO_2 concentrations, and hence climate.

6.3 The Himalayas, Tibet and atmospheric CO_2

Although bordered by impressive mountain ranges, the interior of Tibet is truly a plateau, with a local relief of generally no more than a kilometre or so (Figure 6.16a). The plateau is the catchment area for many of Asia's great rivers, including the Indus, the Brahmaputra (known in Tibet as the Tsangpo), the Yangtze, and the Mekong (Figure 6.17). These rivers have to descend 5 km before they reach the sea, eroding their way through mountain ranges that are steadily being uplifted (Figure 6.16b).

Figure 6.16 (a) The gentle relief typical of much of central Tibet.

Figure 6.16 (b) A deeply incised Himalayan gorge, carved out by a river flowing south from Tibet.

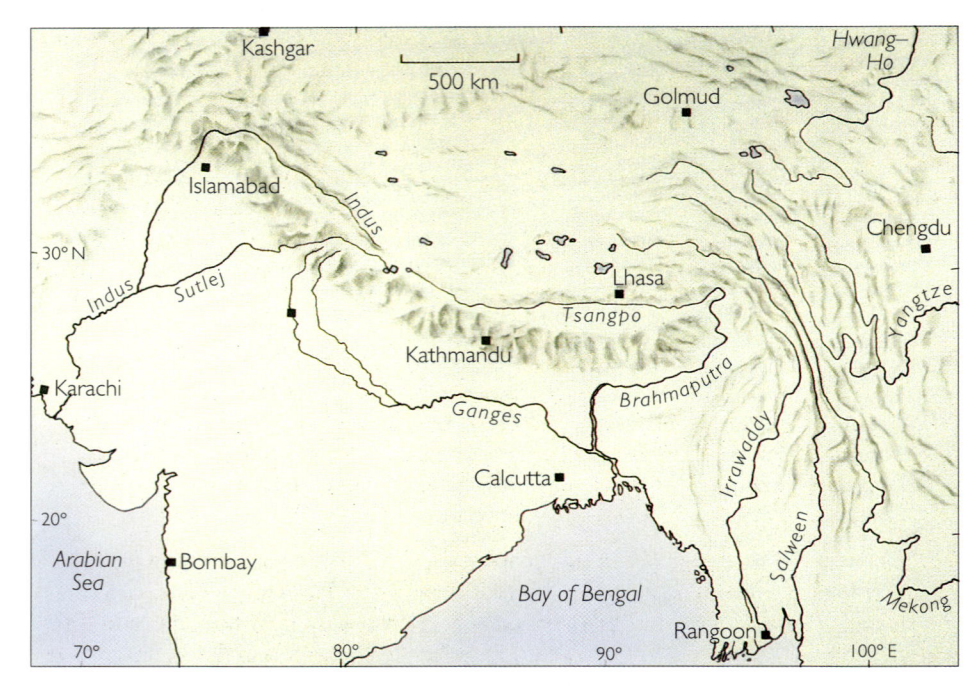

Figure 6.17
The central Asian river system, showing major rivers with sources on the Tibetan Plateau or in the Himalayas.

Question 6.4
Table 6.1 lists the flux of dissolved material of rivers with a source in the Tibet/Himalaya region. The total flux of dissolved material carried by rivers globally is about 3500×10^9 kg yr^{-1}. Given that the catchment area concerned (Figure 6.17) is 5% of the Earth's continental surface, would you say that Table 6.1 suggests the weathering rates in the Tibet/Himalaya region to be unusually high?

Table 6.1 Fluxes of dissolved material in rivers with sources in Tibet or the Himalayas.

River	Flux of dissolved material / 10^9 kg yr^{-1}
Ganges	184
Yangtze	169
Brahmaputra (Tsangpo)	146
Irrawaddy	93
Indus	63
Mekong	61
Salween	46
Hwang-Ho	34

So, rivers flowing down from Tibet and the Himalayas have great erosive power and carry an unusually large load of fragmented rocks (cf. Figure 5.26) and dissolved material (Section 5.4.3).

■ Bearing in mind the high rates of chemical weathering on the southern slopes of the Himalayas (and remembering Section 5.3.3), can you suggest how the formation of the Himalayas and the uplift of the Tibetan Plateau could indirectly affect the concentration of CO_2 in the atmosphere?

■ Weathering of silicate minerals, followed by accumulation/preservation of organic carbon and carbonates in the ocean, results in net removal of CO_2 from the atmosphere (Figure 3.20). So if uplift of the Himalayas and Tibet increased total global weathering rates, it would also have increased the rate at which CO_2 was removed from the atmosphere.

Of course, a change in the concentration of atmospheric CO_2 would affect fluxes into and out of other carbon reservoirs, and it would be some time before a new equilibrium was established. It has been estimated that reduced atmospheric CO_2 levels would begin to be established about one million years after the mountain range had been uplifted. Other factors being equal, this reduction in atmospheric CO_2 concentrations would lead to global cooling.

In climate models proposed in recent years, the role of mountain-building in the global climate system has been treated in various ways. We will consider two approaches that differ in their assumption concerning what primarily determines the CO_2 concentration of the atmosphere, and hence what ultimately drives carbon fluxes between the various carbon reservoirs – the atmosphere, the oceans and the Earth's crust.

The **'BLAG' model** (named after three American scientists – Bob Berner, Tony Lasaga and Bob Garrels) rests on two important assumptions:

1. Global temperatures are determined by the concentration of CO_2 in the atmosphere.
2. The concentration of CO_2 in the atmosphere is determined primarily by the volume of gases emitted from volcanoes.

An *approximate* measure of the global rate of emission of volcanic gas can be obtained from the rate of production of new sea-floor: volcanoes above subduction zones release CO_2 directly into the atmosphere and some of that from hydrothermal vents and underwater eruptions eventually escapes from the ocean (Sections 4.1, 4.6 and 5.3.2). There is evidence to suggest that over the past 100 million years the rate of production of oceanic crust has almost halved, gradually returning to roughly what it was before the eruption of major flood basalts began, about 120 million years ago (Figure 4.22). Less CO_2 was being supplied by volcanism so, according to the BLAG model, there would have been global cooling as a result of lower concentrations of CO_2 in the atmosphere.

An alternative approach has been taken by scientists who suggest that changes in atmospheric CO_2 concentrations are driven primarily by changes in chemical weathering rates. Bill Ruddiman and Maureen Raymo (mentioned earlier) have extended the idea that uplift of the Tibetan Plateau strengthened the South-West Monsoon, and suggested that high weathering rates over a region of steep topography affected by high summer rainfall were at least partly responsible for the global cooling that followed the collision between India and Eurasia. If this is true, uplift of Tibet effectively set the scene for the glacial periods that have characterized the climate in recent geological time.

Distinguishing between these two models is presently an area of active scientific debate. To get an idea of some of the arguments involved, try Question 6.5.

Question 6.5
Figure 6.15 shows an overall downward trend in temperature from about 100 million years ago, but there have been fluctuations and the *rate* of decrease has been variable. On the basis of what you know about the timing of the building of the Himalayas and the uplift of Tibet, and information in Figure 4.22, to what extent do you think the shape of the temperature plot in Figure 6.15 could be used to support (1) the BLAG hypothesis and (2) the Raymo–Ruddiman hypothesis? (*Please read the answer to this question before continuing.*)

While such exercises are useful in assessing the relative merits of the two hypotheses, it is important to bear in mind that diagrams like Figure 4.22 (and Figure 6.15) can only be very rough estimates, so we should be wary of placing too much faith in deductions based on them. Furthermore, supporters of the BLAG model argue that the **Raymo–Ruddiman hypothesis** is a nice idea but is ultimately untestable: until reliable values can be obtained for the carbon fluxes between continents, oceans and atmosphere, the results of their model can really only be described as qualitative.

Let us leave these criticisms on one side for the time being, and explore in more detail whether weathering of the Himalayas and Tibet *could* have caused global climate change over the past 50 million years. The first thing to remember is that weathering only results in the removal of CO_2 from the atmosphere because some of the carbon carried to the ocean in rivers is then *removed* from the oceans and preserved in sediments (Section 3.4): it is shunted out of the intermediate-scale marine carbon cycle and becomes part of the long-term geological carbon cycle (Figure 3.20). As you know, there are two ways in which carbon can be preserved and buried in deep-sea sediments: as carbonaceous sediments (organic carbon) and as carbonates (inorganic carbon).

Let's begin with the organic carbon. As discussed in Section 5.3.4, in connection with oceanic conditions during the Cretaceous (Figure 5.23), one factor that would increase the proportion of organic remains preserved in the deep ocean is a large supply of organic debris in rivers. Before collision of India with Asia, unusually organic-rich sediments were deposited on the continental shelf along the northern margin of the Indian Plate. After collision, the buried sediments were uplifted and exposed at the surface, where they could be weathered and eroded. As a result, much of this organic material would have been carried to the sea, as both particulate and dissolved organic carbon. As discussed in Box 6.1, $\delta^{13}C$ values for organic carbon are lower than those for carbon for other carbon reservoirs. An increase in the flux of organic carbon to the oceans would therefore eventually result in a decrease in the average $\delta^{13}C$ value for sediments (both organic and calcareous) being deposited on the sea-bed. A plot of $\delta^{13}C$ against depth in marine sediments therefore provides a measure of the contribution of organic carbon to the oceans over time. (Note that for reasons we don't need to go into, marine carbonates always have higher $\delta^{13}C$ values than soil carbonates, cf. Figure 6.14.)

Tibet, the Himalayas and the Arabian Sea

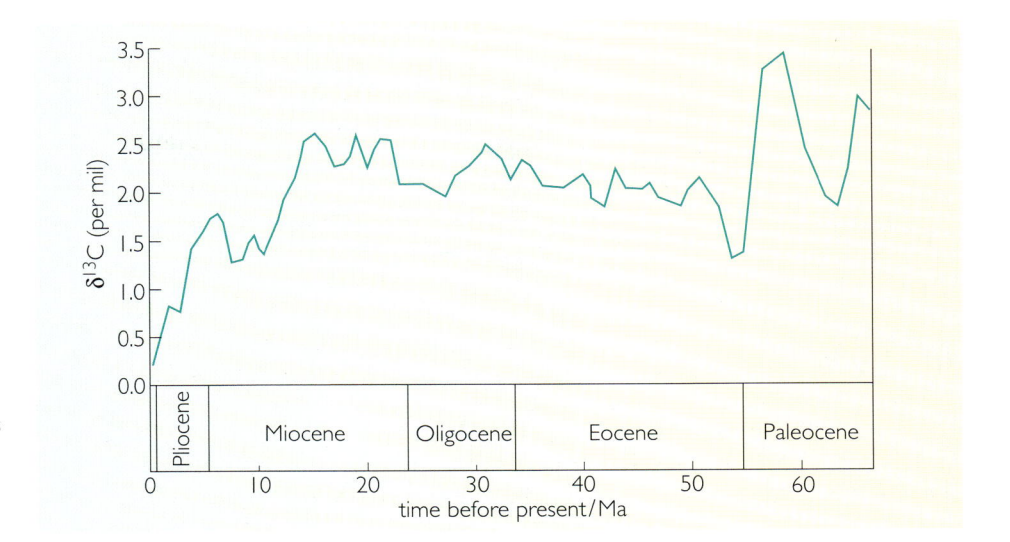

Figure 6.18
The variation of $\delta^{13}C$ in marine carbonates laid down over the past 65 million years or so, i.e. during the Tertiary and Quaternary (the unlabelled division on the far left).

Question 6.6

(a) (i) What would happen to the average $\delta^{13}C$ value of marine carbonates if the global flux of organic carbon from the continents increased significantly?

(ii) According to Figure 6.18, when in the past 70 million years was the largest increase in the flux of organic carbon to the oceans?

(b) Moving on to preservation of *calcium carbonate* in deep marine sediments, why can we say with some confidence that this sink was operating more efficiently during and after the uplift of Tibet than it had been (say) 100 million years earlier?

(c) Bearing in mind the results of the IRONEX II experiment (Section 2.3.2), can you suggest why an increase in the rate of continental weathering might result in an increase in the accumulation and preservation of both inorganic and organic carbon? What other cause, also related to the uplift of Tibet and the Himalayas, would have a similar effect? (*Hint*: Remember Figure 6.12b.)

In fact, many geochemists regard the preservation of organic carbon as more important than preservation of carbonates. However, as your answer to Question 6.6 suggests, it is hard to determine the extent to which the accumulation of organic carbon is driven by weathering rates, especially as marine primary productivity is greatly affected by other influences, notably wind-driven upwelling, and the supply of nutrients generally.

Ironically, the very fact that uplift of Tibet and the Himalayas appears to have such a strong effect on the levels of atmospheric CO_2 could potentially be a problem for supporters of the Raymo–Ruddiman hypothesis, because it seems to leave open the possibility of runaway cooling – something that is not believed to happen, despite the overall decline in global temperature during the past 100 million years. According to the BLAG model, levels of atmospheric CO_2 will be maintained at more or less the same level through time by a negative feedback loop, as follows: if more CO_2 is released into the atmosphere as a result of increased rates of production of sea-floor, global temperatures will rise. This would lead to increasing weathering rates which in turn would *remove* CO_2 from the atmosphere, thus *decreasing* the temperature towards its initial value; and so on (Figure 6.19).

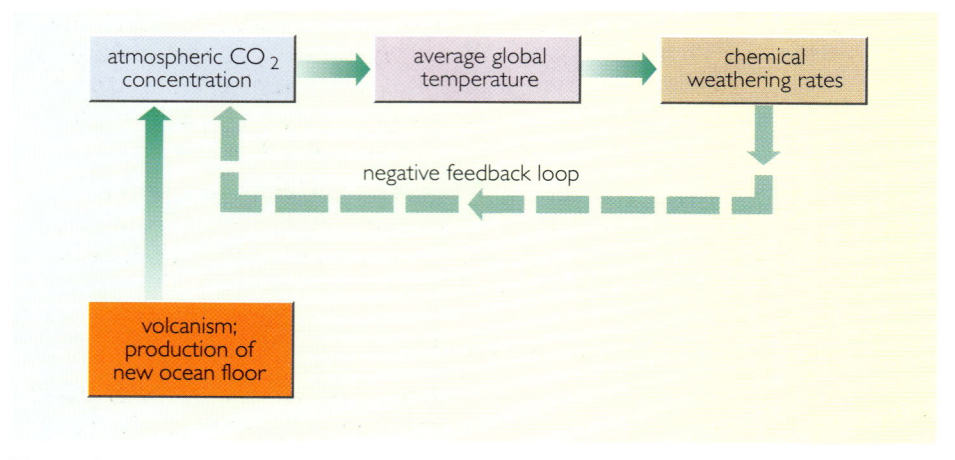

Figure 6.19
The negative feedback loop in the climate cycle derived from the BLAG model. If atmospheric CO_2 concentrations should *rise* because of *increased* rates of sea-floor production, global temperatures will *rise*, leading to *increased* weathering rates which in turn would *remove* CO_2 from the atmosphere, thus *decreasing* the temperature towards its initial value. Similarly, if atmospheric CO_2 concentrations should *fall*, global temperatures will *fall*, leading to *decreased* weathering rates which in turn would *remove less* CO_2 from the atmosphere (there would still be various sources of atmospheric CO_2), thus *increasing* the temperature towards its initial value.

On the other hand, if (as proposed by Raymo and Ruddiman) the concentration of CO_2 in the atmosphere is primarily controlled by mountain uplift, then there is no *direct* link between the rate of operation of the CO_2 sink (preservation and burial of carbon in the oceans) and the rate of operation of the CO_2 source (volcanic emission). As a result, there can be no direct stabilizing feedback loops of the type described above, to prevent runaway cooling. According to the Raymo–Ruddiman estimates of the rate at which weathering of Tibet and the Himalayas has depleted CO_2 in the atmosphere, weathering acting alone would exhaust all the CO_2 in the atmosphere in only a few million years — something that obviously hasn't happened. But we must remember that the processes that eventually lead to mountain-building can at the same time provide a source of atmospheric CO_2.

- Thinking back to Section 5.4.3, can you suggest what the nature of this source is?

- It is *decarbonation* of carbonates mixed with silica (Equation 5.2), both shallow-water carbonates from the continental margins and deep-water carbonates from the slab of oceanic crust descending beneath the collision zone.

Overall, during the lifetime of a particular evolving mountain belt, the CO_2 flux from lithosphere to atmosphere as a result of subduction/collision/volcanism will replenish a significant proportion of the CO_2 lost from the atmosphere through weathering. But it's important to note that although operation of source and sink are related generally through the rates of plate convergence at the destructive plate margin and collision zone, within individual zones of mountain-building decarbonation and weathering are *not* closely coupled: CO_2 gain to the atmosphere will occur early in the history of the mountain belt when carbonates and silicates are first heated; CO_2 loss will predominate later, with the formation of high mountains and rapid weathering.

Yet another source of CO_2 associated with the collision of continents could be the oxidation of organic carbon in the crust. As mentioned above, after the collision of India and Asia, buried carbon-rich sediments were exhumed. Once exposed at the surface, this organic carbon could be oxidized to CO_2 through bacterial activity and by simple chemical reaction with atmospheric oxygen (Section 3.4).

So there are at least two mechanisms that would have counteracted any tendency towards a runaway loss of CO_2 from the atmosphere through weathering of Tibet and the Himalayas: decarbonation reactions in rocks heated beneath the Earth's surface, and oxidation of organic carbon *at* the surface.

6.3.1 Strontium isotopes and the rate of chemical weathering

So far, we have not been able to comment definitively on whether the BLAG model or the Raymo–Ruddiman hypothesis is more appropriate for the estimation of changes in atmospheric CO_2 concentrations over the past 50 million years. When scientists need to decide which of two or more models is 'correct', they usually examine specific outcomes or predictions of the models, and see how they differ. As far as the BLAG and the Raymo–Ruddiman models are concerned, a significant difference arises in the contrasting way in which they link climate change and weathering rates.

- According to the BLAG model, how will a high rate of production of new ocean floor affect the rate at which the continents are weathered?

- According to this model, global temperatures are determined by atmospheric CO_2 concentrations, which in turn are determined primarily by volcanic emissions. Increased rates of production of new ocean floor will therefore lead to higher CO_2 levels and higher temperatures. They will therefore lead to an increase in rates of continental weathering, because chemical reactions occur more rapidly at higher temperatures (cf. Figure 6.19).

Implicit in the BLAG model, therefore, is the assumption that high rates of weathering of continental rocks would result from global warming.

- What would be the relationship between global temperatures and high continental weathering rates if the Raymo–Ruddiman hypothesis were correct?

- According to the Raymo–Ruddiman hypothesis, atmospheric CO_2 concentrations are driven primarily by continental weathering rates. As weathering of the continents removes CO_2 from the atmosphere, *high weathering rates should lead to global cooling.*

The BLAG model and the Raymo–Ruddiman hypothesis – at first sight at least – have contrasting implications for the relationship between global temperature, atmospheric CO_2 concentrations and weathering rates. The BLAG model involves an association between periods of high rates of chemical weathering and *high* global temperatures (high atmospheric CO_2), whereas Raymo and Ruddiman predict an association between high rates of chemical weathering and *low* global temperatures (low atmospheric CO_2). This is a significant point because it means that if we could measure the rate of continental weathering through time and could compare it with the variation in global temperature (Figure 6.15), it might help us decide whether the principal control on climate change (at least over the past 60 million years) has been, on the one hand, the rate of CO_2-release as a result of the production of new ocean floor or, on the other, the formation of high mountain ranges.

Fortunately, geochemists have discovered that certain isotopic ratios in marine sediments go some way towards providing a 'proxy' for rates of weathering of continental rocks. Before we can illustrate this point, it is necessary to explain a little about the isotopes of strontium; see Box 6.2.

Box 6.2 The isotopic composition of strontium

The element strontium (Sr) occurs in nature as several different isotopes. Geochemists are concerned with two of these, ^{86}Sr and ^{87}Sr. The first, ^{86}Sr, is stable and is not the decay product of any other isotope. The other isotope, ^{87}Sr, is known as **radiogenic strontium** because it is the product of radioactive decay of one of the isotopes of rubidium, ^{87}Rb.

Most rocks contain both strontium and rubidium. The strontium includes both ^{87}Sr and ^{86}Sr; the amount of ^{87}Sr is continually increasing due to the decay of radioactive ^{87}Rb, but the amount of ^{86}Sr remains unchanged. In other words, the ^{87}Sr/^{86}Sr ratio will increase with time. As a result, geochemists almost always refer to the ^{87}Sr/^{86}Sr ratio rather than to the absolute concentration of ^{87}Sr because the isotopic compositions of different rocks (or fluids) tell us more about the history of the rocks (or fluids) than the total concentration of Sr.

Rocks that make up the continents have variable ^{87}Sr/^{86}Sr ratios, but these values are all much greater than the ^{87}Sr/^{86}Sr ratio of rocks from the upper mantle.

The ^{87}Sr/^{86}Sr ratio of seawater today represents a combination of two components. The first, with a high average value of about 0.7119, comes from rivers which have entered the oceans after flowing over the continents. The second, with a low average value of 0.7035, comes from hydrothermal fluids which have circulated within the oceanic lithosphere and escaped at vents along ocean ridges; this lower value is similar to that of rocks of the upper mantle.

To explore the relative contributions of different chemical fluxes to a particular reservoir, we must use a mass-balance equation. Let us assume the reservoir is fed from two sources, the first with an Sr flux of F_1 and the second with an Sr flux of F_2. If the ^{87}Sr/^{86}Sr ratios for the two sources are R_1 and R_2 respectively, the ^{87}Sr/^{86}Sr ratio of the reservoir (R_0) is given by:

$$R_0 \times (F_1 + F_2) = (F_1 \times R_1) + (F_2 \times R_2) \qquad \text{(Equation 6.2)}$$

In the case of seawater, its ^{87}Sr/^{86}Sr ratio (R_{sw}) will be determined by two sources: hydrothermal fluids with an Sr flux (F_1) of 137×10^7 kg yr^{-1} and rivers with an Sr flux (F_2) of 290×10^7 kg yr^{-1}.

Question 6.7

According to Equation 6.2, what is the present-day ^{87}Sr/^{86}Sr ratio for seawater (R_{sw})?

The value you have calculated is the present-day ^{87}Sr/^{86}Sr ratio for seawater. The ratio is the same throughout the world's oceans, because they are continually stirred by currents. By measuring the isotopic composition of marine carbonates of known ages, geochemists have calculated what the ^{87}Sr/^{86}Sr ratio of seawater has been at various times over the past 100 million years. Their results are plotted in Figure 6.20.

Tibet, the Himalayas and the Arabian Sea

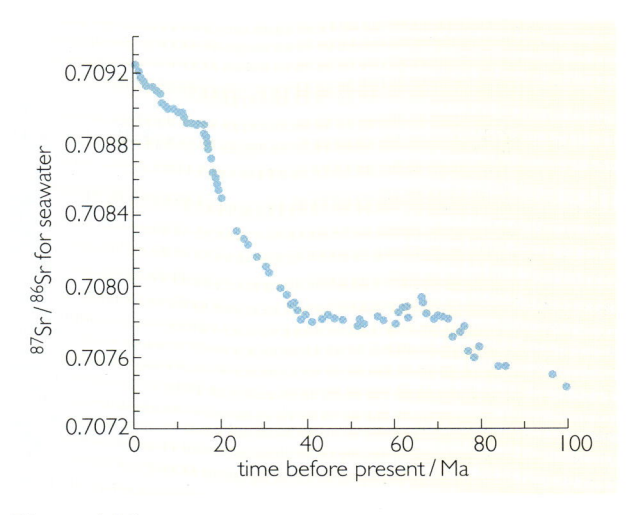

Figure 6.20
The variation in $^{87}Sr/^{86}Sr$ ratio for seawater over the past 100 million years.

- What has been the overall trend in the $^{87}Sr/^{86}Sr$ ratio of the ocean, since the Himalayan collision ~50 million years ago?

- There has been a marked increase.

This means that either the rate of supply of hydrothermal fluids (with low $^{87}Sr/^{86}Sr$ ratios) has decreased, or that the flux of Sr from rivers (with high $^{87}Sr/^{86}Sr$ ratios) has increased, or perhaps a combination of the two. Hydrothermal circulation is driven by the heating associated with sea-floor spreading and other volcanic activity at the sea-bed. As ocean-floor production has been occurring at roughly the same rate over the past 40 million years (Figure 4.22), the trend in Figure 6.20 is unlikely to reflect a marked decrease in hydrothermal circulation. On the basis of the data plotted in Figure 6.20, therefore, it appears that the flux of Sr into the ocean via rivers has *increased*. This in turn could be interpreted to mean that the rate of weathering of continental crust has been increasing since the Himalayan collision.

- If this interpretation is correct, does it support the Raymo–Ruddiman hypothesis?

- Yes it does, because the hypothesis predicts that weathering of silicates results in global cooling, and there has indeed been global cooling over the period in question (Figure 6.15).

However, so far, the evidence linking the uplift of the Himalayas and the Tibetan Plateau with weathering rates via seawater chemistry is entirely circumstantial. The following Activity should establish whether the rivers flowing through the Himalayas in particular have a strong influence on the present-day $^{87}Sr/^{86}Sr$ ratio of seawater.

Activity 6.1

Table 6.2 Fluxes of dissolved strontium and $^{87}Sr/^{86}Sr$ isotopic ratios for some major rivers.

	Flux of dissolved Sr / 10^7 kg yr^{-1}	$^{87}Sr/^{86}Sr$ ratio	$\dfrac{1}{\text{Sr flux }(10^7 \text{ kg yr}^{-1})}$
Amazon	22.18	0.7109	0.045
Brahmaputra	5.61	0.7210	0.178
Ganges	7.11	0.7257	0.141
Indus	7.93	0.7112	0.126
Mackenzie	5.78	0.7110	0.173
Mekong	15.95	0.7102	0.063
Orinoco	2.31	0.7183	0.433
Xingu	0.79	0.7292	1.269
Yangtze	18.48	0.7109	0.054
Yukon	3.10	0.7137	0.323
Zaire	3.91	0.7155	0.256

Table 6.2 provides data for the flux of dissolved strontium and the **Sr-isotope ratio** ($^{87}Sr/^{86}Sr$) for some of the world's major rivers. Several of these rivers drain the Himalayan and/or Tibetan Plateau (see Figure 6.17); of these, two – the Brahmaputra and Ganges – carry to the sea weathering products originating mainly in the Himalayas.

(a) (i) Figure 6.21a is a graph of the Sr flux versus the $^{87}Sr/^{86}Sr$ ratio, on which we have already plotted data for six of the rivers. Plot the data for the remainder of the rivers (the Amazon, Brahmaputra, Ganges, Xingu and Yangtze), and label them.

(ii) Now do the same for Figure 6.21b, a graph of the reciprocal of the Sr flux against the $^{87}Sr/^{86}Sr$ ratio.

(iii) Describe the overall trends in your completed Figure 6.21 *if* you ignore the Brahmaputra and the Ganges.

(b) As Equation 6.2 demonstrates, the effect a particular river has on the $^{87}Sr/^{86}Sr$ ratio of seawater depends on both the $^{87}Sr/^{86}Sr$ ratio of the river water and the flux of dissolved Sr in that river. For each of the Brahmaputra, Ganges, Xingu and Yangtze, discuss the relative effects of these two variables on the $^{87}Sr/^{86}Sr$ ratio of seawater.

Plots such as the ones you have drawn in Figure 6.21 confirm that the Ganges and Brahmaputra rivers are likely to have a significant effect on the $^{87}Sr/^{86}Sr$ ratios of seawater. However, is it really possible that the chemistry of the world's oceans could be so strongly influenced by only two rivers which, even when combined, carry less than 10% of the dissolved material entering the world's oceans? We can test this idea by simple mass-balance calculations, set out in Box 6.3.

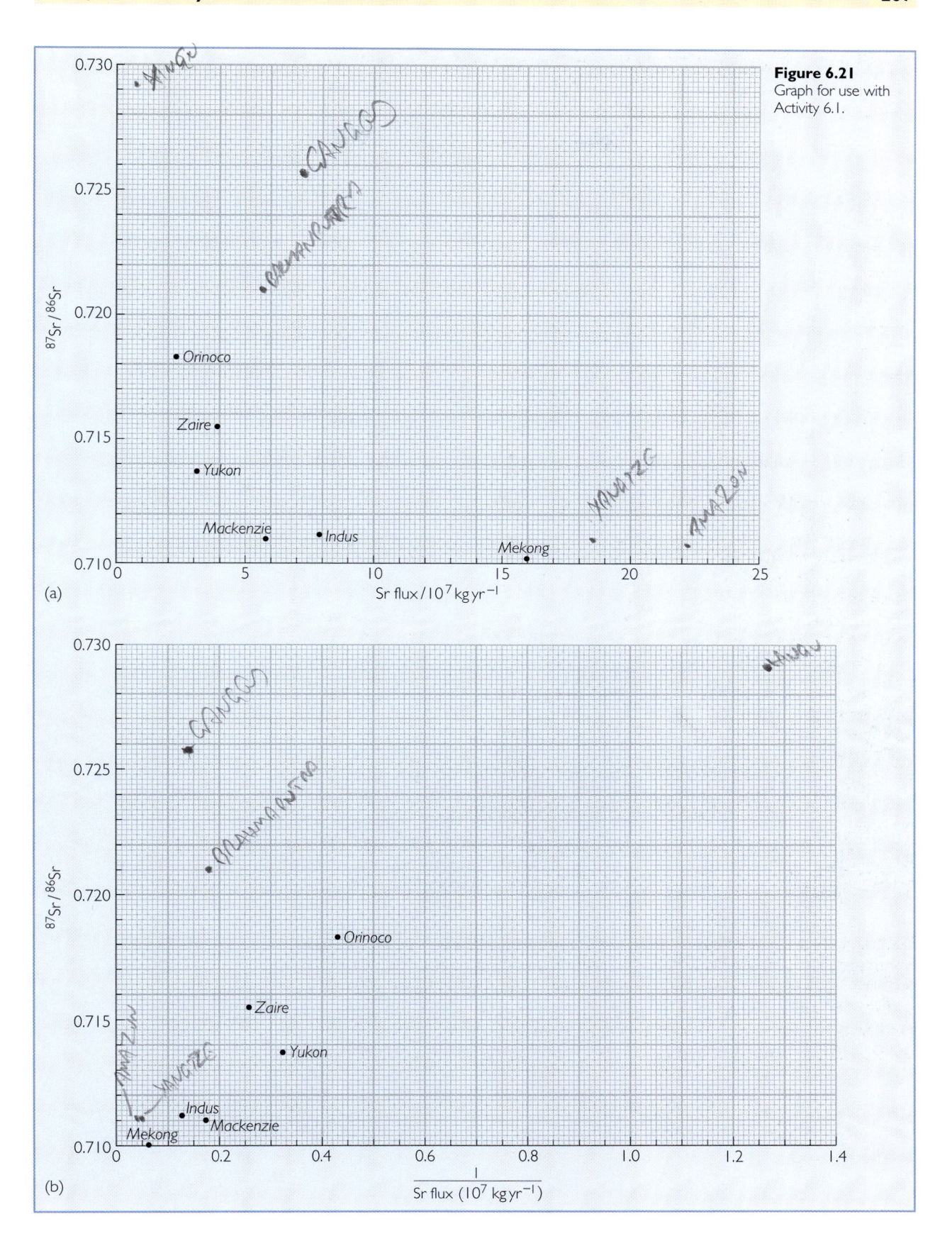

Figure 6.21 Graph for use with Activity 6.1.

Box 6.3 How significant are the Himalayas for the seawater $^{87}Sr/^{86}Sr$ ratio?

In order to assess quantitatively the significance of Himalayan rivers on seawater geochemistry, we need to calculate what the $^{87}Sr/^{86}Sr$ ratio of seawater would be without the contribution of strontium from the Ganges and Brahmaputra rivers. These rivers together contribute a flux of $\sim 13 \times 10^7 \, kg \, yr^{-1}$ of Sr with an average $^{87}Sr/^{86}Sr = 0.7236$. We also need to know the total Sr flux and Sr-isotope ratios for rivers globally. As mentioned in the text, these are $290 \times 10^7 \, kg \, yr^{-1}$ and 0.7119, respectively.

First we must calculate the flux (F^*) and isotope ratio (R^*) of global rivers *without* contributions from the two Himalayan rivers. The Sr flux of global rivers must equal the flux of Himalayan rivers plus that of global rivers *without* the Himalayan contribution. Hence:

$$290 \times 10^7 = (13 \times 10^7) + F^*$$
$$F^* = 277 \times 10^7 \, kg \, yr^{-1}$$

Similarly, we can consider the Sr flux from global rivers to be made up of two components: Sr from Himalayan rivers, and Sr from all other rivers. If F_r and R_r are the flux and isotope ratios of global rivers and F_H and R_H are the flux and isotope ratios of Himalayan rivers, $F_r = F_H + F^*$ and we can write Equation 6.2 as follows:

$$R_r \times F_r = (F_H \times R_H) + (F^* \times R^*)$$

and we can therefore obtain a value for R^*:

$$(290 \times 0.7119) \times 10^7 = (13 \times 0.7236) \times 10^7 + (277 \times R^*) \times 10^7$$

$$R^* = \frac{(290 \times 0.7119) - (13 \times 0.7236)}{277}$$
$$= (206.45 - 9.41)/277$$
$$= 197.04/277$$
$$= 0.7113$$

We have now calculated the average Sr-isotope ratio of all rivers assuming there was no runoff from the Himalayas.

We can now solve Equation 6.2 for R_{sw} using fluxes from hydrothermal sources and from rivers *without* the Himalayan contribution:

$$R_{sw} \times (137 + 277) \times 10^7 = (137 \times 0.7035) \times 10^7 + (277 \times 0.7113) \times 10^7$$

Rearranging:

$$R_{sw} = \frac{(137 \times 0.7035) + (277 \times 0.7113)}{137 + 277}$$
$$= (96.38 + 197.03)/414 = 293.41/414 \approx 0.7087$$

(i.e. 0.0005 less than the present-day value of 0.7092). This value is the Sr-isotope ratio of seawater without the contribution from Himalayan rivers. In effect, it represents seawater in a world without the Himalayas.

The calculation in Box 6.3 shows what a strong influence the Brahmaputra and Ganges river systems have on the strontium geochemistry of the world's oceans. The change obtained by 'removing' these rivers from the world system is greater than the change actually observed over the past 20 million years (Figure 6.20). Interestingly, if all the rivers that rise in Tibet are excluded instead of just the Ganges and Brahmaputra, the isotopic ratio of seawater drops only slightly, to 0.7086. We can see from Table 6.2 that the rivers flowing off the south-east side of the Tibetan Plateau, such as the Yangtze and Mekong (Figure 6.17), have Sr-isotope ratios typical for the fluxes of Sr they transport. Therefore, it is uplift of the Himalayas, not of Tibet, that is particularly important for the strontium isotope composition of seawater. Nonetheless, the important overall conclusion is that uplift of Tibet and the Himalayas has caused a marked increase in the $^{87}Sr/^{86}Sr$ ratio of seawater which indicates an increase in the rate of chemical weathering of the continents.

It is important to emphasize that the Tibetan Plateau and the Himalayan mountains appear to play quite different roles in the climate system. The strengthening of the monsoon seems to be the result of a *large high plateau* being uplifted, as it was this that affected the atmospheric circulation. In contrast, the link between changes in atmospheric CO_2 concentrations and weathering involves increasing weathering rates, and it is in the *Himalayas* that these rates become particularly high, not on the Tibetan Plateau. The plateau itself has little rainfall (of which much evaporates), and has relatively modest relief (see for example Figures 6.6 and 6.16).

So, which is right – the BLAG model or the Raymo–Ruddiman hypothesis? Well – as you may have realized by now – the answer is that they both are! The BLAG model for the carbon cycle is a **steady-state** model. This means that it assumes that the system has attained equilibrium; it assumes the flux of CO_2 into (say) the atmospheric CO_2 reservoir equals the flux out of it. If a such a steady-state system is perturbed, then steady-state conditions no longer apply – such a perturbation is in effect a forcing function – and the system will then adjust until the balance of fluxes is re-established. Uplift of the Himalayas and the Tibetan Plateau can be considered a transient event which disturbed the steady-state carbon cycle, and – it is believed – during this transient event chemical weathering rates rather than rates of production of new ocean floor determined atmospheric CO_2 levels.

So the BLAG model provides a satisfactory mechanism for long-term climate change, i.e. change over hundreds of millions of years, while the Raymo–Ruddiman hypothesis is appropriate for the relatively short-term disturbance to global climate caused by uplift of Tibet and the Himalayas over a period of several million years. Taken *together*, the two approaches explain why global weathering rates are currently high (cf. Figure 6.20) – despite the fact that the climate has been cooling over the past tens of millions of years.

We have been emphasizing the role in the climate system of the uplift of the Himalayas and the Tibetan Plateau so as to provide a focus for our discussion, but (as, hopefully, is now clear) it is highly improbable that a single mechanism could be generally responsible for global climate change. This is not least because the atmosphere is a relatively small reservoir of CO_2 and its size depends on differences between large fluxes. Subtle changes in these fluxes may cause significant climate change. For example, eruption of the Columbia River flood basalts in the western United States around 17 million years ago (Section 4.3) may have been a significant source of atmospheric CO_2 and global warming during the period we have been considering. The best that can be said for any supposed cause of climate change is that such-and-such a mechanism *could* result in an observed trend in the climate record – that is not to say that it was unaided, or even that it was a significant player in such a complex game.

6.4 Cycles in the sediments: climate records in the Arabian Sea

During the previous sections, we have speculated about how the seasonal cycles in the vicinity of southern Asia and the Arabian Sea may have been modified by the rise of the Himalayas and the Tibetan Plateau, over time-scales of millions of years. Now let's briefly consider the effects of forcing factors acting over time-scales of thousands of years, and their record in the sediments of the Arabian Sea.

- What forcing factors would these be, and what climatic fluctuations occur over similar time-scales?

- The forcing factors are the Milankovich cycles and the corresponding climatic fluctuations are glacials and interglacials (Section 5.2).

Since the early 1980s, there have been a number of detailed studies of sediment cores from the Arabian Sea. We are going to look at just a small part of one study, which will nevertheless demonstrate what fascinating information can be gleaned from apparently unpromising material. The drill core we will be concentrating on was taken from the sea-bed at a depth of about 4 km, south-east of the coast of Oman, near to a topographic feature known as the Owen Ridge.

A large proportion of the sediment was found to be biogenic carbonate – shells and skeletons of marine plankton such as coccolithophores and foraminiferans. Figure 6.22a shows the variation in the percentage of CaCO$_3$ with depth in the core, but depth has been converted to *age* using correlations with $\delta^{18}O$ values for a particular species of foraminiferan (Figure 6.22b). Because variations of $\delta^{18}O$ with time are well known (e.g. Figures 5.14, 5.15 and 5.19), they can be used to provide a chronology for sediment cores containing carbonate remains.

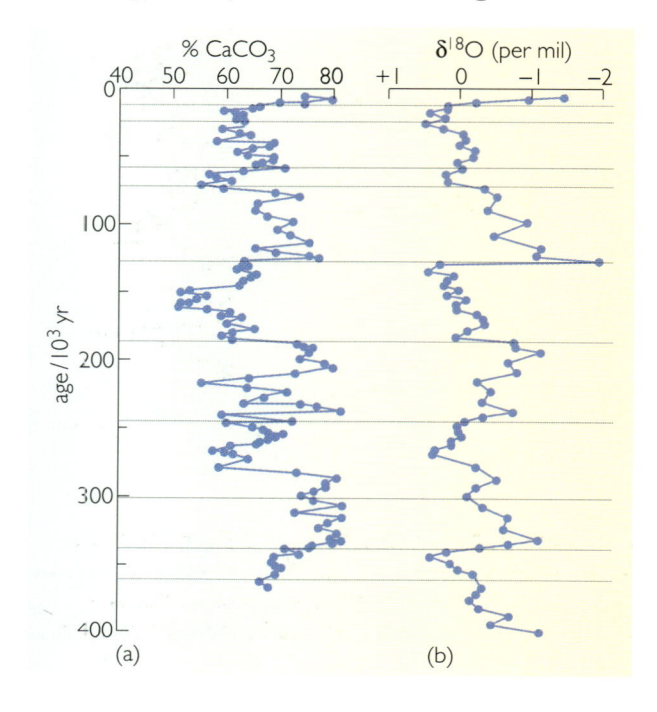

Figure 6.22
(a) Variation with age of the percentage of carbonate in a sediment core from the Owen Ridge (~ 60° E, 17° N), in the Arabian Sea.
(b) Variation in $\delta^{18}O$ of biogenic carbonate from the same site (the species of foraminiferan used was *G. sacculifer*); the fine lines indicate where ages deduced from this plot were use to determine the vertical age scale.

■ With what do the fluctuations in $\delta^{18}O$ correspond?

■ With the peaks of glacials and interglacials (for which the dates are well known).

At first sight, it might seem that the variation in the percentage of carbonate debris in the sediment in the core would be due to variations in primary productivity over time, or perhaps to variations in the carbonate compensation depth. However, a significant proportion of sediment reaching the floor of the Arabian Sea is originally windblown dust, and it seems that the variations in CaCO$_3$ content are mostly determined by the amount of terrestrial dust diluting the biogenic component.

It has been shown that most dust reaching the area to the south-east of Oman is deposited during the South-West Monsoon.

■ In that case, where does it come from?

■ Being carried in south-westerlies (Figure 2.18), it comes from the east coast of Africa, in particular the Somali Peninsula.

Figure 6.23a is the same plot as shown in Figure 6.22a, and part (b) (its mirror image as far as shape is concerned) is the percentage of the sediment carried from the land (100 – % of CaCO$_3$). Part (c) of the diagram shows the variation in the

Tibet, the Himalayas and the Arabian Sea

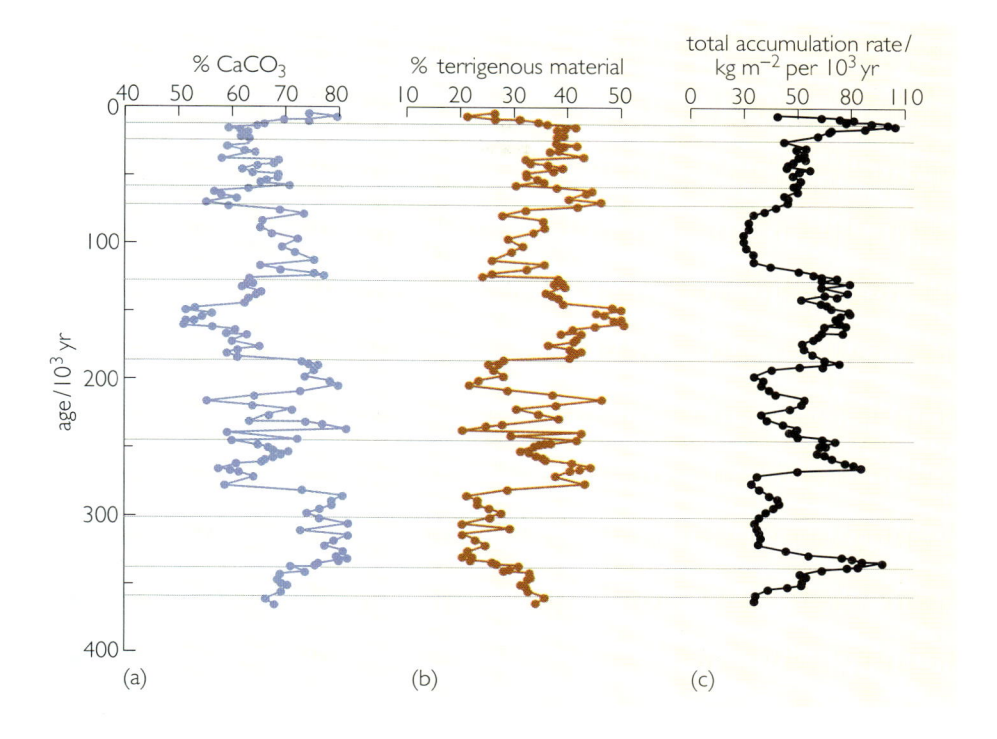

Figure 6.23
(a) Variation with age of the percentage of carbonate in a sediment core from the Owen Ridge (as in Figure 6.22).
(b) Variation with age of the terrigenous component (100 − % of $CaCO_3$) at the same site.
(c) Variation with age of the total rate of accumulation of sediment at the site.

total rate of accumulation of sediment at the site in question (biogenic plus that from rivers plus that carried by wind), expressed in kg m^{-2}. In Activity 6.2, you can use information from Figures 6.22 and 6.23, along with Figure 6.24, to discover more about how the passing of glacials and interglacials has been recorded in the sea-bed sediments of the Arabian Sea.

Activity 6.2

(a) To see more clearly what Figures 6.22 and 6.23 can tell us, begin by using Figure 6.22b to determine which peaks on Figure 6.23 (those to the left or those to the right) correspond to glacial periods and which to interglacials (refer to Figure 5.14 if necessary).

(b) (i) Comparing Figure 6.23b and c, would you say that high sedimentation rates were generally associated with large inputs of windborne dust, or with large inputs of carbonate debris?

(ii) Was (1) the proportion of windblown dust in the sediment, and (2) the total rate of accumulation of sediment, greater or less during glacial periods? What does your answer suggest about the aridity of the African landmass (and presumably other landmasses) during glacial periods? Bearing in mind Figure 2.34, is this what you would have expected?

Another factor that seems likely to have affected the rate of supply of windborne dust is the strength of the winds of the South-West Monsoon. If these winds were stronger during glacial periods, we might expect the proportion of heavy minerals in the windblown dust to have been higher at such times, and indeed this seems to be the case. A further clue to glacial wind strengths might be found in the traces in the sediment of past primary productivity.

▪ Why might this be?

▪ Because stronger winds are likely to lead to more vigorous upwelling, which in turn will lead to higher rates of primary productivity (which may be recorded in biogenic deposits on the sea-bed).

For reasons we don't need to go into, the barium content of sediment is thought to be a good indication of the levels of primary productivity that once occurred in overlying waters, producing organic debris that fell through the water column and was incorporated into the sea-bed.

(c) (i) As in much of the Arabian Sea, primary productivity off the south-east coast of Oman is at a maximum during the South-West Monsoon (i.e. the northern summer); see Figure 6.12b. Bearing in mind the information given above concerning barium content as a proxy for paleoproductivity, does Figure 6.24 indicate that monsoon south-westerlies were indeed stronger during glacial periods?

(ii) What additional factor, itself a consequence of an increased rate of input of windblown dust, might also have tended to increase primary productivity during glacial periods?

(d) Is there any evidence in Figures 6.22 to 6.24 that *more than one* of the component Milankovich cycles have played a part in determining: (i) continental aridity (Figure 6.23b and c); and (ii) primary productivity?

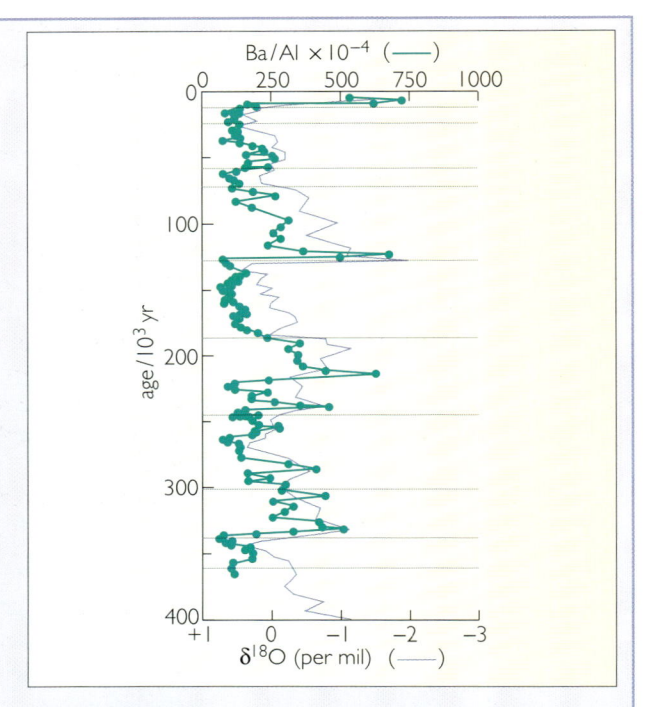

Figure 6.24
Variation with age of the proportion of barium (Ba) in sediments from the Owen Ridge site, expressed in terms of its ratio (by weight) to aluminium (Al) from clay minerals; the fine blue line is the $\delta^{18}O$ plot used in Figure 6.22.

In working through Activity 6.2, you saw how the passage of glacials and interglacials is accompanied by changes in continental aridity, in wind patterns, upwelling and marine primary productivity. Unfortunately, it is difficult to deduce the strengths of past upwellings, and the levels of primary productivity that resulted. This is particularly frustrating because, as you will know from your reading of this and earlier chapters, primary productivity – i.e. life – plays an important role in the carbon cycle, and hence the climate system, on a wide range of time-scales.

In this final chapter of *The Dynamic Earth*, we have considered how the rise of the Himalayas and the Tibetan Plateau may have played a role in climate change, both regionally and globally. We have considered possible mechanisms for climate change over time-scales of hundreds of millions of years and several million years; and, finally, we have briefly looked at some evidence of climatic fluctuations over hundreds of thousands of years. During much of this discussion, we have alluded to the causes and effects of a seasonal phenomenon – the South-West Monsoon.

Tibet, the Himalayas and the Arabian Sea

Explaining climatic change over such a wide range of time-scales is difficult, not least because the different types of change may all be occurring at the same time. However, as first mentioned in Chapter 2, when trying to pin down a forcing mechanism for climatic change it is often helpful to consider the time- (and space-) scale of the climatic change in question. At this point, it seems appropriate to take a final look at our 'bubble diagram', last encountered as Figure 5.6. We've inserted the labels you deduced during Chapter 5 ('Ice Ages' and 'glacial/interglacial') and also added 'mountain climate' (cf. Section 5.4.2), 'cyclical changes in monsoons' (Section 6.3), and 'mountain uplift' (Section 6.2). Note that this diagram is not definitive, and has some drawbacks. Certain processes, that may in the future turn out to be important, may have been omitted. Others may not fit as neatly into 'bubbles' as shown here – for instance, climate change in response to changes in weathering rates (as postulated by Raymo and Ruddiman) should ideally be positioned on the time axis coincident with 'mountain uplift', but (as the effects of changing CO_2 concentrations would be global) at the *top* of the bubble. Generally speaking, however, this type of diagram is a useful device for helping us to come to grips with a tricky scientific problem.

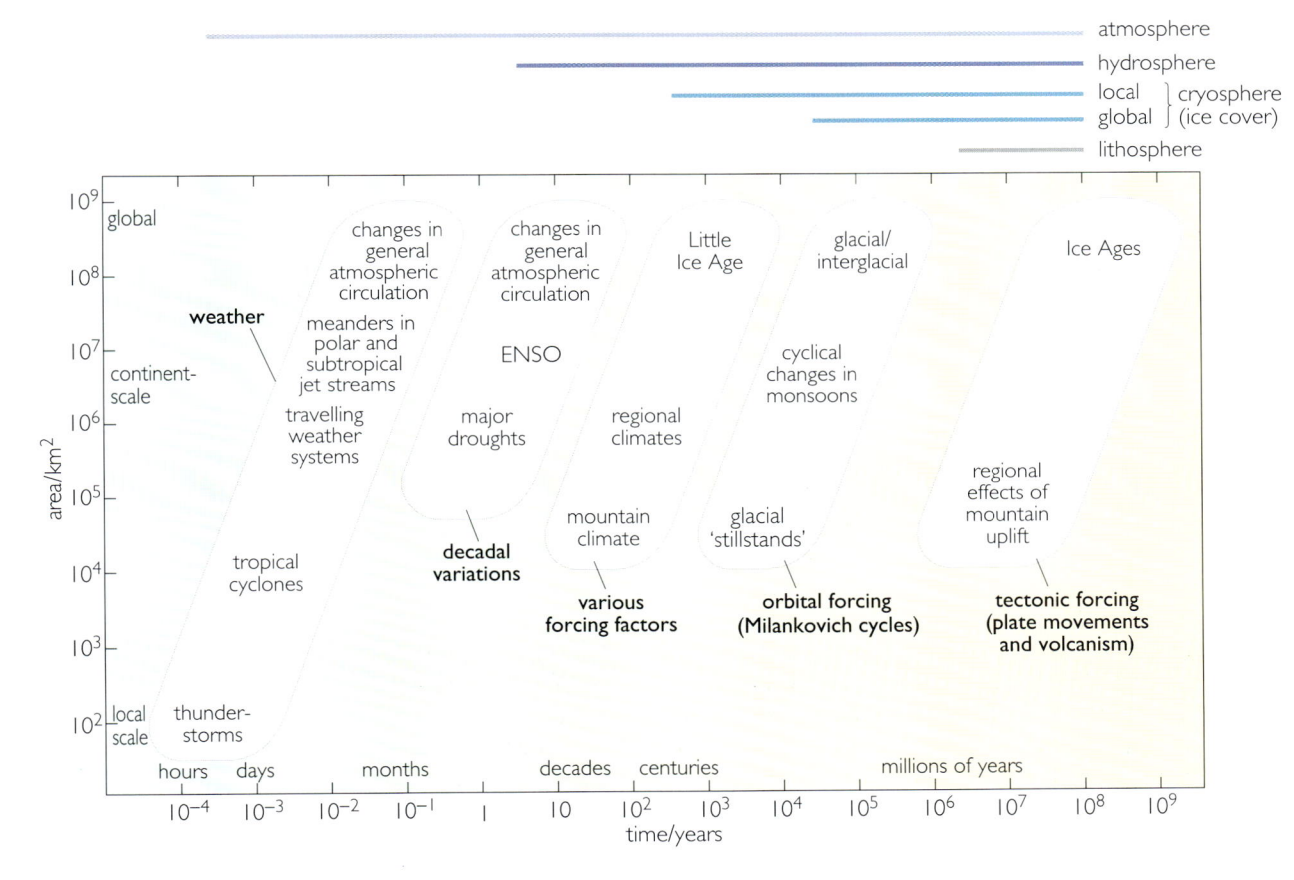

Figure 6.25 Completed version of Figure 2.41 (see text). Included for completeness are *glacial 'stillstands'*, which are times when global ice volumes and sea-level do not change for long periods.

Finally, take a look at the bars along the top of Figure 6.25, intended to illustrate the general principle that while short-term variations are mostly limited to the atmosphere, longer-term changes involve progressively more components of the Earth system. In this chapter, for example, we have considered the effects of the monsoons (primarily atmospheric phenomena) while investigating how uplift of the Himalayas and Tibet (a result of lithospheric plate movements, occurring over very long time-scales) caused changes in the atmosphere, hydrosphere and biosphere. Thus, in studying climatic change – both in southern Asia in particular, and over the dynamic Earth as a whole – we have been forced to appreciate that to have any hope of understanding the climate system properly we must try come to grips with the *interactions* between the lithosphere, atmosphere, hydrosphere and – last but not least – biosphere.

6.5 Summary of Chapter 6

1. Mountain ranges and plateaux affect the climate physically by redirecting air masses around and/or over them. Moisture-laden winds release their precipitation on the windward side while the leeward side (in the case of the Himalayas, the Tibetan Plateau) is dry. The rise of the Himalayas and Tibet is believed to have intensified the strength of the South-West Monsoon by providing a source of heat (in particular, *latent* heat), at a critical position in the atmospheric circulation. (See also item 9 of the Chapter 5 Summary.)

2. The collision between India and Eurasia at about 50 Ma was responsible for uplift of the Tibetan Plateau and the Himalayas. Continued thickening of the lithosphere beneath Tibet led to convective thinning followed by rapid uplift.

3. The rate of uplift of the Tibetan Plateau is not well established. Attempts to measure it are being made by use of fossil plants and dating of normal faults. There is some evidence that the plateau reached its present elevation between 14 and 7 million years ago.

4. The history of climate change in southern Asia is not well established. Attempts to determine it include studies of foraminiferans in the Indian Ocean, studies of $\delta^{13}C$ (to throw light on plant growth) and analysis of the types of sediments eroded off the Himalayas. There is some evidence to suggest strengthening of the monsoon between 9 and 6 million years ago.

5. The Earth's climate has cooled markedly over the past 50 million years. The Raymo–Ruddiman hypothesis postulates that uplift of the Himalayas and Tibet imposed cooling on the global climate, by strengthening the South-West Monsoon and hence increasing rates of chemical weathering: increased weathering of silicates followed by accumulation and burial of carbonates and/or organic carbons in the ocean is assumed to result in the long-term removal of CO_2 from the atmosphere. To prevent 'runaway cooling', the rapid removal of CO_2 by weathering would need to be partially compensated for. Possible mechanisms for replenishing atmospheric CO_2 include decarbonation at subduction zones and the oxidation of organic carbon from exhumed sediments.

6. The steady-state, carbon cycle model (i.e. the BLAG model) interprets the role of mountains as providing a negative feedback loop that stabilizes fluctuations in temperature resulting from variations in volcanism associated with the production of new sea-floor.

7. The isotope ratio $^{87}Sr/^{86}Sr$ of the world's oceans has increased over the past 50 million years. This has been interpreted as indicating increased weathering rates due to the uplift of Tibet and more especially the Himalayas. This interpretation is supported by the Sr fluxes and isotope ratios of rivers currently eroding the Himalayas.

8. Sediments laid down in the Arabian Sea over the past 400 000 years show cyclical variations in the strength of the South-West Monsoon winds, in primary productivity and in aridity of continental landmasses. These

variations are related to glacial/interglacial cycles, i.e. to the Milankovich cycles. The challenge for those wishing to study climatic change is to disentangle the various types of change occurring in response to different forcing factors over a number of different time-scales.

Now try the following questions to consolidate your understanding of this chapter.

Question 6.8
Suggest at least *two* reasons why the effect on climate of the uplift of a plateau the size, shape and elevation of Tibet in the vicinity of one of the poles would not be similar to the effect of the Tibetan Plateau.

Question 6.9
The geological carbon cycle has been modelled as a steady-state system by Berner, Lasaga and Garrels (the BLAG model). Give two reasons why such a model does not provide a satisfactory basis for understanding climate change over the past 50 million years.

Question 6.10
What two alternative scenarios could account for a rapid increase in the $^{87}Sr/^{86}Sr$ ratio in seawater?

Objectives

When you have finished this book, you should be able to display your understanding of the terms printed in **bold** type within the text as well as the following topics, concepts and principles and, where appropriate, perform simple calculations related to them:

1   the main factors that allow the Earth to be hospitable to life;

2   the guiding principle that climatic change on a particular time-scale is likely to be driven by a forcing factor acting on a similar time- (and space-) scale;

3   the causes of variations in incoming solar radiation, both spatially over the surface of the Earth, and in time, on a number of different time-scales;

4   the different types of radiation within the electromagnetic spectrum, and their roles in the Earth's radiation budget; hence why the presence of an atmosphere means that global temperatures are higher than they would otherwise be (i.e. the greenhouse effect);

5   the roles of the atmosphere, surface currents and thermohaline circulation in redistributing heat over the surface of the Earth, and the effect of the configuration of landmasses and oceans on the distribution of temperature at the Earth's surface;

6   the significance of coupling between the atmosphere and the ocean, as manifested in El Niño events and other examples of the operation of teleconnections; the difference between positive feedback, which causes instability, and negative feedback, which leads to stability;

7   the importance of the hydrological cycle, and of clouds (and hence of land plants and phytoplankton), in the climate system; associated with this, the importance of available nutrients for the spatial distribution, and productivity, of plants on land and in the sea;

8   the operation of the 'terrestrial' carbon cycle, the marine carbon cycle and the geological carbon cycle, and the differences between the various reservoirs in terms of size, flux and residence time; the concept of equilibrium;

9   the distinction between organic carbon (i.e. that initially fixed by primary producers) and inorganic carbon, and the different types of dissolved inorganic carbon in the marine carbonate system; the use of $\delta^{13}C$ as an indicator of organic carbon content;

10 the different settings for volcanism, in the context of plate tectonics; also the formation of major flood basalts/submarine plateaux as a result of plumes (or superplumes);

11 the importance of volcanic CO_2 emissions to the atmosphere for climate over the long term (though currently dwarfed by the anthropogenic flux);

12 the importance of volcanogenic SO_2 for climate in the short term, and the various factors that determine the effects of an eruption on climate; the possibility of links between eruptions of flood basalts and mass extinctions;

13 how to use information about historic eruptions (e.g. the Laki Fissure eruption), to deduce simple 'rules of thumb' that can be used to estimate the effects of large eruptions in the geological record;

Objectives

14. the different (and interacting) ways in which the changing positions and configurations of the continents and oceans can affect climate: by changing current patterns, albedo, the hydrological cycle and weathering; and, for a particular landmass, simply changing latitude;

15. the distinction between Ice Ages and glacial periods; the use of $\delta^{18}O$ as an indicator of the amount of ice in the ice-caps, and hence of sea-level; the different causes of eustatic and isostatic sea-level change;

16. the (interacting) effects of changing global temperature, sea-level and carbon preservation (as organic carbon and via deep and shallow carbonate factories);

17. effects of subduction, continental collision and mountain-building on the global carbon cycle; possible effects of mountains on climate, both regionally and globally;

18. why Tibet and the Himalayas are particularly high, and the different methods used to estimate their rate of uplift;

19. likely effects on climate of the uplift of the Tibetan Plateau and the Himalayas, and the various types of evidence for climatic change in southern Asia over the period concerned;

20. why weathering of silicate rocks, followed by preservation and burial of carbonates and/or organic carbon, results in net removal of CO_2 from the atmosphere; hence, the basis of the Raymo–Ruddiman hypothesis, and how it differs from the BLAG model;

21. the use of $^{87}Sr/^{86}Sr$ ratios to assess changes in global weathering rates over time, and the contribution to the ocean of dissolved weathering products from particular rivers.

Answers to Questions

Question 1.1

(a) (i) At 50° N, the incoming solar radiation on 21 March (the spring equinox) is about $1 \times 10^7 \, \text{J m}^{-2}$; it increases up to a maximum of $2.3-2.4 \times 10^7 \, \text{J m}^{-2}$ in late June (around the time of the summer solstice/longest day), then declines to about $0.3 \times 10^7 \, \text{J m}^{-2}$ towards the end of December (around the time of the winter solstice/shortest day), after which it begins to rise again.

(ii) Watts are joules per second, therefore to convert J m^{-2} (incoming solar energy per unit area per day) to W m^{-2} (average solar power) we must divide the contour values by the number of seconds in a day, i.e. by $24 \times 60 \times 60 = 8.64 \times 10^4$. The contour values would therefore range from $(0.4 \times 10^7)/(8.64 \times 10^4)$ to $(2.5 \times 10^7)/(8.64 \times 10^4) \, \text{W m}^{-2}$ or about 46 to 289 W m^{-2}.

(b) Mid-latitudes receive the most solar radiation at any one time: $> 2.5 \times 10^7 \, \text{J m}^{-2}$ in summer; this is because of long day-lengths at a time of year when the noonday Sun is high. (You probably noticed that there is an asymmetry between the hemispheres – unlike the seasons, this asymmetry has nothing to do with the tilt of the Earth's axis; we will come to it next.)

Question 1.2

(a) The main and most obvious reason is that more solar radiation reaches the Earth's surface at low latitudes than at high, for reasons demonstrated by Figure 1.6. The second reason is that of the solar radiation that reaches the Earth's surface, much more is reflected at high latitudes than low, particularly because of the high albedos of ice and snow (Table 1.1).

(b) The Earth–atmosphere system has a net gain of heat between about 40° S and 35° N (Figure A1); poleward of those latitudes, there is a net loss of heat.

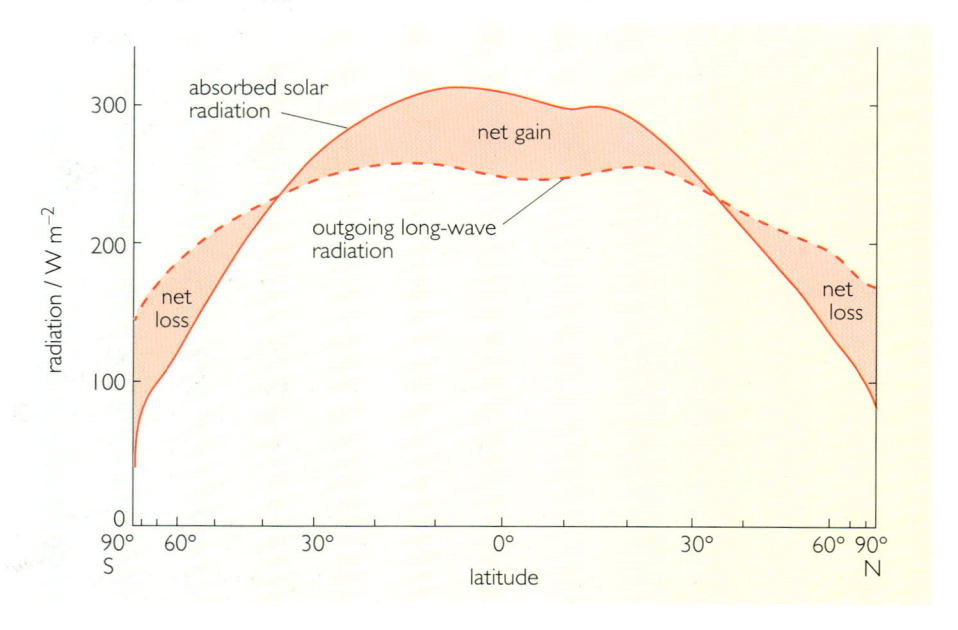

Figure A1 The variation with latitude of the solar radiation absorbed by the Earth–atmosphere system (solid curve) and the outgoing radiation lost to space (dashed curve). Values are averaged over the year, and are scaled according to the area of the Earth's surface in different latitude bands.

Question 1.3
If during the Little Ice Age average global temperatures were about 1 °C lower, and this were attributable entirely to a decrease in solar luminosity, there would have to have been a reduction in incoming solar radiation of about $1 / 0.75 = 1.3$ W m^{-2}. Expressed as a percentage of the present-day average value, 1.3 W m^{-2} is $(1.3 / 343) \times 100 \approx 0.4\%$ below average.

Question 1.4
The martian surface must have an albedo appropriate to bare rock/soil (10–20% according to Table 1.1), and dust must have a relatively low albedo (i.e. closer to that of rock or soil than of clouds or ice). The albedo of Mars as a whole must therefore be significantly lower than the 30% value for Earth. Mars' value is in fact 14%. By contrast, thick cloud is very reflective (albedo up to 80%) so Venus must have an albedo significantly greater than 30%; its value is in fact 77%. This high albedo (coupled with Venus' relatively close proximity to the Earth) is why Venus shines so brightly in the night sky.

Question 1.5
The Tropics are not shown on Figure 1.7 because they are the latitudes at which the noonday Sun is overhead at the summer solstices. They are therefore a consequence of the tilt of the Earth's axis with respect to the plane of the Earth's orbit (which is why the latitude of the Tropics, 23.4°, is the same as the angle of tilt of the axis). If the axis were not tilted – the hypothetical situation illustrated in Figure 1.7 – the whole idea of Tropics would be meaningless.

Question 1.6
(a) (i) The diagram shows the Earth in perihelion and aphelion in two different circumstances: one with the axis tilted so the North Pole is pointing away from the Sun (the northern winter solstice) at aphelion and towards it at perihelion; and the other with the reverse. The cyclical variations involved are therefore those in orbital shape (the 110 000-year eccentricity cycle) and the *direction* of tilt of the Earth's axis (the 22 000-year precession cycle).
(ii) The diagram illustrates Croll's idea that glacial conditions become established in a particular hemisphere when the coldest season (i.e. when the hemisphere is 'away from the Sun') coincides with the time in the orbit when the Earth is farthest from the Sun. This is despite the fact that the hemisphere in question will be experiencing summer when closest to the Sun. The implication of this is that Croll believed that whether ice-sheets persist depends on how low temperatures are in winter, not how high they are in summer.

(b) As shown in Figure 1.8a, at the present time, perihelion occurs during the northern winter, close to the southern summer solstice (i.e. when the North Pole is pointing away from the Sun). This is the situation shown in the second diagram. The two diagrams are half a precession cycle, i.e. 11 000 years, apart – like Figures 1.8a and 1.11 (plus, hopefully, your own sketch).

(c) According to the second illustration, conditions in the Northern Hemisphere (where winter coincides with perihelion) should be 'interglacial', while in the Southern Hemisphere (where winter coincides with aphelion) conditions would be 'glacial'. Present-day conditions in the Arctic might well be viewed as 'interglacial' in comparison with the more extreme 'glacial' conditions in the Antarctic (the Antarctic ice-cap is considerably more extensive than the

Arctic ice-cap, and the Antarctic is colder than the Arctic). In fact, modern climatologists regard the Earth as currently in an interglacial period *within* an Ice Age – we will come back to this topic later.

Question 2.1
(a) Generally, winds blow from regions of high pressure to regions of low pressure; this is most clearly seen in the case of the Trade Winds blowing towards the zone of low surface pressure along the Equator. In addition, winds blow clockwise around high pressure regions in the Northern Hemisphere and anticlockwise around them in the Southern Hemisphere; such flow is referred to as anticyclonic. Winds blow in the opposite direction (i.e. are cyclonic) around low pressure regions.

(b) In regions of high surface pressure, air is sinking; in regions of low pressure, air is rising.

Question 2.2
(a) First, the ITCZ is (mostly) further north in July and further south in January; in other words, it is somewhat shifted into the hemisphere experiencing summer, particularly over large landmasses. This is what we would expect because the ITCZ – an area of vigorous upward convection – tends to be located over the warmest parts of the Earth's surface and, as discussed in connection with Figure 2.1, continental masses heat up much faster than the oceans in summer (and cool down much faster in winter).

(b) In July, when the ITCZ is at its most northerly, the winds over tropical west Africa are mainly south-westerly (i.e. from the south-west). In January, when the ITCZ is at its most southerly, the winds over the region are mainly easterly or north-easterly (from the east or north-east). Such seasonally reversing winds, or *monsoons*, are discussed further in the text.

(c) The north–south shift in the position of the ITCZ is most marked over the northern Indian Ocean and over the western tropical Pacific and South-East Asia.

Question 2.3
(a) Tropical cyclones only form over the ocean because they depend on evaporation from a warm sea-surface to provide the latent heat needed to sustain them. As soon as they move over land, they begin to die out.

(b) If global warming continues, tropical cyclones are likely to become more frequent, as sea-surface temperatures will be higher.

Question 2.4
The value needed from Box 2.1 is the specific heat of water, $4.18 \times 10^3 \, \text{J kg}^{-1} \, ^\circ\text{C}^{-1}$. $1300 \, \text{km}^3$ is $1300 \times 10^9 \, \text{m}^3$ which has a mass of $1300 \times 10^{12} \, \text{kg}$. The total amount of heat given up each day, on average, is given by:

specific heat × mass of water × temperature drop
= $(4.18 \times 10^3 \, \text{J kg}^{-1} \, ^\circ\text{C}^{-1}) \times (1300 \times 10^{12} \, \text{kg}) \times 11 \, ^\circ\text{C}$
= $59\,774 \times 10^{15} \approx 60\,000 \times 10^{15} \, \text{J}$ or $6 \times 10^{19} \, \text{J}$

So, on average, 6×10^{19} joules of heat are released to the atmosphere over the northern North Atlantic every day. Given that $6 \times 10^{19} \, \text{J day}^{-1}$ is about $0.7 \times 10^{15} \, \text{W}$, this is equivalent to the heat output of about half a million large power stations. In reality, of course, this heat loss to the atmosphere would be concentrated in the winter months, so 6×10^{19} joules per day must be an underestimate.

Question 2.5

(a) (i) 10–20% of visible radiation is absorbed in the atmosphere, with absorption at the red end of the spectrum being greater than that at the blue/violet end. The gas responsible is ozone.
(ii) About half of incoming infrared radiation is absorbed by the atmosphere. The gases mainly responsible in this case are water vapour and carbon dioxide.

(b) Outgoing long-wave radiation is absorbed mainly by water vapour and carbon dioxide, with lesser absorption by methane, nitrous oxide and ozone.

Question 2.6

(a) The percentage of the Earth's water presently in the ocean is (1 322 000 / 1 360 0000) × 100 or about 97%; and that in ice-caps and glaciers is (29 300 / 1 360 000) × 100 or about 2%. (Remember that you do no need to worry about the units involved as they are the same both above and below the fraction line.)

(b) Even if all the ice-caps and glaciers melted, the volume of water in the oceans would only increase by about 2%, so if a quarter of them melted it would increase by 0.5%. This may not sound much, but it would mean a rise in sea-level of about 15 m. Given that a large proportion of human habitations are concentrated around the coasts, such a sea-level rise would be disastrous.

Question 2.7

(a) Figure 1.9 shows that in December only low levels of solar radiation reach high latitudes in the Northern Hemisphere – insufficient to allow phytoplankton populations to grow. By May, however, light levels there will have risen considerably; indeed the image in Figure 2.40a shows the burst in primary productivity known as the 'spring bloom'.

(b) The region in question is the centre of the Atlantic subtropical gyre, driven by anticyclonic winds. As discussed in connection with Figures 2.27 and 2.28, such gyres are regions of downwelling, and hence can only support limited primary productivity. (By contrast, the cyclonic subpolar gyres are regions of divergence and upwelling (Figure 2.27b(ii)), which is another reason why primary productivity can be high there once light levels rise in spring.)

Question 2.8

Increased primary productivity would mean bigger plankton blooms, and a greater flux of DMS to the atmosphere. This in turn could lead to more cloud formation, resulting in an increase in the Earth's albedo.

Question 2.9

Reasons you may have thought of are:

1. Being darker in colour (see Figure 1.3), vegetated regions reflect less incoming radiation and so absorb more heat than deserts, i.e. their albedo is significantly less (Table 1.1).

2. Vegetation transpires, and so releases water vapour, and hence latent heat, into the atmosphere. As mentioned in the text, in some ways rainforests behave rather like the tropical ocean as far as the climate system is concerned.

3. Deserts generally have clear skies (see 2). Because clouds trap outgoing long-wave radiation, heat loss from a desert at night is likely to be greater than from a vegetated area (particularly rainforest).

Question 2.10

(a) (i) The total area under the dotted curve is about twice that under the long-dashed curve, so clouds must reflect about twice as much solar radiation as the Earth's surface.

(ii) Yes, we can estimate the Earth's albedo from Figure 2.44 by comparing the sum of the areas under the dashed and dotted curves with the area under the solid curve for incoming solar radiation. Together, the areas under the dashed and dotted curves come to about a third of the area under the solid curve, indicating that about a third of incoming solar radiation is reflected either by clouds or by the Earth's surface. This deduction is consistent with the value for the Earth's albedo, given in the text, of 30%.

(b) The maximum reflectance (particular high albedo) in the vicinity of the Equator is due to the extensive cover by cumulonimbus clouds, generated by atmospheric convection in the Intertropical Convergence Zone.

Question 2.11

(a) Image (a) shows a plume of productive water extending westwards from the islands, presumably carried by prevailing currents. This is consistent with the 'normal' situation in which the Trade Winds are strong and the currents along the Equator are westwards (Figure 2.21). Map (b) shows the high productivity water to the east of the islands, implying that both winds and current flow along the Equator have reversed, or at the very least are weak. In both cases, high productivity would be the result of upwelling in the lee of the wind, as a result of surface water being driven away from the islands.

(b) The iron would have been in dust blown off the islands; in non-El Niño circumstances, it would be carried westwards, and during El Niño events eastwards.

Question 2.12

The statement is partly true, but it is incomplete. The distribution of terrestrial vegetation is largely determined by the availability of water, and to a lesser extent by temperature, but primary production in the oceans is limited by the availability of nutrients. Both on land and in the ocean, plants can only grow when light levels are sufficiently high for photosynthesis to occur.

Question 3.1

(a) Energy is released when carbon is transformed from a reduced to an oxidized state: respiration releases energy, so must be an oxidizing reaction.

(b) Conversely, energy is *used* in reduction of carbon (cf. point 4 at the beginning of this section): photosynthesis uses the energy of sunlight, so must be a reducing reaction.

Alternatively, you could have answered from first principles, on the basis that oxidation involves the addition of oxygen or the removal of hydrogen. On the left-hand side of Equation 1.2 for photosynthesis, each carbon atom has two oxygens (in CO_2), while on the right-hand side, each carbon atom has only one oxygen (in $C_6H_{12}O_6$) *and* has two hydrogens. This demonstrates that photosynthesis must be a reducing reaction. As the respiration equation is the exact reverse of this, *it* must be an oxidizing reaction.

Answers to Questions

Question 3.2

(a) According to Figure 3.3, the amount of carbon in terrestrial plant material is about 560×10^{12} kgC, and we have stated that the net fixation of carbon from the atmosphere (i.e. the net annual input of carbon to the reservoir of living plants) is 60×10^{12} kgC yr^{-1}. The mean residence time for carbon in plant biomass is therefore 560×10^{12} kgC yr^{-1} / 60×10^{12} kgC yr^{-1} ≈ 9 years. The amount of carbon deposited in soil is also 60×10^{12} kgC yr^{-1}; the amount of carbon in the soil reservoir at any one time is about 1500×10^{12} kgC (Figure 3.3). The mean residence time for carbon in soil organic matter is therefore 1500×10^{12} kgC / 60×10^{12} kgC yr^{-1} = 25 years.

(b) Short residence times, because small numbers divided by large numbers give even smaller numbers.

Question 3.3

(a) (i) The annual flux of carbon into tropical rainforests is the total amount of carbon fixed in them annually, i.e. their total net primary productivity (NPP), which is $(17.0 \times 10^6 \text{ km}^2) \times (0.9 \text{ kgC m}^{-2} \text{ yr}^{-1}) = 15.3 \times 10^{12}$ kgC yr^{-1}. (Having now calculated this value, you can complete the last column of Table 3.1.)
(ii) This works out at about $(15.3 / 48.3) \times 100 \approx 32\%$. As you will probably have noticed in working out your answer, this high percentage is partly due to the large area of the globe occupied, even now, by tropical rainforest. (Note, however, that in doing this calculation, we are making the assumption that the area of the globe covered by rainforests is remaining constant, whereas in reality it is decreasing.)

(b) (i) To make a comparison area for area, we must use the column headed 'Mean NPP per unit area'. The value for tropical evergreen forests is 0.900 kgC m^{-2} yr^{-1}, and that for boreal forests is 0.360 kgC m^{-2} yr^{-1}, so tropical evergreen forests are $0.900 / 0.360 \approx 2.5$ times more productive than boreal ones.
(ii) One factor to be considered is rainfall. Tropical forests grow at low latitudes where there is heavy rainfall associated with the ITCZ; however, boreal forests grow in the subpolar regions where precipitation is also fairly high (Figure 2.7) so this cannot be the main reason. Another important difference between the two areas is the amount of light they receive during the year: tropical regions have high daytime light levels all year, but subpolar regions have very low light levels (or are dark) for much of the year (Figure 1.9), and hence a very short growing season (because of length of the growing season, even temperate savannah and tall grassland are more productive than boreal woodland and boreal forest). The most important reason of all, however, is that growth rates are also greatly affected by temperature; indeed, in the world at the present time, this is the factor that limits terrestrial primary productivity at high latitudes (see Section 2.3.2).
(iii) Residence time in living plant material is given by the mass of plant material divided by the net primary productivity (cf. (a)). Therefore, according to Table 3.1, the residence time of carbon in swamps and marshes is $13.6 / 2.2 \approx 6$ years (it is then mostly converted to methane and/or CO_2), while that in boreal forests is $108 / 4.3 = \sim 25$ years (effectively the average lifetime of a tree). (Again, we are assuming that the areal extents of swamps and marshes, and of boreal forests, are remaining constant.)

Question 3.4

(a) The values on the right-hand axis are all negative, and as explained in the caption, negative values correspond to CO_2 concentrations in the atmosphere being higher than those in surface waters. This means that the situation is out of equilibrium, with a concentration gradient across the air–sea interface, and a net flux of CO_2 *into* the ocean (cf. Figure 3.8).

(b) As it was the time of the spring bloom, the phytoplankton were multiplying, fixing carbon (i.e. there was high net primary productivity). Where there were more phytoplankton in the surface water (i.e. chlorophyll concentrations were high), more carbon was being fixed, causing more CO_2 gas to enter surface waters from the atmosphere; where there were fewer phytoplankton, the reverse was true. (As you may have realized, the phytoplankton – and indeed zooplankton feeding on them – would also have been respiring, releasing CO_2 into the water; but as you will see, at times of high primary productivity, sufficient organic debris would be falling out of the surface layers to drive a net flux of CO_2 into the ocean.)

Question 3.5

(a) If the Earth's surface became warmer, evaporation of water (from both the sea-surface and land) would increase, there would be more water in the atmosphere, and so precipitation would increase; in other words, the hydrological cycle would become more active. (To some extent, this is being seen at the present time, in the form of more droughts at lower latitudes and increased rainfall in subpolar latitudes.)

As a result of warmer, wetter conditions, terrestrial biological productivity would increase (consider how you encourage plants to grow). This would fix more carbon from the atmosphere, supply more to soil in organic debris and perhaps increase the rate at which carbon is preserved on land (in swamps, etc.). Marine productivity might also increase as a result of higher sea-surface temperatures (though other factors might counteract this). (There is also some (inconclusive) evidence that an increase in the concentration of atmospheric CO_2 increases net primary productivity.)

Warmer, wetter conditions would also increase rates of chemical weathering, particularly if plant growth were increased. Increased weathering of silicates would result in an increase of the net flux of carbon to the ocean (Figure 3.21). (Do not worry if you only thought of some of these points.)

(b) Increased weathering of silicates and increased primary production on land and in the sea both result in increased rates of sedimentation of calcareous and organic sediments in the ocean. However, although an increased flux of organic-rich material is likely to result in more rapid burial and preservation of carbon, the same cannot necessarily be said for inorganic carbon. This is because an increase in the concentration of CO_2 in the atmosphere and increased primary productivity would both lead to an increase of dissolved inorganic carbon in the deep ocean, and hence an increase in the acidity of deep waters and an increase in the rate of dissolution of calcareous remains (i.e. a deeper carbonate compensation depth). Do not worry if you did not think of this aspect; it illustrates the complexity of the global carbon cycle, and how difficult it is to make predictions about the results of changing fluxes in any part of the cycle.

Answers to Questions

Question 3.6

(a) The fluctuations in atmospheric CO_2 concentration are a result of the uptake of CO_2 by plants during photosynthesis in spring and summer, i.e. removal of carbon from the atmosphere and its fixation in living plant material. Note that it is the lows that correspond to spring and summer, and the highs that correspond to winter.

(b) It is a very evocative description, but rather misleading. Breathing – effectively *respiration* – involves the uptake of oxygen and the release of carbon dioxide (Equation 3.1). Respiration of biomass continues all year round (though more so in the spring and summer), but high rates of primary production and hence *net* primary productivity (production minus respiration) *only* occur in spring and summer. Thus (as discussed in (a)), the pattern is primarily a manifestation of photosynthesis, rather than of respiration.

(c) The pattern is damped in the Southern Hemisphere because primary productivity per unit area in the ocean is much less than on land (compare the columns for 'Mean NPP per unit area' in Tables 3.1 and 3.2), and the Southern Hemisphere is largely ocean.

Question 3.7

(a) The reverse reaction for photosynthesis is that for respiration (Equations 1.2 and 3.1):

$$\underbrace{C_6H_{12}O_6}_{\text{organic matter}} + \underbrace{6O_2}_{\text{oxygen}} \rightleftharpoons \underbrace{6CO_2}_{\text{carbon dioxide}} + H_2O + \text{energy}$$

(b) The reverse reaction for carbonate weathering is carbonate precipitation (Equations 3.6a and 3.7):

$$\underbrace{Ca^{2+}(aq) + 2HCO_3^-(aq)}_{\text{in solution in seawater}} \rightleftharpoons \underbrace{CaCO_3(s)}_{\text{precipitated by organisms}} + H_2O + CO_2$$

Question 3.8

The statement is only partly true. Both limestone and shells/skeletal remains (which eventually become limestones) are forms of inorganic carbon. They are precipitated from dissolved inorganic carbon in seawater (mainly HCO_3^-), and are not composed of large organic molecules made up of carbon, hydrogen and oxygen.

Question 3.9

(a) According to Table 3.2, the total mass of carbon in marine plant material is 1.76×10^{12} kgC, i.e. about 2×10^{12} kgC. This is a small percentage of the standing stock of biomass on land, whether we use the value for this of 560×10^{12} kgC (from Figure 3.3) or 827×10^{12} kgC (from Table 3.1). In the first case, we get ~0.36%, in the second we get ~0.24%.

(b) Residence time is given by: amount in reservoir / flux in (or out). If we use an average of the input and output fluxes shown in Figure 3.19, the residence time for carbon in living phytoplankton will work out as 2×10^{12} kgC / 38×10^{12} kgC yr^{-1} ≈ 0.05 years or about 18 days. By contrast, the residence time of carbon in the terrestrial biomass reservoir is about nine years (cf. Question 3.2), so carbon cycles much faster through the marine biomass reservoir than through the terrestrial biomass reservoir.

Question 4.1
The total amount of CO_2 liberated to the atmosphere during Mauna Loa's construction so far is $6.6 \times 10^{17} \times 0.0032$ kg, or about 2.1×10^{15} kg.

Question 4.2
Your completed Figure 4.13 should look something like Figure A2. There is obviously *broad* correlation – the bigger the eruption, the more aerosol is produced. However, you need to bear in mind that the scales are logarithmic, so that the relationship is not linear, as it appears to be at first.

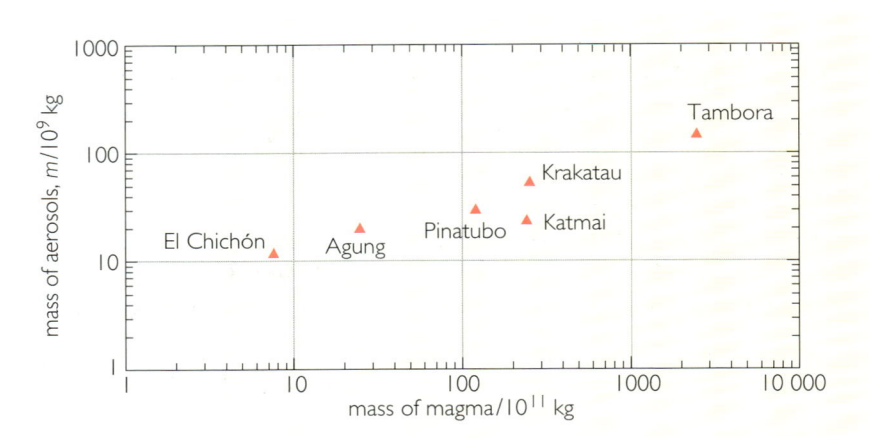

Figure A2
Completed Figure 4.13. Plot of the mass of aerosols injected into the stratosphere against the mass of magma erupted for several major historic eruptions (given in Table 4.3). (The axes are logarithmic.)

Question 4.3
(a) As shown in Figure 2.7, the tropospheric wind systems of the two hemispheres are to a large extent separate, meeting at the Intertropical Convergence Zone. For example, air being carried northwards across the Equator in the Trade Winds is likely to rise at the ITCZ and then be carried southwards again, rather than be carried northwards in the Northern Hemisphere.

(b) As shown schematically in Figure 2.7, the tropopause – the boundary between the troposphere and the stratosphere – is higher at low latitudes than at high latitudes. As discussed in Section 2.2.1, it is generally at a height of about 17 km in low latitudes but only 9–10 km high at high latitudes. Hence a volcano is more likely to be able to send gases (and aerosols) into the stratosphere at high latitudes than at low latitudes.

Question 4.4
(a) According to Equation 4.2, $R_{net} = -(1.43 \times 10^{-10}) \times m$ W m^{-2}. The climate forcing resulting from the Toba eruption would therefore be about $-(1.43 \times 10^{-10}) \times (3.3 \times 10^{12}) = -4.719 \times 10^2 \approx -470$ W m^{-2}.

(b) This is clearly a nonsense because the size of the forcing is more than the total effective solar flux (see Figure 1.5)!

Question 4.5
One molecule of water is 18/64 times the mass of a molecule of SO_2. Thus, if one molecule of SO_2 consumes one molecule of water, the total mass of water consumed would be $2.1 \times 10^{12} \times 18/64$ kg, or 5.9×10^{11} kg. But we are told that *three* molecules of water are consumed for every one molecule of SO_2. On this basis, therefore, $\sim 18 \times 10^{11}$ kg of water would be consumed – more than is present in the entire stratosphere!

Question 4.6

(a) In this case:
$$R_{net} = -1.43 \times 10^{-10} \times 1.5 \times 10^{11} \text{ W m}^{-2}$$
$$\approx -21 \text{ W m}^{-2}.$$

(b) The forcing due to the Laki eruption would therefore have been comparable to that of the great Tambora eruption in 1815, which according to Table 4.4 produced a climatic forcing of -21.4 W m^{-2}.

Question 4.7

(a) Density = mass / volume, so mass = density × volume. If the density of basalt lavas is $2.7 \times 10^3 \text{ kg m}^{-3}$, the mass of rock produced would have been at least $1200 \times 10^9 \text{ m}^3 \times 2.7 \times 10^3 \text{ kg m}^{-3}$, or about 3.2×10^{15} kg. We know that for every kilogram of rock produced, 0.0013 kg of sulfur were released into the atmosphere (see text); multiplying 3.2×10^{15} kg by 0.0013, we get a total mass of sulfur of 4.2×10^{12} kg.

(b) The relative molecular mass of sulfuric acid (H_2SO_4) is $((2 \times 1) + 32 + (4 \times 16))$ = 2 + 32 + 64 = 98. Therefore, 4.2×10^{12} kg of sulfur (relative atomic mass 32) would produce $\frac{98}{32} \times 4.2 \times 10^{12} = 1.3 \times 10^{13}$ kg of sulfuric acid aerosols.

(We have been working from a minimum value for the mass of rock, so the mass of aerosols produced was probably much greater than this.)

Question 4.8

If the gases produced by an eruption do not get into the stratosphere, any aerosols are formed in the troposphere and are soon washed out. Even the biggest explosive volcanic eruptions – such as that of Mount Pinatubo in 1991 – are short-lived, and their aerosols remain in the stratosphere for only a year or two. Thus, although their short-term effects may be profound, the climate forcing they cause is not sustained for sufficiently long for the climate system to respond. For this to happen, feedback loops (e.g. involving growth of vegetation or spread of ice-sheets) have to have time to become established, over periods of tens or hundreds of years. However, flood basalt eruptions may be exceptional, in that they involve the steady effusion of sulfur-rich magmas over periods of decades or more.

Question 4.9

If we assume that 3.8×10^{13} kg of magma were erupted, then $3.8 \times 10^{13} \times 0.0013 \approx 4.9 \times 10^{10}$ kg of sulfur must have been erupted. To obtain the mass of sulfuric aerosol produced (rather than sulfur), we must multiply by the relative molecular mass of sulfuric acid divided by the relative atomic mass of sulfur, i.e. by 98 / 32 (cf. Question 4.7). This gives $4.9 \times 10^{10} \times \frac{98}{32} = 15 \times 10^{10} = 1.5 \times 10^{11}$ kg.

Question 4.10

As discussed in Section 4.2, sulfur dioxide gas in the atmosphere is eventually converted to sulfuric acid (or sulphate) aerosols. These aerosols effectively increase the Earth's albedo, and lead to cooling (Figure 4.9c). Indeed, it is now believed that were it not for the production of these aerosols by industry, the Earth would have experienced more global warming than has in fact occurred; this hypothesis is supported by the fact that greenhouse warming seems to have had least effect over the most industrialized regions of the world, i.e. North America and Europe.

Question 5.1
We have been assuming that the Earth's angle of tilt has been more or less the same over the time involved. If it were significantly different, it would not be possible to refer so confidently to, for example, 'polar' or 'equatorial' conditions.

Question 5.2
(a) Given that they occur at intervals of 100s of millions of years (and the whole globe is affected), Ice Ages would plot in the upper part of the right-hand bubble in Figure 2.41. (You should now add Ice Ages into the last bubble, if you have not already done so.)

(b) Because if major ocean basins open and close on time-scales of 100–200 million years, then the time-scale on which the disposition of continents over the surface of the globe changes significantly is of the same order as the time-scale at which Ice Ages come and go.

Question 5.3
(a) Features you may have recognized as being fundamental to a rotating Earth are: anticyclonic subtropical gyres; cyclonic subpolar gyres; poleward-flowing western boundary currents like the Gulf Stream; westward-flowing North and South Equatorial Currents. (Note that only configurations (1) and (3) have eastward-flowing Equatorial Counter-currents, because these are a result of westward flow in the vicinity of the Equator being deflected back *along the Equator* (where the Coriolis force is zero) by the western boundary (which is not present in configuration 2).

(b) No, the temperature distribution would not be symmetrical in any of the configurations shown. As discussed in connection with Figure 2.1, over much of an ocean, water warmed at low latitudes flows polewards along the western side, while water cooled at high latitudes flows equatorwards in the eastern part of the ocean. This results in the western sides of oceans being generally warmer than eastern sides (though, at the present day, the north-easterly flow of the Gulf Stream causes the north-eastern North Atlantic to be warmer than the north-western part).

(c) The second configuration would lead to the warmest ocean overall. Surface water may flow uninterrupted around the planet at low latitudes, becoming significantly warmer than would be possible in the other two configurations. Moreover, as with the other two configurations shown in Figure 5.10, there is no barrier to heat transfer by current flow between low and high latitudes. As a result, the 'extra' heat gained by the water circulating for longer in regions of high levels of incoming solar radiation is redistributed efficiently to higher latitudes. The overall effect is that contrasts in temperature between low and high latitudes will be much less in continent–ocean configurations that permit circum-equatorial currents. (Don't worry if you did not come up with this answer.)

Question 5.4
The label 'glacial/interglacial' plots towards the top of the bubble labelled 'Milankovich cycles, orbital forcing', indicating that glacial–interglacial cycles are at least partly controlled by such forcing. We can be even more precise than this, because the period of the cyclicity is very close to 110 000 years, the period of the eccentricity cycle, which is therefore likely to be the main forcing factor, at least over the time covered by Figure 5.14.

Question 5.5

(i) Changing sea-levels may open or close oceanic gateways to surface (or even deep) currents. This could influence how much heat is carried poleward in currents, or – in the case of gateways permitting circum-equatorial flow – allow surface water to remain longer at low latitudes, becoming warmer, before flowing to higher latitudes. This latter situation could result in increased average temperatures for the ocean and warming of the climate as a whole.

(ii) Generally speaking, during periods of high sea-level, the Earth's average albedo is reduced, as sea-surfaces generally have a lower albedo than land (Table 1.1). More solar radiation is absorbed, which would reinforce any global warming trends. Conversely, falling sea-level exposes more higher-albedo land, which would encourage global cooling. Note, however, that this argument is somewhat oversimplified, as both land and (especially) sea have a greater albedo for low Sun elevations (i.e. at high latitudes).

Question 5.6 *Note: You were not expected to come up with all the details given below.*

(a) (i) During periods of low sea-level, larger areas of continental crust (notably silicate minerals) are exposed to weathering processes, which will tend to increase the rate of removal of atmospheric CO_2, further reinforcing any tendency towards global cooling. (As discussed later, the upper parts of mountains are sites of vigorous *physical* weathering. However, the weathering products are carried away (leaving more rock surface to be weathered) and are deposited on the lower slopes and on coastal plains, where they are subjected to chemical weathering – particularly if there is abundant vegetation.)

When sea-levels are high, the area of coastal plains available for (chemical) weathering is less, the rate of removal of atmospheric CO_2 will tend to be less, so that the concentration of CO_2 in the atmosphere would eventually begin to rise again, favouring global warming. (Note, however, that unless carbon added to the ocean through weathering is preserved in sediments (carbonaceous deposits and limestones), it will be returned to the atmosphere in 1000 years or so.)

(ii) and (iii) As mentioned in Box 5.3, because continental shelves have a flat topography, relatively small changes in sea-level can result in relatively large areas being flooded or exposed. Assuming sea-level is starting from a low level, a rise in sea-level would initially result in a significant increase in the area of low-lying, swampy coastal land, where organic remains could be preserved (continuing sea-level rise could eventually mean a decrease in the area of low-lying coastal regions, as the sea flooded steep terrain – as in the lochs of Scotland and the fjords of Norway). Similarly, a rise in sea-level would increase the area of shallow seas, where shallow-water carbonate-secreting organisms (such as certain algae, bivalves and corals) could flourish. As discussed in Chapter 3, both preserved organic carbon (peat, coal, etc.) and carbonate remains (i.e. accumulations of inorganic carbon) are long-term sinks for carbon, and so remove carbon dioxide from the atmosphere on geological time-scales. Thus, an increase in sea-level could lead to a decrease in the concentration of atmospheric carbon dioxide, and global cooling. In this case, negative feedbacks would be brought into play.

A decrease in sea-level could expose both organic remains and carbonate rocks, allowing them to be oxidized and eroded, and so returning carbon dioxide to the atmosphere and encouraging global warming. On long (geological) time-scales, the organic carbon and carbonates that are respectively oxidized and eroded following a fall in sea-level would not necessarily be the same ones that accumulated in the previous period of high sea-level, as much tectonic activity might have occurred in between.

The interactions of sea-level and climate are very complex, and if you thought of all these points you were doing very well.

(b) Considered in isolation, the effect of sea-level change on climate via weathering (i) is one of positive feedback. As discussed in Section 2.2.3, positive feedback mechanisms lead to instability in the climate system. By contrast, the effects described for (ii) and (iii) (both of which involve parts of the biosphere) are mechanisms for negative feedback, and therefore act so as to bring about stability. Both of the effects outlined in Question 5.5 (via albedo and current patterns) are positive feedbacks, leading to instability.

Question 5.7

(a) (i) As you read in Section 3.3.2, the CCD is depressed beneath areas of high productivity, i.e. it lies deeper, so that a greater proportion of production in the deep-water carbonate factory is preserved. (ii) If large amounts of extra CO_2 are introduced into the oceans, a substantial proportion must dissolve, because of the high pressure and (even in Cretaceous times) relatively low temperatures prevailing in the deep ocean. This must inevitably make the deep water more acid and corrosive to calcium carbonate remains, and so the CCD will rise, i.e. it will become shallower.

(b) Figure 5.22 shows a gradual rise in the CCD during the Cretaceous (up to ~ 100 Ma), which is consistent with an increased acidity (lower pH) in the deep oceans. Thereafter, it gradually deepens again, which would be broadly consistent with the spread of pelagic organisms secreting calcareous shells and skeletons – though there must have been other factors contributing to this pattern.

Question 5.8

(a) Tillites are fossil glacial deposits and so are indicative of cold conditions (Section 5.1), whereas remains of shallow-water carbonates including stromatolites would be expected to have accumulated in warm conditions (Section 5.3 3). At first sight, the two types of deposit are unlikely to be found in close proximity, and in particular we would not expect to find ice-sheets at low latitudes.

(b) We know nothing about the tilt of the Earth's axis during the Precambrian. As discussed in the text, if it were greater than 54°, changes in temperature over the course of the year would be greater than today, *and* climatic zones would be reversed – so allowing ice-sheets to form at the Equator.

For large angles of tilt, low latitudes would experience several (cool/cold) months of twilight alternating twice a year with (hot) months when the Sun would rise high in the sky during the day. If temperatures dropped sufficiently low during the cold season of the year, ice-sheets could become established and – because of their high albedo – survive the hot season. (Meanwhile at high latitudes, months of permanent twilight would alternate with periods of permanent night and periods of permanent day.)

Incidentally, it may not be necessary to explain the close association of tillites and stromatolites in this way. It has now been realized that stromatolites may form in dry regions that are cold – recently formed carbonate sediments, including stromatolites, have been discovered (in association with evaporites) in Antarctica (remember that while subpolar regions experience high precipitation, very high latitudes, beneath the polar atmospheric high, are *dry*).

Answers to Questions

Question 6.1
No. The normal fault studies suggest that the plateau reached its maximum elevation between 8 and 14 Ma. The data on Figure 6.5 suggest that it reached its present elevation much more recently, i.e. within the past million years.

Question 6.2
Condensation of moisture to form rain releases latent heat originally taken up from the ocean as latent heat of evaporation. Thus, air that is rising because it has been warmed by the underlying continent (Figure 2.19b) has an extra heat source, and rises all the more, so intensifying the low pressure over the region and drawing in air from the south yet more strongly.

Question 6.3
(a) There have been a number of large fluctuations in the relative abundance of *Globigerina bulloides*, but the main change occurred between about 8 and 9 million years ago when the relative abundance increased rapidly from <15% to more than 50%. In more recent times, the proportion has fluctuated, but has never been reduced to less than it was more than 8.5 million years ago (though it was very low at about 5.5 Ma).

(b) The increase in the relative abundance of *G. bulloides* 8–9 million years ago could well indicate an increase in upwelling of nutrient-rich water. As shown in Figure 6.12b, at the present day, upwelling and high levels of primary productivity occur during the South-West Monsoon. It is therefore reasonable to interpret Figure 6.13b as an indication of a strengthening of the winds of the South-West Monsoon 8–9 million years ago.

Question 6.4
The total flux of dissolved solids from the rivers listed in Table 6.1 is 796×10^9 kg yr^{-1} which is $\frac{796}{3500} \times 100 \approx 22.7\%$ of the global total. Hence, nearly a quarter of the dissolved products of weathering come from 5% of the available land area. This certainly suggests unusually high weathering rates in the Tibet/Himalaya region.

Question 6.5
According to Figure 6.15, there was a period of relatively fast cooling starting at about 85 Ma, another period with a high rate of cooling starting just before 50 Ma, a sharper increase in the rate of cooling starting about 15 Ma, and another perhaps even sharper decline starting at about 5 Ma.

1. As discussed earlier, Figure 4.22 indicates that the rate of production of oceanic crust gradually decreased between 120 Ma and 80 Ma, and then deceased more sharply over the next 20 million years. After that, it rose slightly, but has generally remained about the same ever since. As mentioned in the text, the general decline in sea-floor production over the course of the past 100 million years or so is consistent with the BLAG hypothesis. In addition, the hypothesis is also supported by the coincidence of the fairly sharp temperature drop ~85–70 million years ago with the decrease in ocean-floor production over roughly the same period. However, according to Figure 4.22, the decrease in temperature over the past 50 million years *cannot* be accounted for on the basis of the BLAG hypothesis as the rate of production of ocean floor has remained fairly constant.

2   The Himalayas and the Tibetan Plateau began to rise some time after 70 Ma, and probably only after 50 Ma, but presumably high rates of weathering would not begin to be established until some time after this, when the topography had become quite high and steep. The phase of global cooling over the past 50 million years is therefore consistent with the Raymo–Ruddiman hypothesis.

Question 6.6

(a) (i) If the flux of organic carbon from the continents increased significantly, the average $\delta^{13}C$ value of marine sediments would decrease.

(ii) The largest increase in the flux of organic carbon to the ocean will be indicated by the largest decrease in $\delta^{13}C$ in marine carbonates, which according to Figure 6.18 occurred about 58–56 million years ago. (We can assume that the $\delta^{13}C$ plot in Figure 6.18 is determined mainly by the flux of organic carbon from *land*, because that from marine primary and secondary production is very small by comparison.)

(b) Because the deep-water carbonate factory only began to be established about 100 million years ago, when the coccolithophores and foraminiferans evolved (Section 5.3.3). (Incidentally, as discussed in Section 5.3.3, a high production of carbonate debris depresses the CCD, allowing more calcareous sediments to accumulate. Furthermore, a large flux of HCO_3^- into the oceans, as a result of high rates of continental weathering, would – if accompanied by excess anions – also depress the CCD. On the other hand, consumption and decomposition of large amounts of organic material in the deep ocean, as a result of the increase in the riverborne flux of organic carbon discussed in (a), would have increased the acidity of the deep sea, so *increasing* the amount of inorganic carbon that would dissolve. Rest assured that we do not expect you to have come up with these extra details.)

(c) Amongst the constituents whose concentration in seawater could be increased by increased chemical weathering are the nutrients, dissolved nitrate, phosphate and silica, plus – perhaps more importantly – micronutrients such as iron. An increase in the supply of nutrients to the oceans could result in increased primary productivity of phytoplankton, both with and without calcium carbonate hard parts. This would result in increased accumulation of both organic carbon and (especially if the carbonate compensation depth were depressed, see (b)) inorganic carbon.

An increase in primary productivity in response to increased nutrient supply would also result from increased upwelling. As discussed earlier, vigorous upwelling occurs during the South-West Monsoon, so if the uplift of the Himalayas and Tibet caused the monsoon to strengthen, it could at the same time have enhanced the rate of preservation of carbon in the deep sea.

Question 6.7

The working is as follows:

$$R_{sw} \times (137 + 290) \times 10^7 = (137 \times 0.7035) \times 10^7 + (290 \times 0.7119) \times 10^7$$

Rearranging:

$$R_{sw} = \frac{(137 \times 0.7035) + (290 \times 0.7119)}{137 + 290}$$

$$= (96.380 + 206.45) / 427 = 302.83 / 427$$

$$= 0.7092$$

Answers to Questions

Question 6.8
Reasons you may have thought of are:

1. A polar plateau could not possibly interact with the atmospheric circulation in the same way as Tibet, particularly as at high latitudes the atmospheric circulatory patterns are slantwise or horizontal, rather than more or less in the vertical plane like the Hadley circulation. Being at high latitudes, the polar plateau would not experience such intense summer heating; and even if a low pressure region did develop, air drawn into it would *not* have passed over a very warm sea-surface (comparable to the Arabian Sea) and would therefore contain much less moisture, and could not add to the warming effect through release of latent heat.

2. Following on from (1), if there were no high summer rainfall, there would not be extensive chemical weathering, particularly as temperatures would be much lower. Less chemical weathering would result in less CO_2 being removed from the atmosphere and, other things being equal, less cooling.

3. The plateau could not affect the path of the subtropical jet stream. However, it might interfere with the polar jet stream – any effects that this might have are outside the scope of this Course!

4. Such a plateau would not change the proportion of the Earth's surface covered by ice, so there would be no increase in the global albedo.

Question 6.9
First, the BLAG model does not predict decreasing abundances of CO_2 in the atmosphere and hence decreasing global temperatures within the past 50 million years. Secondly, the model assumes that weathering rates will *decrease* as global temperatures cool, but in fact over the past 50 million years weathering rates (as measured by $^{87}Sr/^{86}Sr$ in seawater) have actually been *increasing* during a period of global cooling.

Question 6.10
The $^{87}Sr/^{86}Sr$ of seawater will increase if:

(1) There is a decrease in sea-bed volcanism, i.e. in the rate of production of new ocean floor (hydrothermal fluids have a low $^{87}Sr/^{86}Sr$ ratio).

(2) There is an increase in the rate of chemical weathering of continental rocks (which have a high $^{87}Sr/^{86}Sr$ ratio).

It could also be argued that even if chemical weathering rates stay the same, if continental rocks with unusually high $^{87}Sr/^{86}Sr$ ratios are exposed to weathering, the $^{87}Sr/^{86}Sr$ ratio for rivers draining the region will increase, so increasing the $^{87}Sr/^{86}Sr$ ratio of seawater.

Comments on Activities

Activity 1.1
(a) Figure 1.9, which takes into account absorption in the atmosphere, gives values for daily summer sunshine levels at 65° N of $\sim 2.0 \times 10^7\,\text{J}\,\text{m}^{-2}$. According to the left-hand end of the curve in Figure 1.14, current daily summer sunshine levels at this latitude are nearly $3.9 \times 10^7\,\text{J}\,\text{m}^{-2}$; this plot must therefore be of the amount of solar radiation *reaching the top of the atmosphere*, not that reaching the Earth's surface after absorption in the atmosphere.

(b) The curve for high southern latitudes would not be the same as that in Figure 1.14, because the (changing) ellipticity of the orbit means that when the seasons (i.e. the effect of the tilt) are very intense in one hemisphere they are moderated in the other. It is difficult to predict what the curve would look like, but it certainly would not be a mirror image either; for one thing, the main effect of the changing shape of the orbit is to change the amount of solar radiation *reaching the Earth as a whole,* so increasing or decreasing incoming solar radiation by the same amount in both hemispheres.

(c) According to Figure 1.9, at 50° N the difference between the solar radiation reaching the Earth's surface at mid-summer and mid-winter is about $(2.3 - 0.3) \times 10^7 = 2.0 \times 10^7\,\text{J}\,\text{m}^{-2}\,\text{day}^{-1}$. According to Figure 1.14, the most rapid changes in solar radiation resulting from Milankovich cycles (at 65° N) are of the order of $0.6 \times 10^7\,\text{m}^{-2}\,\text{day}^{-1}$ (i.e. about a third of the current difference between summer and winter values for the southern British Isles) over a period of $10 \times 10^3 = 10\,000$ years. This would be the change at the top of the atmosphere – at the Earth's surface, the change would be even less. We can therefore confidently say that the Milankovich cycles do not result *directly* in climatic changes detectable on human time-scales.

Activity 2.1
1. Tropical cyclones are transient phenomena, each affecting a fairly small area of the Earth, and so plot towards the lower part of the 'weather' bubble.

   The polar and subtropical jet streams both encircle the globe, and you know that wave motions in the polar jet stream develop over a month or so, so these plot near the top of the first bubble.

2. El Niño–Southern Oscillation (ENSO) events occur every 4–10 years, and affect a large part of the globe (though not all of it), and so plot in the middle (or towards the top) of the second 'bubble'.

   Don't worry if your answer was not exactly the same as ours, as long as the phenomena were plotted in the right 'bubble'.

Comments on Activities

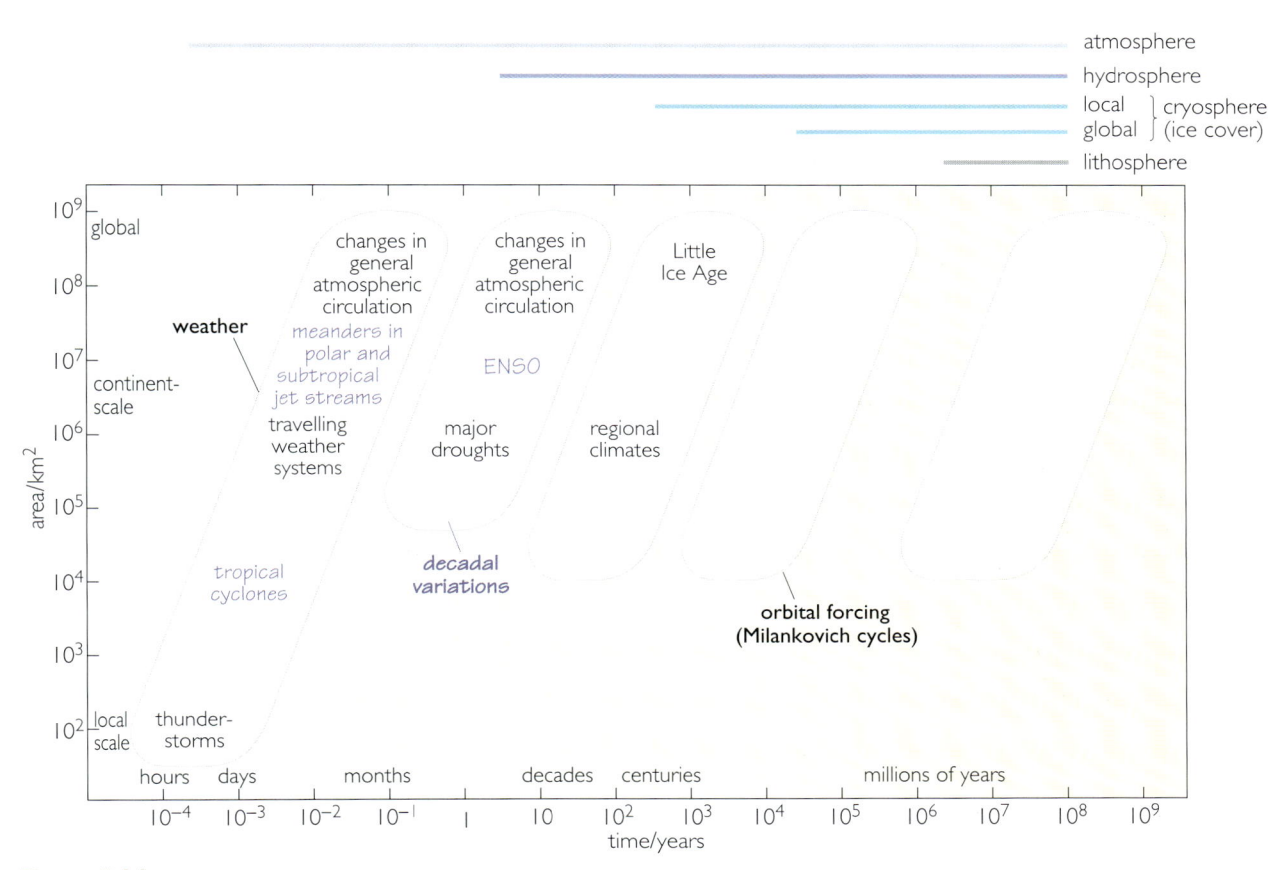

Figure AA1
Answer to Activity 2.1.

Activity 3.1

(a) In the north-eastern North Atlantic (as well as at high southern latitudes), the net flux of CO_2 is *into* the ocean. This is a region where deep-water formation is occurring as a result of intense cooling of surface water (and brine-rejection) (Figures 2.29, 2.30). Cold water can take up a relatively large amount of carbon dioxide before becoming saturated, and on sinking down from the surface would allow yet more CO_2 to dissolve.

(b) At both seasons of the year there is a flux of CO_2 into the ocean in the central North Atlantic, partly because surface waters converge and sink there (Figure 2.28). However, by April–June, the water is being warmed and so the region where there is a net flux of CO_2 *out* of surface water is beginning to spread northwards. Also at this time of year there is high net primary productivity in the northern part of the ocean (Figure 2.40a), which also contributes to a net flux of CO_2 from air to sea.

(c) In low latitudes, water is upwelling to the surface and warming. Both the decrease in pressure and the increase in temperature cause CO_2 to come out of solution, so that there is a net flux from sea to air (despite any high primary productivity along the Equator which would tend to draw CO_2 down into the ocean).

(In studying Figure 3.17, you may have noticed that in the northern Pacific and northern Indian Ocean, there is a net flux out of the ocean, particularly noticeable in January and March. The reason for this will be explained in the text shortly.)

Activity 4.1

(a) (i) The average annual contribution of CO_2 to the atmosphere from the Deccan Traps would have been $5.8 \times 10^{15} \times$ kg / 500 000 years, or about 1.16×10^{10} kg yr^{-1}. If the total mass of CO_2 in the atmosphere was 2.8×10^{15} kg, then the amount of CO_2 erupted annually would have been only a very small fraction of it ($1.16 \times 10^{10} / 2.8 \times 10^{15} \approx 0.000\,004$).

Note: we have assumed for convenience that the mass of CO_2 in the atmosphere at the time was not that different from what it is today. In fact, as you will see, the CO_2 concentration in the atmosphere around the end of the Cretaceous was probably somewhat higher than it is today (and the impact of the eruption would have been proportionately less).

(ii) Some of the 'extra' CO_2 would be taken up by other parts of the Earth system, i.e. it would be dissolved in the ocean and taken up in increased primary productivity, mainly on land (see Question 3.5). We would therefore not expect the mass of CO_2 in the atmosphere to increase annually by the amount calculated in (i).

(b) (i) Doubling the CO_2 in the atmosphere would mean the addition of an extra 330 p.p.m. An increment of 75 p.p.m. is only 75 / 330 of this, or about a quarter. One would expect this to result in a climate forcing of around $+4.4$ W m^{-2} / $4 \approx +1.0$ W m^{-2}.

(ii) Assuming that the Earth's climate sensitivity is about 0.75 °C per W m^{-2} of forcing, the extra CO_2 from the eruption of the Deccan Traps would result in a warming of only 0.75 °C. However, we need to take into account that the duration of Deccan volcanism was long enough for some feedback processes to 'kick in', and perhaps amplify any warming trend. On the other hand, the warming would almost certainly have been offset by cooling due to reflection of incoming solar radiation by aerosols (Figures 4.8 and 4.9).

(iii) The residence time of carbon in the atmosphere (a small reservoir with relatively large fluxes in and out of it) is very short – of the order of a few years, so any changes in the fluxes will be 'noticeable' effectively instantaneously. This means that while the volcanism was occurring spasmodically, the concentration of CO_2 in the atmosphere would have been fluctuating, as would the size of the resulting 'greenhouse effect' (cf. Section 3.4).

Activity 6.1

(a) (i) and (ii) The completed graphs should look like Figure AA2.

(iii) There is a general trend of decreasing Sr flux (increasing 1 / Sr flux) with increasing $^{87}Sr/^{86}Sr$. This is seen most clearly in Figure AA2b where the relationship is almost linear.

(b) Rivers that have the strongest effect on the $^{87}Sr/^{86}Sr$ ratio of seawater must deliver a large Sr flux with high $^{87}Sr/^{86}Sr$ ratios. This is not true for either the Xingu river (a tributary of the Amazon), which delivers small fluxes of Sr with high $^{87}Sr/^{86}Sr$ values, nor for the Yangtze, which delivers large fluxes of Sr but with low $^{87}Sr/^{86}Sr$. However, both the Ganges and the Brahmaputra rivers lie outside the general trend and have both high Sr concentrations and high $^{87}Sr/^{86}Sr$ ratios. These two rivers (particularly the Ganges) will therefore have a strong influence on the $^{87}Sr/^{86}Sr$ ratio of seawater. Comparison between the two plots illustrates the advantage of plotting the reciprocal of the Sr flux rather than the actual value. There is, in fact, a linear relationship between ($^{87}Sr/^{86}Sr$) and 1 / Sr flux for most rivers (cf. (a) (ii) above), and points lying off this trend are much more apparent in plot (b) than plot (a).

Comments on Activities

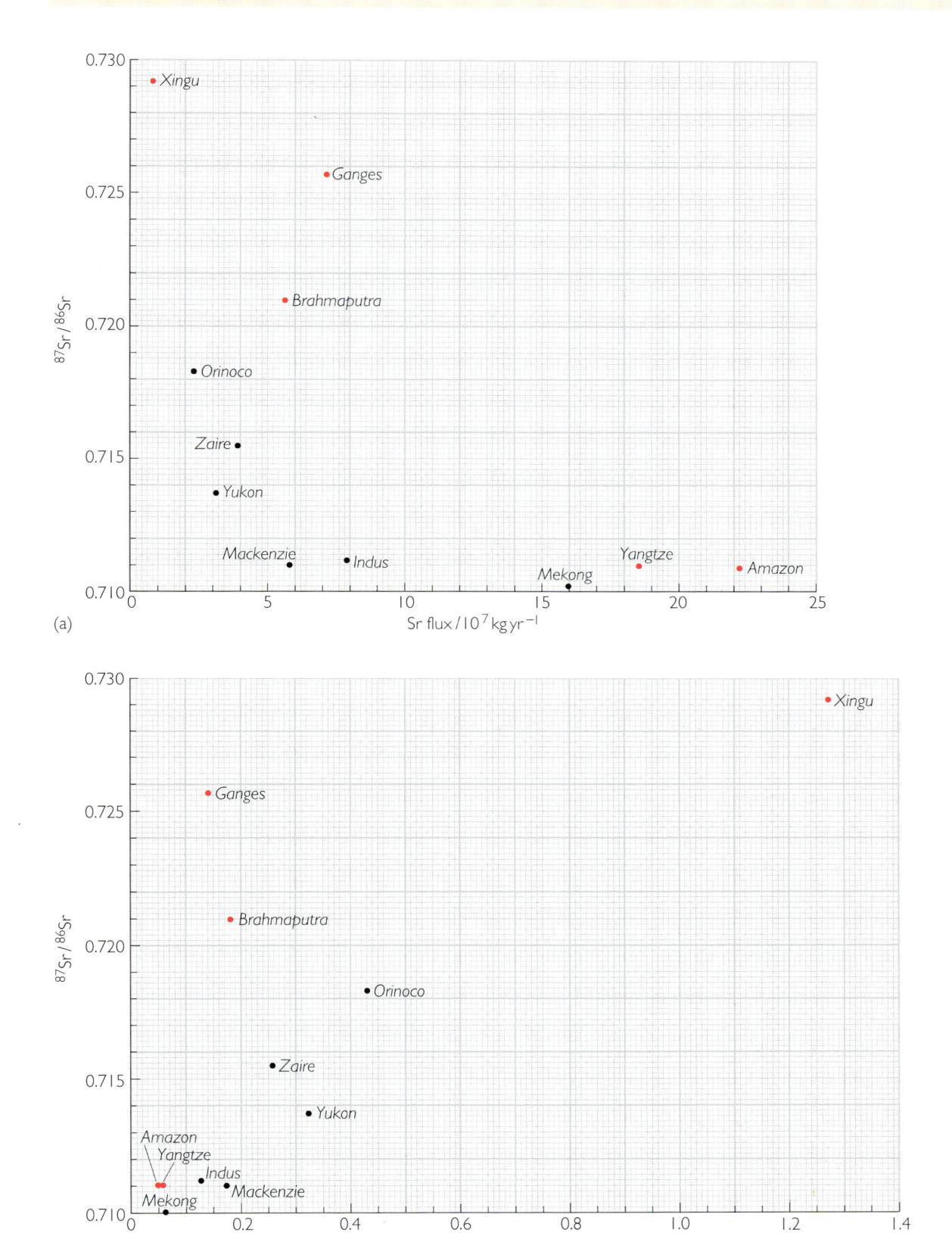

Figure AA2
Completed versions of Figure 6.21a and b.

Activity 6.2

(a) The higher $\delta^{18}O$ values correspond to peaks of glacials and the lower (more negative) values to the peaks of interglacials (cf. Figure 5.14). Even if you couldn't remember the relationship between $\delta^{18}O$ and ice volume (global temperature), you could have tackled this by remembering that we are currently in an interglacial, so the low $\delta^{18}O$ values at the top of the core must correspond to interglacial conditions.

(b) (i) Peaks in Figure 6.23b and c correspond, so the high sedimentation rates were generally associated with large inputs of windblown dust; in other words, the increased rates of sedimentation were largely due to increases in the input of windblown dust from the land.

(ii) Both the proportion of windblown dust and (not surprisingly, given (i)) the total accumulation rate were higher during glacial periods. This suggests that the source area of the dust (and presumably landmasses in general) were more arid during glacial periods. This is what we would expect, as during glacial periods the hydrological cycle is less active and much of the water that is cycling though the Earth system as water vapour or liquid water during interglacials is locked up in the ice-caps.

(c) (i) Perhaps surprisingly, no it doesn't. In fact, it seems to show the opposite, that primary productivity off the coast of Oman has been highest at the peaks of interglacials (including the present one). (The reason for this is not known, but it may have more to do with ocean current patterns than with wind strength.)

(ii) The addition to surface waters of more of the micronutrient iron could also have increased primary productivity. (However, if it did, the effect must have been counteracted by some other factor; cf. (c)(i).)

(d) The 110 000-year cycle is clearly the most dominant cycle as far as (i) continental aridity and (especially) (ii) primary productivity are concerned. (In the case of (i), this is perhaps easiest to see by comparing the shape of the plots on Figure 6.23b and c with the $\delta^{18}O$ plot in Figure 6.22b.) However, in Figure 6.23b and c, the plots are more jagged than the $\delta^{18}O$ plot, and there seems to be a peak every 40 000 years. You might also be able to convince yourself that you can see one every 22 000 years!

In fact, sophisticated spectral analysis of the various plots shows that both continental aridity *and* primary productivity are affected to some extent by *all three* Milankovich cycles, not just the 110 000-year eccentricity cycle, but also the 40 000-year tilt cycle and the 22 000-year precession cycle.

Acknowledgements

The Course Team wishes to thank the following: Professor Bill Chaloner, Dr. G. Shimmield and Frank McDermott, the external assessors, for providing helpful advice on the content and level of this book; also Fran Van Wyk de Vries, Peter Daniels, Margaret Deller, Jim Grundy and Colin Whitmore, and the tutor reader Cynthia Burek, for their comments.

Grateful acknowledgement is made to the following for permission to reproduce material in this book:

Figures

Figures 1.3a, 4.2b, 4.12, 6.12 NASA; *Figure 1.3b* © T. Van Sant/GeoSphere Project, Santa Monica; *Figure 1.4* NSF/NASA-sponsored US Global Ocean Flux Study Office, Woods Hole Oceanographic Institution, with Goddard Space Flight Center, University of Miami, University of Rhode Island; *Figure 1.13* J. Imbrie *et al.* (1984) in Berger, A. *et al.* (eds) *Milankovich and Climate*, Kluwer Academic Publishers; *Figure 1.16* J.F.B. Mitchell (1989) 'The greenhouse effect and climate change', *Reviews of Geophysics*, **27**(1), American Geophysical Union; *Figure 1.17* T. Vander Haar and V. Suomi (1971) *Journal of Atmospheric Science*, **28**, pp. 305–14, American Meteorological Society; *Figure 1.18* Rutherford Appleton Laboratory; *Figure 1.19a* National Optical Astronomy Observatories/NSO, Sacramento Peak; *Figure 1.19b* Lockheed Palo Alto Research Laboratory/Japanese Institute of Space & Astronautical Science; *Figures 2.1, 2.40* InterNetwork Inc., NASA/JPL and GSFC; *Figures 2.17, 2.21* A. Strahler (1973) *Earth Sciences*, Harper & Row; *Figure 2.18* A.H. Perry and J.M. Walker (1977) *The Ocean–Atmosphere System*, Addison–Wesley; *Figures 2.20b, 5.16a, 5.20* T. Waltham, Geophotos; *Figure 2.24* A.E. Gill and E.M. Rasmussen (1983) 'The 1982–83 climate…', *Nature*, **306**, Macmillan; *Figure 2.28* L. Xie and W.W. Hsieh (1995) 'The global distribution of wind-induced upwelling', *Fisheries Oceanography*, **4**, pp. 52–67; *Figure 2.30* M. Brandon; *Figure 2.32* J.F.B. Mitchell (1989) 'The greenhouse effect and climate change', *Reviews of Geophysics*, **27**(1), American Geophysical Union; *Figure 2.35* Mrs. G. Rozario; *Figure 2.39* N.T. Nicoll; *Figures 2.41, AA1, 5.6, 6.25* P.F. McDowell *et al.* (1995) 'Long term environmental change', in Powell, T.M. (ed.) *Ecological Time Series*, Chapman & Hall; *Figure 2.44* W.D. Sellers (1965) *Physical Climatology*, University of Chicago Press; *Figure 3.4* B. Cox *et al.* (1989) *The Atlas of the Living World*, Marshall Editions; *Figure 3.5* Nancy Dise; *Figure 3.10* Mike Dodd; *Figures 3.11, 3.12* P. Williamson/ NERC, BOFS Project; *Figure 3.13* Bob Spicer; *Figure 3.14a* A. Alldredge; *Figure 3.14b–d* Southampton Oceanography Centre; *Figure 3.16a* A. McIntyre, Lamont–Doherty Geological Observatory; *Figure 3.16b* I. Joint, Plymouth Marine Laboratory; *Figure 3.16c* D. Breger, Lamont–Doherty Geological Observatory; *Figure 3.17* N. Lefevre, Plymouth Marine Laboratory; *Figure 3.22* K.W. Thoning *et al.* (1994) 'Atmospheric CO_2 records…', in Boden, T.A. *et al.* (eds) *Trends '93: A Compendium of Data on Global Change*, ORNL/CDIAC-65, Oak Ridge National Laboratory; *Figure 3.23* Conway *et al.* (1988) 'Atmospheric carbon dioxide…', *Tellus 40B*, pp. 81–115; *Figures 4.2a, 4.6, 4.16* P. Francis; *Figure 4.10* NASA/AVHRR; *Figure 4.11* A. Kreuger/NASA/Goddard; *Figures 4.15, 4.18* P. Francis (1993) *Volcanoes: A Planetary Perspective*, Clarendon Press; *Figure 4.19* H. Sigurdsson (1982) 'Volcanic pollution and climate…', *EOS*, **63**, pp. 601–2, American Geophysical Union; *Figure 4.21* M.F. Coffin and O. Eldholm (1993) 'Exploring large subsea igneous provinces', *Oceanus*, **36**(4), Woods Hole Oceanographic Institution; *Figure 4.22* R. Larson (1995)

'The mid-Cretaceous superplume episode', *Scientific American*, Feb., by permission of Scientific American Inc., all rights reserved; *Figure 5.1* K.W. Glennie (1990) *Introduction to the Petroleum Geology of the North Sea*, Blackwell; *Figure 5.2* J. Watson; *Figure 5.3* Natural History Museum; *Figure 5.4a* Chris Wilson; *Figure 5.5* I.P. Martini (1996) *Late Glacial and Postglacial Environment Changes*, Oxford University Press; *Figure 5.10* T.H. van Andel (1985) *New Views on an Old Planet*, CUP; *Figure 5.11b–d* B.U. Haq (1984) 'Paleoceanography ...', in Haq, B.U. and Milliman, J.D. (eds) *Marine Geology and Oceanography of Arabian Sea and Coastal Pakistan*, Van Nostrand Reinhold; *Figure 5.12* Kennett, J.P. *et al.* (eds) (1974) *Initial Reports of the DSDP*, Vol. 29, US Govt Printing Office; *Figure 5.13* J.W. Valentine and E.M. Moores (1970) 'Plate-tectonic regulators...', *Nature*, **228**, p.106, Macmillan; *Figure 5.15 Cairngorms – A Landscape Fashioned by Geology*, British Geological Survey; *Figure 5.16b* Kevin Church; *Figure 5.16c* John Wright; *Figure 5.17* R.A. Davis (1994) *The Evolving Coast*, copyright © 1994 Scientific American Library; *Figure 5.21* Scholle, P.A. and James, N.P. (eds) SEPM 'Photo CD-1 and Photo CD-2 ...', Society for Sedimentary Geology; *Figure 5.24* S. Groom, NERC, BOFS Project; *Figures 5.26, 6.6–6.8, 6.10, 6.11, 6.16* Nigel Harris; *Figure 6.1* US National Geophysical Data Center; *Figure 6.5* M. Raymo *et al.* (1988) 'Influence of late Cenozoic...', *Geology*, **16**(7), Geological Society of America; *Figure 6.13a* R.W. Jordan and W. Smithers, University of Surrey; *Figure 6.13b* D. Kroon *et al.* (1991) 'Onset of monsoonal-related...', *Proc. Ocean Drill Program Sci. Results*, **117**, pp. 257–63; *Figure 6.14* J. Quade *et al.* (1989) 'Development of Asian monsoon...', *Nature*, **342**, Macmillan; *Figure 6.1* D.M. Kerrick and K. Caldeira (1994) 'Metamorphic CO_2 degassing...', *GSA Today*, **4**(3), March, Geological Society of America; *Figure 6.20* F. Richter *et al.* (1992) 'Sr isotope evolution...', *Earth & Planetary Sci. Lett.*, **109**, Elsevier; *Figures 6.22–6.24* G.B. Shimmield *et al.* (1990) 'A 350 ka history...', *Trans Roy. Soc. Edin: Earth Sci.*, **81**, Royal Society of Edinburgh.

Tables
Tables 3.1, 3.2 R.H. Whittaker and G.E. Likens (1973) 'Carbon in the biota', in Woodwill, G.M. and Pecan, E.V. (eds) *Carbon and the Biosphere*, National Technical Information Service, Washington.

Index

Note: Entries in **bold** are Glossary terms. Indexed information on pages indicated by *italics* is carried wholly in a figure or table.

AABW *see* Antarctic Bottom Water
absorption of solar radiation *20*, *21*, 32
acid rain 130
adiabatic 30, 31, 42, *47*, 177
advection 31
aerosols 65
   condensation around 32
   volcanism 133–4, 136
      climatic change 119, *120*, *121*, 122–3, *124*, 125–8
Africa
   African Plate *112*, 113
   agriculture 50
   drought 39
   dust from 210
   upwelling 51
agriculture 50
   drought 38–9
   ruined by volcanism 130–2
Agulhas Current 45
Agung eruption 124–5, 128
air masses 31, 38–9, 41
see also atmosphere; pressure; transport of heat; winds
Alaska
   carbon dioxide fluctuations 107
   Current 45
   sea-level changes and climate 163
   volcanism 124–5, 128, 130
albedo 10, 57
   continental drift and climate 150
   heat transfer 32
   mountain-building and climate 177
   sea-level changes 166, 168, 174
   solar radiation 10, *18*
   Tibet and Himalayas 190
   volcanism 120, 123
albite 88
algae 9, 63
   blooms *174*
   carbon cycle 91, 94, *95*
   mountain-building and climate 178
   sea-level changes and climate 170, *171*
   see also coccolithophores

altitude
   orographic precipitation 42, *43*, 188, *189*
   pressure, air *30*, 176–7
   temperature changes *30*, 31, *33*, 42, *47*, 185–6, 190
   see also mountains; uplift
altocumulus and altostratus 32
Amazonia and Amazon river 109, 179, 206, 207
see also rainforest
animals
   carbon cycle *92*, 93–4, *95*
   continental drift and climate *156*, 157, *158*, 159–60
   El Niño 48, 50
   feeding 75
   insignificant carbon reservoirs 77, 78
   sea-level changes and climate 172
   Tibet and Himalayas *186*, 191, 195
   see also biomass; foraminiferans; zooplankton
anorthite 88
anoxic environments **78**, 144, 170, *173*
Anscom, William *120*
Antarctic
   carbon dioxide concentration 105, 116
   deep waters and currents 44, 45, 52, 53, *70*, *97*, 151, *155*, 156
   ice 128, 147, 167
   Plate 113
   see also polar region; Southern Ocean
Antarctic Bottom Water 52, 53, *70*, *97*, *155*
Antarctic Circle *3*
Antarctic Circumpolar Current 44, 45, 70, 151, 156
Antarctic Divergence *52*, 70, *97*, 98
Antarctic Intermediate Water 52–3
Antarctic Polar Frontal Zone (Antarctic Convergence) *52*
anticyclones *34*, *37*, *38*, 44
anvil head 32
aphelion 14
Aptian extinctions 135
Arabian Plate 113
Arabian Sea
   climate change and phytoplankton 191–2, *193*, 194–5
   records in sediments of 209–14
Arctic 53
   Circle *3*

Ocean 53, 147, 151, 153
    see also polar region
argon 55, 62
arid areas *see* deserts
ash, volcanic 119
Asia
    monsoons 41–2, 50, 70
    see also Deccan Traps; Tibet and Himalayas
Atlantic Ocean
    currents 44–5
        see also Gulf Stream
    deep circulation *52*, 53, *54*
    fluxes across air–sea interface 96, *97*
    gyres 44, 153, *155*
    oscillations 50
    plate tectonics 153, *155*, 156
    primary production and chlorophyll *65*, *90*
    temperature 27, 44
        thermohaline conveyor *54*, 98
    upwelling 51
    winds 40
atmosphere
    chemistry of 55–8
    convection *29*, 31, 36, 50
    gases in 55, *62*, 74
    see also carbon dioxide; methane; nitrogen; oxygen; water vapour
    heat transported by 30–43
        coupling with ocean 46–52
    as support system for life 58–66
    see also air masses; atmospheric carbon; climate; time- and space-scales
atmospheric carbon in carbon cycle 81–2
    diagrams 76, 98, 101, 108
    see also carbon dioxide
atmospheric window 57
Australia *49*, 156
autotrophic organisms **75**
axis of rotation of Earth *12*
    see also tilt
Azores 5

back radiation *57*
    see also long-wave radiation
bacteria 63, 64, 66, 92, 203
Bahamas *171*
Bajocian extinctions 135–6
balanced system, carbon cycle as 103–9

Baltic Sea 164
basalt *see* flood basalts; oceans
Benguela Current 44, 45
benthic organisms 64, 170, 196
    carbon cycle 91, *92*
    continental drift and climate 157, *158*, 159–60
Berner, Bob *see* BLAG
bicarbonate ion (HCO_3^-) 83, **84**, *85*, 87, 88
biogeochemical cycle 58, **76**
biolimiting 64
biological pump 92
biomass 59
    see also animals; biota; plants
biomes 79, *80*, *81*, 83
biosphere 10
biota and carbon cycle *76*, 98, 101, 108
    see also animals; plants
birds 49
bivalves 94, 170, 178
black body
    Earth as 56
    Sun as *19*, 56
BLAG model 199–200, 201, *202*, 203, 209
blooms 65, *174*
boulder clay *see* tillites
Brahmaputra (Tsangpo) river 197, 198, 206, 208
Brazil Current 45
brine-rejection 53
Britain
    continental drift and climate *141*, 142, *143*, 144
    sea-level changes and climate 161, *162*, *171*, 172
    volcanism, effect of 130, 131, 132
Brito–Arctic flood basalt province 135
Burren *60*
C4 and C3 plants 195
calcareous sediments 100, *101*
calcium (Ca^{2+})
    in lava 117
    in rocks 88
    in water 60, 61, 64, 88, 89
calcium carbonate 201
    in Arabian Sea sediments 210, 211
    and silica (decarbonation) 177–8, 202
    see also carbonate; shells and skeletons
California Current 45

Index

Cambrian plate tectonics 157, 158, 165
Canada 53, 130
Canaries Current 45
Cancer, Tropic of *12–13*, 15
cap world, polar *149, 150*, 151, 169
Capricorn, Tropic of *12–13*, 15
carbohydrates 74, 75
carbon
    compounds *74*
    fixing *see* primary production
    isotopes 193, *194*
    in oceans 62–3, 89–100, *96*
        organic factors controlling flux 89–94
        physical factors controlling flux 94–6
carbon cycle 73–110, 200
    as balanced system 103–9
    climate and 75–6
    life and 73–5
    summary diagrams *76, 89, 98, 101, 108*
    see also geological; marine; terrestrial
carbon dioxide *74*
    in atmosphere
        carbon cycle 75–6, 98, 101, 108
        climate 55, 56, 57, 75
        controls on 101-3
        decrease, seasonal *82*, 102
        fluctuations 106
        flux across air–sea interface *96*
        increased *105*, 106–9, *116*
        long-term controls on 101–3
        photosynthesis 9, 74
        phytoplankton counteracting 65
        proportion 62
        reservoir 77, 81–2
        sea-level change and climate 168–74
        seasons *82*, 102, *106*, 107
        subduction 177–9
        temperature 8
        volcanism 117–18, 119, *120*, 168, 177, 199, *202*
        see also Tibet and Himalayas
    in soil 87
    in water 83–7, *84–5*
        carbonate system 83–5
        fluxes *86, 89*, 90, *96–7*
        oceans 62–3
        air–sea interface *96*
        rain 60
carbonaceous sediments 100, *101*, 200
carbonate
    Arabian Sea sediments *210, 211*
    in carbon cycle 88, 101, 102, 108
    equilibrium system 83, *84–6*, 87, *89*, 90
    soil *194*
    in water 83, 84, *85*, 87, *201*
        deep oceans 172, *173*, 174
    weathering 87–8, 99, 101–2, 179
    see also calcium carbonate
carbonate compensation depth (CCD) 94, 110, 172, *173*
carbonate ion (CO_3^{2-}) 83, **84**, *85*, 87
carbonate system 83, *84–5*, 87
carbonic acid
    in forests 87
    in water 60, 61, 83, *84, 85*, 87, 88, 89, 94
Carboniferous 83, 165
    continental drift and climate 141, 142, *143, 145*, 149, 153; 158
Caribbean Plate 113
CCD *see* carbonate compensation depth
Cenomanian extinctions 135–6
Cenozoic plate tectonics 141, 147, 155–6, 158
CFCs (chlorofluorocarbons) 109
chalk 17, 101, *171*
chemical equilibrium 84
chemical weathering 60
    carbonate 87–8, 99, 101–2, 179
    mountain-building and climate 178–9
    sea-level changes and climate 169
    silicates 88, 99, 102, 169, 178–9, 199
    Tibet and Himalayas 195, 199, *202*
    see also strontium
Chichón, El 124–5, 128
Chicxulub impact 134–5
Chile *50*
China Plate 113
chlorides 60, 61, 89, *120*
chlorofluorocarbons 109
chlorophyll 9, *10*, 64, *65*, *90*, *192*
see also photosynthesis
Christmas Island 49
cirrocumulus and cirrostratus 32
cirrus 32
clay *171*
see also tillites

climate and climatic change 55–68
   Arabian Sea sediment records 209–14
   atmosphere 55-8
      –ocean–Earth as support system for life 58–66
   carbon cycle 75–6
   defined 11
   feedback 47, 65, 67–8, 202, *202*
   forcing function 123–4, *126*
   oceans 58–66, 151–8
   see also sea-level change
   sunspots 66–8
   volcanism *see* aerosols; Deccan Traps; flood basalts
   weather systems 67, 148, 213
   see also aerosols; atmosphere; continental drift; flood basalts; fog; greenhouse; precipitation; temperature; Tibet and Himalayas; winds; mountain-building; plate tectonics

clouds 32, *57*, *190*
   in cyclones *47*
   formation 30, 31–3, 46, 159
      phytoplankton and 65

coal measures 142, *143*, 144, *145*, *146*, 153

coccolithophores 94, *95*, *171*, 172, 178, 210
   bloom *174*

Cocos Plate 113

collision of continents *176*, 178, 202
   India–Asia *183*, *184*, 185

Columbia River Province flood basalts *129*, 133–4, *135*, 209

combustion *see* fossil fuels

compounds, carbon's ability to make 73, 74

compression 30
see also pressure

computer modelling 149, 151
see also forcing function; GCM

condensation *see* clouds

conduction 20, 31

conservative margins (transform faults) *112*, 113

constructive margins (mid-ocean spreading ridges) *112*, 113, 114, 119

continental climate 70

continental drift and climate 141–61
   glacials and interglacials 158–61
   ocean currents 151–8
   rearranging continents 144–51

continental shelf 91, 157, 200
   plate tectonics 164, 166, 170, 172
   shelf seas 146, 164, 172

convection 29
   in atmosphere *29*, 31, 36, 50
   clouds 32
   in Earth's interior *115*, *139*, 167
      thinning *184*
   in oceans *29*, 44

convective thinning *184*

cooling
   volcanism 123, *124–6*, 127, 130–1, *132*, 133
   weathering rates 202, 203
   see also adiabatic; temperature

copepods *92*, 93

corals 170, *171*, 178

Coriolis effect 34, *35–6*, 37, 45, 51

coupling 47

Cretaceous
   continental drift and climate 141, *154*, 155–6, 158
   runaway greenhouse prevented 169–70
   sea-level changes and climate 165, 166, 168, 169–74
   Tibet and Himalayas 196
   volcanism 134, 137–8

Croll, James 16
see also Milankovich

crustal thickening *175*, *184*
see also convective thinning

cryosphere *see* time- and space-scales

cumulonimbus 31, *32*, 36, *47*

cumulus 31, *32*

currents *see* oceans

cyanobacteria *171*

cycles *see* biogeochemical cycle; carbon cycle; hydrological cycle; Milankovich; Sun

cycling of elements, global *62*

cyclones *34*, *37*, *38*
   downwelling 50
   El Niño 48
   stages in formation *46*
   tropical (hurricanes/typhoons) 46, *47*, 148, 213

$\delta^{13}C$ 193–5, **194**, 200–1
$\delta^{18}O$ 159–60, 193, *194*, 195, 196, *200*, *201*, 210
dating 195
   isotope *186*, 188

decarbonation reaction 177, 178, 202

Deccan Traps *129*, 135
   carbon dioxide and greenhouse warming *136*, 137
   mass extinctions 135–6

Index

decomposition
    carbon cycle 77, 78, 82, 87, 92, 97
    see also bacteria

deep oceans
    carbon cycle 90, 91, 97–100
        diagrams 76, 98, *101*, 108
            sediments 76, 98, 99, 100, *101*, 108, 109, 195, 200, *201*
    carbonate production 172, *173*, 174
    circulation and currents 44, 45, *52*, 53, *54*, 55, *70*, 97, *151*, 152, *155*
    plate tectonics 170
    trench 131, 133
    see also benthic

deforestation 43, 78, 107, 173

deserts and arid areas 80
    continental drift and climate 142, 143
    Tibet and Himalayas 188, *189*
    see also evaporites

destructive margins (subduction zones)
    mountain-building and climate 175, *176*, 177–8
    Tibet and Himalayas 183–4, 202
    volcanism *112*, 113, *115*, 138

detritus in carbon cycle 77, 78–9, 82, 92, 193, 210

Devonian plate tectonics 141, 142, 158, 165

diatoms 178

DIC *see* dissolved inorganic carbon

dimethyl sulfide from phytoplankton 65

dinosaurs, extinction of 135

dissolved inorganic carbon (DIC) 84, *87*, *89*, 92

dissolved organic carbon (DOC) 79, *89*, 92, 99, 101, 200

DNA and RNA 73

DOC *see* dissolved organic carbon

downwelling *50*, 51–2, *70*
    in carbon cycle diagrams 98, 101, 108
    winds *50*

Drake's Passage 156

drought 38–9, 67, 148, 213

drowned valley 161, *162*

ductile flow *115*

dunes 142, *143*, 144

dust 210

Dzulfian extinctions 135–6

Earth Radiation Budget Experiment 122–3

earthquakes 186

easterlies 33, *39*

eccentricity of Earth's orbit *14*, *15*, *17*

ecosystems 48, 50, 79, *80*, *81*, 83

eddies 45

El Niño 47, *48–9*, 50, 68

electromagnetic spectrum and radiation *19–20*, **19**, 55, *56*

elliptical orbit of Earth *see* eccentricity of Earth's orbit

endothermic reactions **73**, 92–3
see also carbon cycle; photosynthesis

energy
    for plate tectonics 115
    reactions releasing *see* exothermic
    reactions requiring *see* endothermic
    stored and released by carbon 73, 75
    thermal *19–20*
    see also solar radiation

ENSO (El Niño–Southern Oscillation) 48

Eocene
    extinctions 135-6
    plate tectonics *155*, 156
    Tibet and Himalayas 185, 196, 201

epeirogenic sea-level changes *see* isostatic

Equatorial Countercurrent 45

Equatorial Divergence *50*, 51, *52*

equilibrium system
    carbonate *84–6*, 87, *89*, 90

equinoxes 12, *13*, 15, *17*

erosion 166, 179, *197*, 198
see also weathering

estuaries 91, 99

Ethiopian flood basalt province 135–6

Etna, Mount 118

Eurasian Plate 113, *182*, 183

Europe
    volcanism, effect of 130–2
    see also Britain

eustatic sea-level changes 163–4, 165, 166, 167–8

evaporation 31, *59*
    latent heat of 28

evaporites 142, 144, *145*, *146*, 153

evapotranspiration 31

exothermic reactions **73**, 92–3
see also carbon

Falklands Current 45
faults
   normal 186, *187*, 188, 195
   transform (conservative margins) *112*, 113
fauna *see* animals
feedback
   carbon cycle 104
   climate 47, 65, 67–8, *202*
   sea-level changes and climate 166, 168–9
fires 78
fisheries 48
flood basalts and climatic effects 129–36, **129**, 177
   dates of eruption *135*
   mass extinctions 134, *135*, 136
   see also Laki Fissure eruption
flora *see* plants
Florida Current 45
fluorine 130–1
fluxes
   carbon cycle *77*, 78, 82
      see also geological; marine; terrestrial
   carbon dioxide in water *86*, *89*, 90, *96–7*
fog 32, 130–1
foraminiferans
   Arabian Sea sediments *210*
   carbon cycle *92*, 93–4, *95*
   continental drift and climate *156*, 158, 159–60
   mountains and climate 178, 192, *193*
   sea-level changes and climate 172
forcing function 23, 24, 67
   climate 123–4, *126*
   see also glaciation; Milankovich; plate tectonics
forest 80, 188, *189*, 195
   carbon sink 109
   continental drift and climate 142, *143*, 144, 146
   deforestation 43, 78, 107, 173
   fossil 161, *162*
   reservoir size 83, 108
   weathering 87
   see also rainforest; temperate forest
fossil fuels *101*
   combustion 105, 107, 108, 173
fossils 142, *143*, 159, *186*
   plants 161, *162*, 185, *186*
   teeth 195
   till *see* tillites
fractionation of isotopes 194
Franklin, Benjamin 130, 131

frazil ice *53*
fronts, atmospheric 37, *39*
fusion, latent heat of 28

Gaia hypothesis 65, 169
Galápagos Islands 50, 64–5
gamma rays 19
Ganges river 198, 206, 208
Garrels, Bob *see* BLAG
gas hydrates 109
gases
   volcanism 117–18, 119, *120*, *123*, 127, 130;
   see also atmosphere; carbon dioxide; methane;
   water vapour
gastropods *171*
gateways, ocean 152, 169
GCM (general circulation model) 189
see also Raymo
general circulation model *see* GCM
geological carbon cycle 100–3, 200
   short-circuiting 105–6
   see also fossil fuels; rocks
Gilbraltar, Straits of *155*, 156
glaciation 209, 212
   continental drift and climate 153, 156, 158–61, *159*
   deposits *see* tillites
   not triggered by volcanism 128
   see also ice; Ice Ages
global warming 161, 165, 168, 203, 209
   runaway greenhouse prevented 169–70
   see also greenhouse
Globigerina bulloides 192, *193*
glucose *74*
glycine *74*
Gondwanaland 153
GPP *see* gross primary production
grasslands 80, 144, 188, 195
gravitational attraction of planets 15
grease ice *53*
greenhouse effect and greenhouse gases **57**, *74*
   carbon cycle 75–6, 104, 109
      increased *105*, 106–9, *116*
   climate 55, *56*, 57, 75
      sunspots 66
   see also carbon dioxide; methane; nitrous oxide;
   ozone; water vapour

Index

Greenland 147, 167
Greenland Sea *53*, 151
gross primary production (GPP) 74, 77, *82*
groundwater *59*
Guinea Current 45
Gulf Stream 44, *45*, 70
    deep circulation 53–4
    mesoscale eddies *46*
    oscillations 50
    speed and flow *45*, 50
gypsum 142
gyres 44, 45, 153, *155*, 192

Hadley circulation (Hadley cells) 33, **37**
hailstorms *190*
halite 142
Hawaii
    atmosphere and carbon dioxide concentration 105, 107, *116*
    flux 118
    orographic precipitation and rain shadow *43*
    volcanism 112, *116*, *117*, 138
heat
    latent 28, 30, 46
    sensible 31
    thermal energy and electromagnetic spectrum *19–20*
    thermal inertia 28
    thermocline 44, 52, 161
    thermohaline circulation 53, *54*, 98, 170
    thermosphere 33
    transport *see* transport of heat
    see also global warming; temperature
heterotrophic organisms **75**
high pressure areas *33*, *34*, 40, *70*
see also anticyclones; pressure
Himalayas *see* Tibet and Himalayas
historic eruptions *114*
    aerosols and climate change 119, *120*, *122–3*, 124, *125*, *126*, 127, *128*
    flood basalts and climatic effects 132–3, 134
    see also Pinatubo; Tambora; Vesuvius
hospitable planet, Earth as 7–25
    contrast with other planets 7–10
    Sun, energy from *see* solar radiation
hot spots 112, 113, *116*, 138
humus 88
hurricanes *see* cyclones

Hwang-Ho river 198
hydrogen chloride and hydrochloric acid *120*
hydrogen fluoride 130–1
hydrological cycle 58, *59*, 60–1
hydrosphere *see* time- and space-scales; water
hydrothermal vents 61, 62, 168

ice
    -caps and glaciers
        albedo 150
        density 28
        water in 58, *59*
        see also polar region
    continental drift and climate 153
    -cores 128
    cryosphere *see* time- and space-scales
    melting 161, 167
    sea-ice *53*
    see also glaciation; snow
Ice Ages *see* glaciation; Little Ice Age; Quaternary
Iceland 50, *174*
see also Laki Fissure eruption
igneous rocks 141–2, 164, 175
    plateaux 130, 135-6, *137*, 138, 167–8
    see also flood basalts
India 146
    present-day climate 188–9
    northward migration *182*, *183*, *184*, 185
    see also Deccan Traps
Indian Ocean
    currents 45
    deep circulation *54*
    gyres 192
    monsoons 42, 188–90, 192–5
    oscillations 50
    phytoplankton 191, 192
    plate tectonics 156
    upwelling 51, 192
    winds 40
Indian Plate 113, *182*, 183
Indonesia 48, 142
    volcanism *see* Agung; Krakatau; Tambora; Toba
Indonesian Low *49*, 50
Indus river 197, 198, 206, 207
infrared radiation *19*, *20*, 56
inorganic carbon in marine sediments 170–4
see also carbonate; dissolved inorganic carbon; marine carbon cycle

intensity of solar radiation 10, *11*, 12–14, 18, 22
interglacial, present 147, 157, 161
intermediate time-scale carbon cycle *see* marine
Intertropical Convergence Zone (ITCZ) 36, 70
    continental drift and climate 142
    El Niño 48
    heat transfer *29*, 36–7, 40, *42*
    monsoons 189
    precipitation at *59*
inversion
    temperature 33
ions 83–4, *85*, 87, 89
iron 88, 117
    as a micronutrient 64–5, 201
IRONEX II 64–5, 201
Irrawaddy river 198
island arcs and chains 115, *116*
see also Hawaii; Philippines
isostasy 164
isostatic (epeirogenic) changes in sea-level 164, 165, 175
isotherms 26
isotopes
    carbon 193, *194*
    dating *186*, 188
    see also strontium; oxygen
ITCZ *see* Intertropical Convergence Zone
Japan 132, 133
Jefferson, Thomas 131
jet streams 37, 38, *39*, 148, 191, 213
Jupiter 7, 15
Jurassic plate tectonics 141, *154*, 155, 158, 165, *171*

Kamchatka eruption *114*
kaolinite 88
Karakoram Mountains *179*, 182
Katmai eruption 124–5, 128
Keeling, C.D. 105
kelp forest *91*
Kerguelen flood basalt plateau 135–6, 137–8
kerogen 100
Kilauea eruptions *116*, 118
Klyuchevskaya eruption *114*
Krakatau eruption *114*, 119, *120*, 124–5, 128
Kuroshio *45*

Labrador Current 45
Labrador Sea 53
lakes and inland seas, water in 59
Laki Fissure eruption *130–2*, 133–4
Lasaga, Tony *see* BLAG
latent heat 28, 30, 46
latitude
    changing *see* paleolatitudes; solar radiation *21*; temperature *26*, 28
lava *116*, *117*, 175
see also Laki; flood basalts; magma
Le Chatelier's principle 85
lignin 78
limestone 101, 142, 172
    pavement *60*
    soil in 88
lithification *see* rocks
lithosphere 114, 164
    root 184
    thickened at continental margins *176*, *184*
    volcanism *112*, 113–15
    see also plate tectonics; time- and space-scales
Little Ice Age 24, 67, 148, 213
long time-scale carbon cycle *see* geological carbon cycle
long-wave radiation, outgoing 20, *21*, *57*, 58, 123
    balanced with solar radiation 120, *121*, 122
Lovelock, James 65
low pressure areas 33, *34*, 40, *70*
see also cyclones; Intertropical Convergence Zone; pressure
Lyell, Charles *149*

Maastrichtian extinctions 135–6
Mackenzie river 206, 207
macro-algae *91*
Madagascar flood basalt province 135–6
magma *124–5*, 132, 175
    see also lava
magnesium 61, 117
magnesium–iron olivine 88
magnetics
    magnetic field reversal 139
    see also paleomagnetism
manganese 117

Index

mantle plumes 112
see also superplumes

marine carbon cycle 83–100
  diagrams 76, 98, 101, 108
  organic factors controlling 89–94
  physical factors controlling 94–6
  see also marine sediments

marine sediments
  Arabian Sea, climate records in 209–14
  deep-sea 76, 98, 99, 100, *101*, 108, 109, 195, 200, *201*
  inorganic carbon in 170–4
  organic carbon in 169–70
  sea-level changes and climate 167, 168, 169–74
  see also marine carbon cycle

marine snow 92, *93*, 94, 99

Mars 7, *8*

mass extinctions 134, *135*, 136

mass spectrometry 159

Mauna Kea *116*

Mauna Loa 105, *116*, *117*, 118

Maunder Minimum 24

Mediterranean Sea 155, 156, 170

Mekong river 197, 198, 206, 207, 208

Mercury 7

mesopause 33

mesoscale eddies 45, *46*

mesosphere 33

Mesozoic 135

methane 55, *56*, 57, *74*, 75
  increased 109

Mexico
  Chicxulub impact 134–5
  volcanism 124–5, 128

micronutrient, iron as 64, 201

microwaves 19

mid-ocean spreading ridges *see* constructive margins

Milankovich cycles (Milankovich–Croll) **16**, 128, 148, 209, 213
  solar radiation *15*, 16, *17*, 18, 23, 24

Miocene 156, 196, 201
  extinctions 135

molecules, collisions between *20*

monsoons 42, 70
  heat transfer 41–2, *42*, 45, 50
  and uplift of Tibet 188–91, 193, 195, 199, 208–10, 213

mountain-building and climate 164, 175–9
  subduction and atmospheric carbon dioxide 177–9
  see also Tibet and Himalayas
  mountain climate 176–7

mountains *9*, 80
  volcanoes *see* volcanism
  see also altitude; mountain-building

NADW *see* North Atlantic Deep Water

Namibian flood basalt province 135–6

Nazca Plate 113

nearest living relative approach *185*, 186

negative feedback 65, 104, 169, 201, *202*

Nepal *197*

Neptune 7

net primary production (NPP) 75, 77, *81*, *91*, 108

nimbostratus 32

Niña, La 49

Niño, El 47, *48–9*, 50, 68

nitrates 63, 64, 88

nitrogen 55, 62, 74

nitrous oxide 55, *56*, 57

normal faults 186, *187*, 188, 195

North America
  flood basalt province 135–6
  Plate 113
  plate tectonics 155, 156
  see also Canada; United States

North Atlantic Deep Water 52, 53–4, *70*, *97*

North Atlantic Drift 45, 151;
see also Gulf Stream

North Equatorial Current 45

North Magnetic Pole 141

North Pacific Current 45

North Sea 142, 164

North-East Monsoon 41

North-East Trades 33, 36

Norway 130

Norwegian Sea 151

NPP (net primary production) 75, 77, *81*, *91*, 108

nuclei, condensation 32

nutrients 59, 88
see also micronutrient, iron as

ocean
    and atmosphere, coupling 46–52
    basin changes 141, 167–8, *176*
    carbon cycle *97*
    see also marine carbon cycle
    carbon in *see* carbon
    chemical composition of seawater 60, *61*, 63–5
    crust formation 137, *138*, 139, 164, 199
    currents and circulation *29*, 44, *45*
        continental drift and climate 151–8
        deep *see* deep oceans
        heat transfer 52, 53, *54*
        sea-level changes and climate 166, 168, 169
        winds over 40, 43, 44
        see also downwelling; upwelling
    deep *see* deep oceans
    gases in *62*, 63
    global cycling of elements *62*
    gyres 44–5, 153, *155*, 192
    heat transport 28–9, *30*–43, 44–5
        coupling with atmosphere 46–52
    iron in *see* iron
    mid-ocean spreading ridges *see* constructive margins
    pH of seawater *87*
    plants in 9, *10*
    see also alga
    sediments *see* marine sediments
    strontium in seawater 204, *205*, 206, *208*
    as support system for life 58–66
    temperature *8*, *26*, 27
        and coupling with atmosphere 46–52
        and deep circulation 52–5
        see also thermocline; thermohaline; heat transport
    see also Atlantic Ocean; Indian Ocean; Pacific Ocean; plankton; sea-floor spreading; sea-level change; Southern Ocean
Oligocene *155*, 156, 196, 201
olivine 88
Ontong–Java flood basalt province 135
orbit of Earth 7, 14, *15*, *16*, *17*
    see also Milankovich cycles
Ordovician plate tectonics 141, 157–8, 165, 166
organic carbon
    in marine sediments 169–70
    Tibet and Himalayas 200–3
    see also carbonaceous sediments; dissolved organic carbon; particulate organic carbon

organisms, living 75, 169
    atmosphere-ocean-Earth as support system for 58–66
    carbon cycle 73–5
    controlling carbon flux in oceans 94–6
    and plate tectonics 157–8
    see also animals; benthic; organic carbon; pelagic; plankton; plants; primary production; respiration
Orinoco river 206, 207
orographic precipitation 42, *43*, 188, *189*
orthoclase 88
oscillations between pressure systems 48, *49*, 50
Ontong–Java Plateau 137–8
Owen Ridge 209, *210*, *211*–12
oxidation of organic carbon 203
oxygen
    in atmosphere 55, 56, 57, 62, 74
    isotopes and climate record 158, *159*–*60*, *161*, 166, *196*
    in lava 117
    in oceans 62
    in photosynthesis 9, 74, 75
Oyashio 45
ozone 55, 56, 57, 109
    and sunspots 66

Pacific Ocean
    currents 44, 45
    deep circulation *54*
    El Niño 47–9
    fluxes across air–sea interface 96, *97*
    gyres 44, 153
    iron added to 64–5
    oscillations 50
    plate tectonics 112, 113, *115*, 153, 156, 167
    sea-level changes and climate 167
    temperature 27, 44
        thermohaline conveyor *54*, 98
    upwelling 51
    volcanism 112, 113, *115*, *122*–*3*, 124–6, 128, 134, 177
    winds over 40, 43
    see also El Niño
Pacific Plate 113, 114
Pakistan 193, *194*
Paleocene 156, 196, 201
paleoclimatic indicators 144, 146
paleolatitudes, changing 141–2
see also continental drift

Index

paleomagnetism *141*, 144, *145*, 166
Paleozoic 135
Pangea, break-up of 153–7, *158*, 167, 170
Panthalassa 153, *155*, 167
particulate inorganic carbon (PIC) 89
particulate organic carbon (POC) 79, *89*, 99, 101, 200
pelagic organisms 64, 92
perihelion 14
Permian 135, 165
   continental drift and climate 141, 142, *143*, 146, 157, 158
   extinctions 158
Peru
   El Niño 48
   upwelling of *50*
Peru Current (Humboldt) 44, 45
pH 87
Phanerozoic *165*
Philippines *115*
   Plate 113
   see also Pinatubo
phosphates 63, 64, 88
phosphorus 74, 117
photic zone 63, 91, *93*, 157
see also surface of oceans
photosynthesis 9, 73–4, 77, 194
   C4 plants 195
   in oceans 63
   see also chlorophyll
physical factors controlling carbon flux in oceans 89–94
physical weathering 179, 195
phytoplankton 63, 64–5, 90–1, *191*, *192*
PIC *see* particulate inorganic carbon *89*
Pinatubo, Mount *122–3*, 124–6, 128, 134
planets
   contrast with Earth 7–10
   gravitational attraction 15
   temperatures *8*
plankton 63, 64, *156*, 159–60, 178
see also phytoplankton; zooplankton
plants and plant biomass *10*
   belts *9*, *80*, 144
   carbon cycle *80*
      diagrams *76*, *98*, *101*, *108*
      see also reservoirs

drought 38–9
evapotranspiration 31
fossil 161, *162*, *185*, *186*
heat transfer 50
hydrological cycle *59*
plate tectonics 142, *143*, 144, 146, 170
reservoirs and carbon cycle 83, 88, 91, 100–1
   size of *77*, 79, *80*, *81*, 83, 108
Tibet and Himalayas 191, 194–5
see also algae; phytoplankton; forest; photosynthesis
plate tectonics 112–16
   effect on climate 141–81, 202, 213
      see also continental drift; mountain-building; sea-level change
   India 182, 183
   volcanism 112–13, *114–15*, 116, *139*
plateaux, igneous 130, 135–6, *137*, 138, 167–8
see also Tibet
Pliensbachian extinctions 135–6
Pliny the Younger 119
Pliocene 135, 156, 196, 201
plumes, mantle and superplumes 112, 138, *139*, 168
Pluto 7
POC *see* particulate organic carbon
polar air 37, *38*
polar front 37, 39, *70*
polar jet stream 37, *38*, *39*, 148, 213
polar region *9*, 147
   solar radiation at *14*
   temperature *26*
   see also Antarctic; Arctic
Pomona lava flow *129*
positive feedback 47, 68
potassium 61, 88, 117
Precambrian 177
precession 15, *17*
precipitation 46, *59*
   acid rain 130
   chemical composition 60, *61*
   orographic 42, *43*, 188, *189*
   plate tectonics 159
   volcanism *131*
   see also monsoons
precipitation of carbonate 102–3
pressure, air *33*, *34*, 40, *70*
   altitude *30*, 176–7
   oscillations between systems 48, 49–50

solubility of gases 62–3
see also air masses; anticyclones; cyclones; monsoons; winds
primary production 9, 74–5, 108
    carbon cycle 69
        marine *90*, *91*, 92, 93
        terrestrial 77, *81*, *82*
    climate 64, *65*, 66
    continental drift and climate 157
    measuring *82*
    and upwelling 48, 192
    see also chlorophyll; gross primary production; net primary production; organisms, living; photosynthesis

Quaternary Ice Age *201*
    plate tectonics 147, 148, 156, 157, *165*

radiation budget 28
see also electromagnetic; solar radiation
radio waves 19
radiogenic strontium 204
radiolarians 178
radiometer 26
rain *see* precipitation
rain shadow 42, *43*
rainforest, tropical 43, 80
    carbon sink 109
    continental drift and climate 142, *143*, 144, 146
    size of reservoir 83
    South Asia 188, *189*
    weathering 87
    see also deforestation
raised beach 161, *162*
Rajmahal flood basalt province 135
Raymo, M. and Ruddiman, B. 189, 199–200, 201–2, 203, 209, 213
Raymo–Ruddiman hypothesis 203
recrystallized minerals 188
reflection of solar radiation *see* albedo
relative sea-level change 165
reptiles, age of 135
re-radiation *see* back radiation
reservoir 58
    carbon 83, 88, 91, 100–1
        animals insignificant 77, 78
        diagrams *76*, *98*, *101*, *108*

    residence time 77, **78**, 98–9, 101, 109
    size of *77*, 79, *80*, *81*, 83, 108
        time-scales *see* geological; marine; terrestrial
    carbon dioxide 77, 81–2
residence time 58, 78
    aerosols 134
    carbon 77, **78**, 98–9, 101, 109
respiration 75, 77, 78, 87
reverse weathering 94
Rhaetian extinctions 135–6
ridges, ocean *see* constructive margins
ring world, tropical *149*, *150*, 151, 169
rivers
    carbon cycle 79, 87–9, 99
    chemical composition 60, *61*
    Tibet and Himalayas 195, *197*, *198*, 200
    strontium in *206–7*, 208
rocks 31
    formation (lithification) *101*, 108, 169, 170
    see also geological carbon cycle; igneous; sedimentary
Rossby waves 37, 50
rotation of Earth *see* Coriolis effect
Roza lava flow 129, 133–4
run-off and carbon cycle 79, 99

Sahel 39
St Helens, Mount 125, 126
salinity 53
    deep circulation of oceans 53–4
    salts *see* evaporites
    solubility of gases 62–3
Salween river 198
sandstones 142, 143, 144
Saturn 7, 15
savanna 80, 142
sea *see* oceans
sea-ice *53*
sea-floor spreading 114, 141, 167–8
sea-level change and climate 161–74
    atmospheric carbon dioxide 168–74
    basin size and shape changes 167–8
    contributions to 163–6
    eustatic 163–4, 165, 166, 167–8
    isostatic 164–5, 175
    measurement *166*
    water volume changes 167

Index

seamounts (sea-floor volcanoes) 177, 199
seasons
    carbon dioxide levels *82*, 102, *106*, 107
    cause of *12–13*, *14*
    precipitation *see* monsoons
    temperature *26*, 147
seaweed *91*
sedimentary rocks *60*, *88*, 101, 142, 172, 185
    continental drift and climate 141–2, *143*, 144
    sea-level changes *166*
sediments *see* marine sediments; sedimentary rocks
sensible heat 31
sensitivity 23
Serra Geral flood basalt province 135
shelf seas 146, 164, 172
shells and skeletons, calcium carbonate 159, 178, 210
    carbon cycle 92, 94, 100
    sea-level changes and climate 169, 170, *171*, 172
short time-scale carbon cycle *see* terrestrial
short-circuiting geological carbon cycle 105–6
short-wave radiation *see* solar radiation
Siberia 130, 135–6
silica
    and calcium carbonate (decarbonate reaction) 177–8, 202
    in oceans 60, 61, 63, 64
    precipitated 94, 117
    in rocks 88, 100, 102, 103
    in water 60, 61, 88, 89
silicates, weathering of 88, 99, 102, 169, 178–9, 199
silicon 117
Silurian plate tectonics 141, 142, 158, 165
sinking water in oceans *see* downwelling
skeletons *see* shells and skeletons
slice world *150*, 151, 152
snow 159, 176, 177, 190
see also ice
sodium 117
    in water 61, 64, 88, 89
soil
    anoxic 78
    carbon cycle 77, 82
        diagrams 76, 98, 101, 108
    carbon dioxide in 87
    carbonates in *194*
    erosion 79
    fertility 59, 88
    lacking 60
    nutrients in 88
    water in 31, 59–60
solar flux 10, *11*, 23–4, 67
solar radiation 10-24, 56, *57*, 58
    absorption *20*, *21*, 32, 122–8
    balance with outgoing radiation 120, *121*, 122
    continental drift and climate *159*, 160
    reflection *see* albedo
Solar System 7, 21
solstices *12–13*, 15
solubility 73
    of gases 62–3
    see also carbon dioxide
South Africa 135–6
South America
    plate tectonics 155, 156
    upwelling off 50, 51
    volcanism *115*, 124
South American Plate 113
South Asia 188, 193, *194*
see also Tibet and Himalayas
South Equatorial Current 45
South Pacific High *49*, 50
South Pole *see* Antarctic; Southern Ocean
South-East Trades 33, 36
    El Niño 48, *49*
    upwelling 51
South-West Monsoon 41, 188, 189–90, 191, 193, 195, 210, 213
Southern Ocean
    carbon cycle *97*
    currents 44, 45, 70, 151, *155*, 156
    deep circulation 53
    iron deficiency 54
    temperature 27, 44
    water temperature *156*
    winds 40
Southern Oscillation 48, *49*, 50
specific heat of water 28
spectrum *see* electromagnetic spectrum
Sr-isotope ratio 203–8, 206
Sri Lanka 188
steady-state model 209
see also BLAG
Stefan–Boltzmann Law *19*
steppes 188

Stiles, Ezra 131
stillstands, glacial *213*
stomata 74
stratocumulus 32
stratopause 33
stratosphere
   solar radiation 56, 57
   temperature changes *30, 33*
   winds 33
stratus 32
stromatolites *171*, 181
strontium isotopes and chemical weathering 203, *204–6*, 207, *208*, 209
subduction 115, 177–9
   see also destructive margins
submarine plateaux 130
subtropical highs 33, 37
subtropical jet streams *39*, 148, 191, 213
subtropical highs 33
subpolar lows 33
sulfates in natural waters 61
see also aerosols
sulfur 74, 132
sulfur dioxide 32
   and DMS 65
   from volcanic eruptions 117, 118, 119, *120, 123*, 127, 130
   see also aerosols
sulfuric acid aerosols *see* aerosols
Sun 8
   energy from *see* solar radiation
   sunspots, 11-year cycle of 21, *22*, 23
      climatic change 66–8
      length of solar cycle *68*
   see also orbits; planets
supercontinents
   break-up of *see* Pangea
   formation of 153, 168
superplume 138, *139*, 168
surface of oceans (air–sea interface)
   carbon cycle 90, 91, *96-7*
      diagrams 76, 98, 101, 108
   sea-level changes and climate 161, 170–2
   see also photic zone
swamps 142, *143*
Sweden, volcanism, effect of 131–2
Syria 130

Tambora 124–5, 126, 127–8
   contrasted with other eruptions 132–3, 134
Tasman Ridge 156
tectonics *see* plate tectonics
teleconnections 48, 49, **50**
temperate forest 80, 83, 87, 108
temperate regions 39
temperature (mainly changes) *8, 20*, 23
   atmosphere *8*
      altitude *30*, 31, *33*, 42, *47*, 185–6, 190
   average surface *27, 57*
   carbon dioxide 8
   continental drift and climate 149, *150*, 156
   distribution over Earth's surface 26–8, *27*
   sea-level changes and climate 165, 167
   seasons *26*, 147
   solubility of gases 62–3
   sunspots *68*
   vertical structure *33*
   volcanism 123, *124–6*, 127, 130–1, *132*, 133
   water 28
   see also global warming; heat; oceans
Tennyson, Alfred Lord 119–20
terrestrial carbon cycle 77–83
   diagrams 76, *98, 101, 108*
Tertiary 134
   plate tectonics 156, 165
   Tibet and Himalayas and climate change *186*, 191, 195–6, *201*
Tethys Ocean 155, 156
thermal energy *see* heat
thermocline 44, 52, 161
thermohaline circulation **53**, *54*, 98, 170
thermosphere 33
Tibet and Himalayas 182–214
   Arabian Sea, climate records in sediments of 209–14
   atmospheric carbon dioxide 197–209
   formation 182–8
      reasons for height 183–5
   see also uplift of Tibet and Himalyas
tillites 144, *145*, 146
tilt of Earth's axis 12, 15, *16, 17*
   changes *see* Milankovich
   glaciation 147, 148, 160
time- and space-scales *67, 148, 213*
time-scales of carbon cycle 77–103
   intermediate *see* marine
   long *see* geological
   short *see* terrestrial

Index

Tithonian extinctions 135–6
Toba 127–8
Trade Winds 33, 36, *37*
    and El Niño 48, *49*
    upwelling *50*, 51
transform faults (conservative margins) *112*, 113
transpiration 31, 59
transport of heat 28–55, 152
    atmosphere 30–43
    atmosphere–ocean coupling 46–52
    in oceans 44–6
    see also heat; temperature
trench, deep sea 131, 133
Triassic 135
    plate tectonics 141, 158, 165
    sea-level 166
trilobites, extinction of 135
tropical air 37, *38*
tropical jet streams *39*
Tropics of Cancer and Capricorn *12–13*, 15
tropopause *33, 39*
troposphere 30, **33**
Tsangpo *see* Brahmaputra
tundra 80, 109
typhoons *see* cyclones

ultraviolet radiation *19, 20*
    absorbed 56
    ozone produced by 66
United States
    droughts and crop failures 38–9
    orographic precipitation 42–3
    sea-level changes and climate *163*, 164, 171
    volcanism *131*
        flood basalts *129*, 133–4, 135, 209
uplift of Tibet and Himalayas 185–8
    climate change related to 188–96, 197, 199, 200, 201, 208–9
    evidence for 191–6
upwelling 48, *70*
    carbon cycle 98, 101, 108
    relating to continental drift and climate 154, *155*, 157
    heat transfer 48, 50, 51–2
    net primary production *91*
    winds *50*, 51
Uranus 7

vegetation *see* plants
Venus 7, *8*
vertical temperature structure of atmosphere *33*
Vesuvius 119
visible radiation *19, 20*, 56
volcanic winter 127
volcanism 60, 112–40
    carbon dioxide in atmosphere 117–18, 119, *120*, 168, 177, 199, *202*
    eustatic sea-level changes 167–8
    global view 137–9
    mountain-building and climate 175, 177–8
    plate tectonics and hot spots 112–16
    see also aerosols; Deccan Traps; flood basalts

water
    properties important for climate *28*
    residence time 58
    in soils 31, 59–60
    total world supply *59*
    see also groundwater; hydrological cycle; oceans; rivers; time- and space-scales; water vapour; carbon dioxide
water masses 52
water vapour
    in atmosphere 55, 56, 57
        transport of 30–2, 36, 41
    condensation *see* clouds
    from volcanic eruptions 117, 127
wavelengths of electromagnetic radiation *19, 20*
weather systems 67, 148, 213
weathering 94
    physical 179, 195
    see also chemical weathering; erosion
West African flood basalt province 135–6
West Wind Drift *see* Antarctic Circumpolar Current
westerlies 33, 38, *39*, 48
western boundary current *see* Gulf Stream
White, Gilbert 131
winds *29*, 31, 34, 151
    directions *39*, 40–1
    downwelling *50*
    hypothetical *33, 34, 35*
    oceans and currents 40, 43, 44
    plate tectonics 153, 157
    polar jet stream 39
    stratospheric 33
    on water-covered Earth *33*
    see also easterlies; Trade Winds; westerlies
Wrangellian flood basalts 135–6

Xingu river 206
X-rays 19, *22*

yaks *186*
Yangtze river 197, 198, 206, 208
Year without a Summer (1816) **126**, 132
Younger Dryas 54

Yucatan impact 134–5
Yukon river 206, 207
Zaire river 206, 207
zonal temperature distribution 28
zooplankton 63, 191, 192
 role in carbon cycle 92, 93–4, *95*